Advances in Plant Disease Management

Advances in Plant Disease Management: Volume I: Fundamental and Basic Research is an invaluable compilation for researchers/students/stakeholders/policymakers in agriculture. This book aims to offer the latest understanding of fundamental and basic research fronts toward managing crop plant diseases. After clearly explaining the updated knowledge on the host immune system, and pathogen's interplay with the host as unraveled through genomics, bioinformatics, and molecular studies, this book equips readers with the knowledge to confidently account for them during the formulation of management strategies for major crop plant diseases.

This book offers comprehensive coverage of the research advances in plant disease management, including:

- Newer insight into the host–pathogen interaction, including effector-driven pathogenesis in different host–pathogen systems
- Updates on plant defense pathways leading to resistance to pathogens
- Use of novel molecules, antagonists, and genome-editing tools toward manipulating host resistance
- Plant protection policies that support the agricultural production system from a global perspective

Advances in Plant Disease Management

Volume I: Fundamental and Basic Research

Edited by

Pranjib K. Chakrabarty, Kalyan K. Mondal, Mahender S. Saharan, J. Kumar, and Charudatta D. Mayee

CRC Press
Taylor & Francis Group
Boca Raton London New York

CRC Press is an imprint of the
Taylor & Francis Group, an **informa** business

Designed cover credit: © Shutterstock

First edition published 2024
by CRC Press
2385 NW Executive Center Drive, Suite 320, Boca Raton FL 33431

and by CRC Press
4 Park Square, Milton Park, Abingdon, Oxon, OX14 4RN

CRC Press is an imprint of Taylor & Francis Group, LLC

Library of Congress Cataloging-in-Publication Data
Names: Chakrabarty, Pranjib, editor. | Mondal, Kalyan K., editor. |
Saharan, M. S., editor. | Mayee, C. D. (Charudatta Digambarrao), 1946- editor. |
Kumar, J. (Jagdish), editor.
Title: Applied and strategic research advances in plant disease management.
Volume 1 / edited by Pranjib Chakrabarty, Kalyan K. Mondal, M. S. Saharan,
C. D. Mayee, and J. Kumar.
Description: First edition | Boca Raton, FL : CRC Press, 2024 |
Includes bibliographical references and index.
Identifiers: LCCN 2023036696 (print) | LCCN 2023036697 (ebook) |
ISBN 9781032516066 (hardback) | ISBN 9781032516196 (paperback) |
ISBN 9781003403142 (ebook)
Subjects: LCSH: Plant diseases. | Plants, Protection of. | Agricultural innovations.
Classification: LCC SB601 .A68 2024 (print) | LCC SB601 (ebook) | DDC 571.9/2—dc23/eng/20231018
LC record available at https://lccn.loc.gov/2023036696
LC ebook record available at https://lccn.loc.gov/2023036697

ISBN: 9781032516066 (hbk)
ISBN: 9781032516196 (pbk)
ISBN: 9781003403142 (ebk)

DOI: 10.1201/9781003403142

Typeset in Times
by codeMantra

Contents

Foreword

Global agriculture has noticeably been threatened by pest and diseases which adversely affect agricultural productivity, profitability, and livelihood of the farmers. The changing global climate and human activities have impacted the production and productivity of food crops. Plant Protection is crucial to food and nutritional security not only in India but all over the world. The task of protecting crops from biotic stresses, particularly under the changing climatic conditions becomes highly challenging. The alarm is further exacerbated due to globalization of trade, introduction of invasive pests and diseases, and emergence of new and aggressive pests and pathogens. Advancement in science and technology during the last decade has opened several avenues for novel and strategic deployment of technologies to efficiently address the avoidable losses. Together with traditional technologies and enabling policies, the capabilities of crops to defend against intruding plant pathogens can be greatly enhanced. Coupling artificial intelligence for biotic stress monitoring and diagnosis can result in greater precision on stress surveillance. Future research may target novel and innovative methods contrived after deciphering molecular mechanisms of host-biotic stress interactions. The next generation of climate resilient crop protection strategies would focus more on the use of smart formulations based on antimicrobial peptides, recombinant antibodies, dsRNA, gene silencing, CRISPR/CAS editing, etc. Host-Induced Gene Silencing (HIGS) and Spray-Induced Gene Silencing (SIGS) through the application of miRNA and siRNA offer best alternatives, that could cut down the use of ecologically hostile chemo-therapeutants for the management of biotic stresses in agricultural and horticultural crops. However, many of the potential technologies developed using advanced tools have inherent regulatory limitations for deployment. A basic tenet for effective management lies in understanding various limitations so that rational ways and means can be developed with enabling policies for deployment of such technologies in a field ready condition.

The book *Advances in Plant Disease Management: Volume I: Fundamental and Basic Research (two volumes)* is a compilation work on plant health protection, technology advancements, and future research direction and development. Editors of this book are renowned scientists, well-acclaimed in their respective areas of specialization and have been engaged in framing policies governing plant protection. I am confident that this book will serve as a ready reckoner for scientists, policymakers, and students for future research into the realms of Plant Protection. The editors deserve appreciation for bringing out the book which will be used as a textbook in agriculture and allied sciences. I compliment the editors for their all-round efforts in bringing out this publication.

(Himanshu Pathak)
Dated the 4th December, 2023
Dr. Himanshu Pathak,
Secretary (DARE) & Director General (ICAR),
Government of India
New Delhi

Preface

The Indian economy largely depends on agriculture. Over a period of time, India has shown its potential by transforming from a food deficit to a food sufficient to finally food surplus nation. This transformation was made possible through scientific intervention in food production and food protection technology. Plant protection technologies played a key role in this transformation. However, despite new technologies and tools, threating pests and diseases of crop plants is of serious concern for today's researchers working on plant protection. There is precedence wherein newer insight into the basic understanding of host–pathogen interaction, including effector-driven pathogenesis, updates on plant defense, and newer eco-friendly molecules fit for organic agriculture, enabled us to contain the emerging pathogens in a better way. However, all these developments have not been documented systematically, which can serve as a database for future generations or provide research leads to manage the disease more appropriately. In this context, this book entitled *Advances in Plant Disease Management: Volume I: Fundamental and Basic Research* **is an attempt to offer a hand-ready reference on the research developments that changed our understanding as well as provided nature-friendly options for protecting crop health.**

The primary aim of this book is to comprehend the updates on fundamental and basic research fronts of plant protection that not only provide deeper insight into the subject but also deliver scope for exploring novel strategies of plant health management and thereby improving agricultural productivity. This book will serve as a reference book for stakeholders associated with agriculture, including students, scientists, researchers and policymakers of crop production and protection technology. Volume I of this noteworthy compilation contains 14 chapters written by globally acclaimed scientists on different areas of plant protection sciences. Briefly, the current volume discusses the updated insight into host–pathogen interaction and disease management aspects using omics/genomic tools, T3SS-effectors, signal transduction, mycotoxins, diagnostics, small RNAs, and CRISPR-Cas editing. Besides, the fundamental policies on crop biosecurity, invasive species/**pest risk analysis,** and their effective implementation toward protecting plant health **have been discussed. The editors thank all the contributors for their valuable and timely contributions and time-to-time co-operation as and when required during the publication process. The editors put their best effort into shaping and bringing out this book with minimum error and would look forward to learning from the esteemed readers about its practical utility.**

Editors:
Pranjib K. Chakrabarty, Kalyan K. Mondal,
Mahender S. Saharan, J. Kumar, and Charudatta D. Mayee

Editor Biography

Dr. Pranjib K. Chakrabarty has served ICAR in various capacities as Scientist (Plant Pathology), Principal Scientist (Biotechnology), and Head of the Division of Crop Improvement at the Central Institute for Cotton Research (CICR) Nagpur. Before joining ICAR, he served as Assistant Director General (Plant Protection & Biosafety) of Krishi Bhavan, New Delhi. He extensively studied and did pioneering research on general and molecular plant pathology with special reference to host–pathogen interaction, signal transduction for resistance signaling, functional genomics of pathogenicity genes, molecular diagnostics, and transgenic research. He did his post-doctoral research at the University of Florida, Gainesville and at the University of Missouri-Columbia in Columbia, USA. He is also an adjunct faculty member at UF, Gainesville. He visited several countries including Australia, the UK, Canada, Brazil, Ecuador, Switzerland, the Netherlands, China, Malaysia, Singapore, Nepal, Bhutan, Ethiopia, and Bangladesh, representing national and international causes on plant protection and biosafety. Besides his assignments as ADG plant protection in ICAR, he was the nodal officer on GM Biosafety and Biosecurity; Member of the CIBRC, Ministry of Agriculture; Member of RCGM, DBT; Chairman of Technical Committee of Central Sector Scheme on Monitoring Pesticide Residue at National Level, DAC, MOA; Member of Task Force (Agricultural Biotechnology), DBT; Member of Task Force, DRDO, MEA; Chairman of the Board of Institute of Pesticides Formulation Technology, Ministry of Chemicals and Fertilizers, GOI; and Member of Governing Body and Executive Body of NIPHM, Hyderabad. With his direct association with four National Plant Protection Institutes of ICAR and seven all India coordinated research projects on plant protection, as ADG (PP), besides policymakers in the Ministry of Agriculture, he has been associated with the development of policies and guidelines for enabling and ensuring crop biosecurity.

Since April, 2019, he has been working as a Member (Plant Sciences), ASRB, New Delhi, in the Ministry of Agriculture and Farmers Welfare, GOI, New Delhi. He was the President of Indian Society of Mycology and Plant Pathology for two consecutive years, from 2018–2019 to 2019–2020, and also of the Indian Phytopathological Society during 2020–2021. He is a Fellow of NAAS, New Delhi, besides several other professional scientific societies.

Dr. Kalyan K. Mondal received both his BSc (Ag) Hons and his MSc (Ag) in Plant Pathology from the Bidhan Chandra Krishi Viswa Vidyalaya, West Bengal, in 1991 and 1993, respectively. He obtained his doctoral degree in Plant Pathology from ICAR-IARI, New Delhi, in 1999. After a stint (2001–2006) as Scientist at CSKHPKV, Palampur, Dr. Mondal joined the Division of Plant Pathology, ICAR-IARI, New Delhi, as Senior Scientist (2006–2012) and continued as Principal Scientist (2012–2023). Presently, Dr. Mondal is working as Joint Director (Research) at ICAR-NIBSM, Raipur, Chhattisgarh, India (since August, 2023). He was trained at Stanford University, California, USA, as a visiting scholar to gain insight into the role of bacterial Type-3-Secretion-System (T3SS) effectors in the modulation of plant defense. He deciphered the first-time effector repertoire of two important bacterial pathogens, *Xanthomonas oryzae* pv. *oryzae* (Xoo) and *X. axonopodis* pv. *punicae* (Xap), infecting rice and pomegranate, respectively. He functionally demonstrated the role of crucial effectors of both the pathogens during disease development in respective crops; he generated the full genome sequence of Xoo race 4, a virulent population from India; and he was invovled in the development and commercialization of seven blight-resistant rice varieties namely Pusa 1592, Pusa 1718, and Pusa 1728, Pusa 1847, Pusa 1885, Pusa 1886 and Samba 1853 having *R*-genes *xa13* and *Xa21*. Dr. Mondal became Fellows of NAAS (2023), NABS (2020), ARRW (2021), IPS (2013), SARP (2020), and a member of the National Academy of Sciences India (2013). He is the recipient of many prestigious awards, including Nanaji Deshmukh ICAR Award (2021), Dr, BP Pal Memorial NABS Best Scientist Award (2022), Prof. MK Patel Memorial Young Scientist Award (2008), Prof. MJ Narasimhan Academic Merit Award (2001), and Prof. JP Verma Memorial Award (2016). He serves as a guest reviewer for journals of international repute, including *BMC Microbiology*, *Plant Disease*, *Frontiers of Microbiology*, *MPMI*, *Microbial Pathogenesis*, *Scientific Reports*, *PMPP*, and *Plant Pathology*. He served as a member of important academic and research committees including Task Force DBT (Agricultural Biotechnology), Project Reviewer of DST-SERB, Member of IMC, ICAR-NBAIM (2016–2019), and Editor (2017–2019) for *Indian Phytopathology*, and is presently Jt. Secretary of the Indian Phytopathological Society. He published >50 research papers in journals of international repute and authored a very popular text book, *Plant Bacteriology*, published by Kalyani Publishers. He has delivered lead talks and guest lectures at international conferences across the world, including in the Philippines, Vietnam, China, Thailand, and the USA.

Dr. Mahender S. Saharan received his PhD in Plant Pathology (Gold Medalist) from CCS HAU, Hisar, Haryana. He is presently Head, Division of Plant Pathology, ICAR-Indian Agricultural Research Institute, New Delhi. Earlier till 2016, he served in different capacities like Scientist, Senior Scientist, Principal Scientist, and Principal Investigator, Crop Protection at ICAR-IIWBR, Karnal. He contributed significantly to wheat pathology research and actively participated in strategy implemenation for managing wheat rusts and Karnal bunt in main wheat belt of the country. He characterized *Fusarium* spp. as causing a head scab of wheat. He was also associated with the draft genome sequencing of *Tilletia indica* and *Bipolaris sorokiniana*, two major pathogens of wheat. He has rich experience in the genetics of resistance and the management aspects of plant diseases. He contributed significantly to the development and release of 16 wheat varieties in the country, besides registering 23 wheat genetic stocks for rusts, Karnal bunt, head scab, and spot blotch resistance. He worked on many international collaborative research projects, viz., ISCB, ACIAR, CIMMYT, BBSRC, etc. He is a widely traveled scientist and has visited Mexico, Switzerland, Russia, Japan, China, Bangladesh, Nepal, Kenya, Turkey, Malaysia, Australia, Italy, and the USA to participate in and present invited talks at international conferences. He is the recipient of prestigious BGRI Gene Stewardship Award and is also a Fellow of National Academy of Agricultural Sciences, New Delhi, besides Indian Phytopathological Society, New Delhi, and Society for Advancement of Wheat and Barley Research, Karnal.

Dr. J. Kumar received his Ph.D. from Dr. YSPUH&F, Solan, H.P., in Mycology and Plant Pathology and is an Alexander Humboldt Fellow of Germany. He served ICAR in several capacities, such as Principal Scientist, IIWBR, Karnal; Head of IARI Regional Research Stations at Wellington (T.N.) and also at Shimla (H.P.); Joint Director, School of Crop Health Biology Research, ICAR-NIBSM, Raipur, Chhattisgarh (2015); and Director of ICAR – National Institute of Biotic Stress Management, Raipur, Chhattisgarh – a deemed university. He has vast experience in developing plant disease forecasting models. He worked on biochemical basis of host resistance in cereals, besides earning specialization in breeding for disease resistance in wheat. He is associated with the development of 12 wheat varieties resistant to various diseases in India. He has rich experience in research in plant pathology and molecular biology through international exposure during his research stay at CIMMYT, Mexico, and Justus Liebig University, Germany. He registered five genetic stocks with ICAR-NBPGR. Dr. Kumar visited Germany, Mexico, Hungary, Belgium, China, and Nepal as part of his research assignments. Besides having several awards, he is a fellow of Indian Phytopathological Society, New Delhi, and Society for Advancement of Wheat and Barley Research, Karnal.

Dr. Charudatta D. Mayee is a renowned Plant Protection Scientist and is currently serving as the President of the South Asia Biotechnology Centre (SABC), New Delhi, and Chairman of Agricultural Finance Corporation of India, Mumbai. Born in a farming family in Maharashtra, Dr. Mayee obtained his PhD from IARI in Plant Pathology. He commenced his career in Plant Pathology Research at IARI and worked in various capacities at Central Rice Research Institute, Cuttack; Punjab Agricultural University, Ludhiana; the Ford Foundation, Delhi; and Maharashtra Agricultural University (MAU) Parbhani for nearly 30 years. The research, teaching, and extension experiences led him to serve as Vice Chancellor-MAU Parbhani; Director-Central Institute of Cotton, Nagpur; and Agriculture Commissioner, Government of India, New Delhi, before superannuating as the Chairman, Agricultural Scientists Recruitment Board, Ministry of Agriculture and Farmers' Welfare, Government of India.

Though specialized in Plant Pathology, Dr. Mayee committed himself to the overall growth of Indian agriculture. He guided 20 PhD and more than 38 MSc students, wrote several books and monographs, published over 200 scientific publications in journals of repute, and served the cause of Agriculture Cop Protection through development of the subject. He is Fellow of several scientific committees and associations, including Alexander Humboldt Fellow of Germany and National Academy of Agricultural Sciences (NAAS). He is also a member of several international societies, including ISAAA and ABNE, Africa. Biotechnology is his passion, and he was instrumental in the first commercial release of Bt cotton in India in 2002, a great feat that revolutionized cotton production in the country. He is the recipient of several honors and awards for agriculture development, including the lifetime Achievements Awards and honorary Doctors of Philosophy from many universities. Currently, Dr. Mayee is a member of the Board of Management of many organizations and universities. His super administrative capabilities and practical knowledge on plant protection issues and rational policy decisions in agriculture, including those related to crop biosecurity, make him one of the most sought-after scientists in academia and among policymakers. He has widely traveled and visited more than 50 countries around the world.

List of Contributors

Rashmi Aggarwal
Emeritus Scientist, Division of Plant Pathology
ICAR-IARI
New Delhi, India

Anjali, V. A.
University of Horticultural Sciences
Bagalkot, India

Aparna, E.
ICAR-Indian Institute of Rice Research
Hyderabad, India

Archana, T. S.
School of Agriculture
Lovely Professional University
Phagwara, India

Awaji, S. M.
ICAR-National Rice Research Institute
Cuttack, India

Barbadikar, K. M.
ICAR-Indian Institute of Rice Research
Hyderabad, India

Dipankar Barman
Division of Plant Physiology
ICAR-Indian Agricultural Research Institute, Pusa
New Delhi, India

Bishnu Maya Bashyal
Division of Plant Pathology
ICAR-IARI
New Delhi, India

Satish C. Bhatla
Bio-stimulant Cell
Ministry of Agriculture and Farmers Welfare
New Delhi, India

Pranjib K. Chakrabarty
ASRB
New Delhi, India

Supriya Chakraborty
Molecular Virology Laboratory, School of Life Sciences
Jawaharlal Nehru University
New Delhi, India

V. Celia Chalam
Division of Plant Quarantine
ICAR-National Bureau of Plant Genetic Resources
New Delhi, India

Viswanathan Chinnusamy
Division of Plant Physiology
ICAR-Indian Agricultural Research Institute, Pusa
New Delhi, India

Kavya Dashora
Indian Institute of Technology Delhi
New Delhi, India

Devanna, B. N.
ICAR-National Rice Research Institute
Cuttack, India

Divya Bharathi, R.
Division of Plant Physiology
ICAR-Indian Agricultural Research Institute, Pusa
New Delhi, India

Divya, K.
ICAR-Central Tuber Crops Research Institute
Thiruvananthapuram, India

Kavita Gupta
Division of Plant Quarantine
ICAR-National Bureau of Plant Genetic Resources
New Delhi, India

Manjunath Hubballi
University of Horticultural Sciences
Bagalkot, India

Deeba Kamil
Division of Plant Pathology
ICAR-IARI
New Delhi, India

Katara, J. L.
ICAR-National Rice Research Institute
Cuttack, India

Fatima Nazish Khan
Division of Plant Physiology
ICAR-Indian Agricultural Research Institute, Pusa
New Delhi, India

Ravinder K. Khetarpal
Asia-Pacific Association of Agricultural Research Institutions (APAARI)
Bangkok, Thailand

Chaitra C. Kulkarni
University of Horticultural Sciences
Bagalkot, India

Jagdish Kumar
ICAR-National Institute of Biotic Stress Management
Raipur, India

Kumari, M.
ICAR-National Institute for Plant Biotechnology, Pusa
New Delhi, India

Madhav, M. S.
ICAR-Indian Institute of Rice Research
Hyderabad, India

Makeshkumar, T.
ICAR-Central Tuber Crops Research Institute
Thiruvananthapuram, India

Ashish Marathe
ICAR-National Institute of Biotic Stress Management
Raipur, India

Kalyan K. Mondal
Joint Director (Research)
ICAR-National Institute of Biotic Stress Management
Raipur, Chhattisgarh, India

Enrique Moriones
Instituto de Hortofruticultura Subtropical y Mediterránea "La Mayora"
Consejo Superior de InvestigacionesCientíficas
Málaga, Spain

Senthilkumar K. Muthusamy
ICAR-Central Tuber Crops Research Institute
Thiruvananthapuram, India

Shivani Nagar
Division of Plant Physiology
ICAR-Indian Agricultural Research Institute, Pusa
New Delhi, India

Nakkeeran, S.
Tamil Nadu Agricultural University
Coimbatore, India

Hanu R. Pappu
Department of Plant Pathology
Washington State University
Pullman, Washington

Parameswaran, C.
ICAR-National Rice Research Institute
Cuttack, India

Patra, B. C.
ICAR-National Rice Research Institute
Cuttack, India

Deepak Pawar
ICAR - Directorate of Weed Research (DWR)
Jabalpur, India

Pramesh, D.
University of Agricultural Science
Raichur, India

Prashant, S. H.
ICAR-National Rice Research Institute
Cuttack, India

Jyoti Priya
Division of Plant Physiology
ICAR-Indian Agricultural Research Institute, Pusa
New Delhi, India

Raghu, S.
ICAR-National Rice Research Institute
Cuttack, India

Rajashekara, H.
ICAR-Vivekananda Parvatiya Krishi Anusandhan Sansthan
Almora, India

Soham Ray
Division of Plant Physiology
ICAR-Indian Agricultural Research Institute, Pusa
New Delhi, India

Sujoy Saha
ICAR-National Research Centre for Grapes
Pune, Maharashtra, India

Mahender S. Saharan
Division of Plant Pathology
ICAR-IARI
New Delhi, India

Samantaray, S.
ICAR-National Rice Research Institute
Cuttack, India

S. Selvakumaran
ASRB
New Delhi, India

Shashi Bhooshan Sharma
School of Veterinary and Life Sciences
Murdoch University
South Perth, Australia
and
Plant Biosecurity, Department of Primary Industries and Regional Development
Baron – Hay Court
South Perth, Australia

Sharma, T. R.
Division of Crop Science
Indian Council of Agricultural Research
New Delhi, India

Palaiyur Nanjappan Sivalingam
ICAR-National Institute of Biotic Stress Management
Raipur, Chhattisgarh, India

Aditi Tailor
Scientist B, Genetics and Tree Improvement Division
Arid Forest Research Institute
Jodhpur, Rajasthan, India

Ratna U. Thosar
ICAR-National Research Centre for Grapes
Pune, Maharashtra, India

Archana Watts
Division of Plant Physiology
ICAR-Indian Agricultural Research Institute, Pusa
New Delhi, India

Pragya Yadav
Division of Plant Physiology
ICAR-Indian Agricultural Research Institute, Pusa
New Delhi, India

Priya Yadav
ICAR-National Bureau of Plant Genetic Resources
New Delhi, India

Pranjal Yadava
Division of Plant Physiology
ICAR-Indian Agricultural Research Institute, Pusa
New Delhi, India

Fauzia Zarreen
Molecular Virology Laboratory, School of Life Sciences
Jawaharlal Nehru University
New Delhi, India

Ying Zhai
Department of Plant Pathology
Washington State University
Pullman, Washington

1 Nucleic Acid Interventions in Plant Disease Management

Fauzia Zarreen and Supriya Chakraborty
Jawaharlal Nehru University

1.1 INTRODUCTION

In his writing "An Essay on the Principle of Population," the 18th-century British economist Thomas Robert Malthus theorized that the geometrical growth of the human population would outpace the arithmetical increase in food production, eventually triggering a population die-off. Steady advances in agricultural practices and the ushering in of the green revolution did prevent the Malthusian catastrophe, but hunger remains a challenge even in the 21st century. From the earliest days of farming, plant disease and pest pose a serious threat to food security, causing an average of 20%–40% yield loss globally (https://www.weforum.org/agenda/2021/06/climate-change-insects-pests-crops agriculture/).

Losses from diseases have been minimized in the past using an integrated approach combining best agricultural practices, agrochemical use and plant breeding. The benefits of an integrated approach can be seen in the management of rust of wheat, a disease that caused huge losses across the world. Cultural practices such as removal of the sexual stage of this pathogen, chemical control such as foliar fungicides, and breeding programmes involving screening of local cultivars for resistance against rust have enabled containment of this disease. However, these efforts are not bereft of challenges. Pesticides have an undesirable effect on the environment, and pathogens may develop resistance to them. Furthermore, it is challenging and labour-intensive to select for genetic resistance against multiple diseases simultaneously. Dwindling agricultural land area, global warming, increased global trade that facilitates spread of the pathogen, and rapid emergence of new pathogens compound the challenge. The world population is projected to increase to 10 billion by 2050, and world crop production needs to increase by at least 60% to meet the nutritional needs (United Nations, Department of Economic and Social Affairs, Population Division, 2017). These observations reinforce the need for more research on plant–pathogen interactions and to develop efficacious, affordable and sustainable approaches to plant disease management.

Genetic improvement has been a central pillar of improved agricultural productivity. The selection of plants with desirable characteristics such as yield, nutrition and disease resistance has resulted in our modern crop varieties. Conventional plant

DOI: 10.1201/9781003403142-1

breeding and genetic engineering are two main approaches to producing crops with improved characteristics (Figure 1.1). Both of these approaches entail altering the genetic composition of plants to enhance their characteristics.

Disease-resistant crop varieties have been developed by plant breeding methods in the past. This process usually involves screening of varieties for resistance, followed by crossing the resistant variety with the elite variety, and selection of offsprings with the desired combination of characteristics of the two parents. Technological advances such as whole-genome sequencing and marker-assisted selection have practically revolutionized the screening process by enhancing the accuracy of breeding practices and saving time. However, genetic crosses and selection of segregant progeny remain labour-intensive and time-consuming processes. The non-availability of a source of genetic resistance for use in plant breeding programmes and inter-specific incompatibility can be another challenge. Often, undesirable traits are introduced along with the resistant trait. Furthermore, the introgression of a single gene for resistance via conventional breeding could facilitate pathogen evolution to resistance (Figure 1.1).

As opposed to conventional breeding, genetic engineering, which is the direct alteration of an organism's genetic material via biotechnology (Christou, 2013), has several advantages. First, genetic engineering allows for insertion, removal and modification at a precise location in the genome, facilitating the production of crops with desired agronomic qualities in fewer generations as compared to conventional breeding. Second, it enables the cross-species exchange of genetic material. As a result, the raw genetic materials that can be used in this procedure are not limited to the genes found within the species. Third, genetic engineering permits the introduction of new genes into vegetatively propagated (e.g., cassava) and polyploid crops with complex genetics (e.g., alfalfa and potatoes). To accelerate the introduction of improved or novel traits into important crop plants, we rely on the conventional transgenic approach or the more recent genome-editing technology. In conventional transgenic methods, a genetically modified (GM) crop variety is created by identifying a piece of DNA that confers the desired trait, such as a virus resistance gene; cloning the DNA into

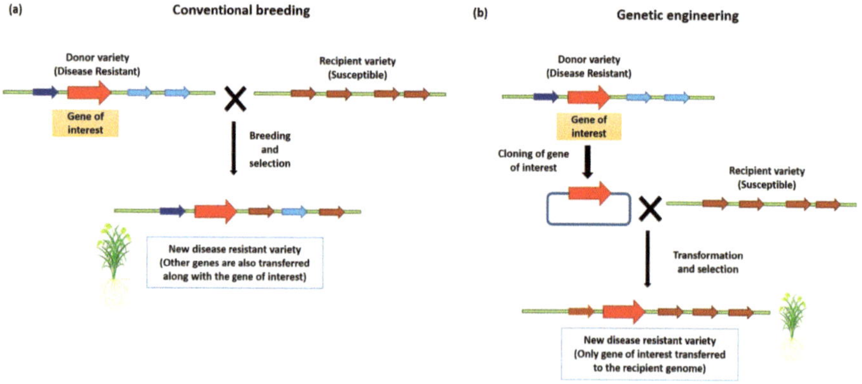

FIGURE 1.1 Strategy used for generating disease-resistant varieties using (a) conventional breeding and (b) genetic engineering.

the carrier or vector plasmid; delivering the DNA to the target plant, usually using *Agrobacterium*-mediated transformation; and generating modified plants with the desired trait (Lorence and Verpoorte, 2004). However, these GM crop varieties could contain undesired foreign DNA, such as the gene for antibiotic resistance. Genome editing, on the other hand, allows for alterations to endogenous plant DNA, such as deletions, insertions and substitutions of DNA of varying lengths at specific locations (Barrangou and Doudna, 2016). These precision genome-editing platforms include the clustered regularly interspaced short palindromic repeat (CRISPR)/CRISPR-associated protein (Cas), transcription activator-like effector nuclease (TALEN), Zn-finger nuclease and meganuclease methodologies. The genome-edited plants may or may not contain foreign DNA, depending on the sort of alterations used, and the ones that do not contain foreign DNA are not regulated as genetically modified organisms (GMOs) in various parts of the world. Furthermore, the development of genome-editing protocols that do not require transfer DNA (T-DNA) for the transfer of the genome-editing cassette into the crop plant and its removal via segregation has now enabled the development of transgene-free plants (He and Zhao, 2019). In addition to genome editing, technological breakthroughs such as speed breeding (Ghosh et al., 2018; Watson et al., 2018) next-generation genotyping (Nguyen et al., 2019), and phenotyping (Watt et al., 2020) have revolutionized crop improvement and plant disease management.

To suppress plant defence, pathogens produce elicitors called pathogen/microbe-associated molecular patterns (PAMP/MAMP), which include peptides, metabolites, cell wall components, enzymes and toxins defence (Boller and Felix, 2009; Dodds and Rathjen, 2010; Giraldo and Valent, 2013; Wirthmueller et al., 2013). The wounded host creates damage-associated molecular patterns (DAMP), which include plant signal molecules, after being attacked by a pathogen (Boller and Felix, 2009). Pattern recognition receptors (PRRs) present on the plasma membrane recognize these elicitors (PAMP/MAMP/DAMP), triggering downstream genes as the first line of defence, also known as PAMP/pattern-triggered immunity (PTI) (Jones and Dangl, 2006; Niks et al., 2015; Dodds and Rathjen, 2010). Particular microbes produce race-specific intracellular elicitors called effectors, encoded by specific avirulence (AVR) genes that suppress PTI. The effectors are perceived by specific plant receptors (R proteins), encoded by R genes. Recognition of pathogen effectors results in effector triggers in the effector-triggered immunity (ETI), which is basically manifested as induced programmed cell death at the infection site, known as the hypersensitive response (HR), and thus locally limits pathogen spread (Jones and Dangl, 2006; Spoel and Dong, 2012). ETI also induces the production of mobile immune signals, such as methyl salicylic acid (MeSA), azelaic acid and glycerol-3-phosphate (G3P), which are transported to uninfected tissues. These signals induce the accumulation of salicylic acid (SA), and SA mediates systemic acquired resistance (SAR) in the uninfected tissue, leading to massive transcriptional programming and activation of defence genes (Spoel and Dong, 2012; Fu and Dong, 2013). Plants also employ RNA interference (RNAi) to protect them from viral infections. Viral double-stranded RNA (dsRNA) is processed by Dicer-like (DCL) proteins, producing viral siRNAs. Viral siRNAs are loaded with Argonaute (AGO) proteins to guide viral RNA cleavages, thus restricting viral replication (Rosa et al., 2018). To counteract this, plant viruses produce viral suppressors of RNAi (Zamore, 2004; Csorba et al., 2015). Several aspects of host-microbial interactions provide genetic engineering opportunities for disease resistance (Dangl et al., 2013). For example, reprogramming important defence hub regulatory genes and susceptible host factors, engineering plants to encode antimicrobial peptides, natural/engineered immune receptors and inducing the host RNAi machinery can be used to fine-tune defence responses.

In this chapter, we discuss the genetic/nucleic acid alteration tools and solutions currently available for generating resistance against bacterial, viral, fungal and other pathogens. We evaluate these solutions in light of their efficacy and the benefits they offer. We highlight the success stories and challenges of the potential genetic engineering tool and attempt to describe the trend for the future.

1.2 GENETIC ENGINEERING APPROACH FOR PLANT DISEASE MANAGEMENT

1.2.1 RNA Interference

Experimental results have established that ectopic production of genes encoding wild-type or mutant viral proteins could disrupt the viral life cycle and confer tolerance to the virus and related strains of the virus (Lomonossoff, 1995), and this immunity is mediated by a natural gene regulation mechanism of the host, RNA interference (RNAi). RNAi plays a major role in antiviral defence in plants but is also used as a defence mechanism against other plant pathogens as well, examples of which are discussed later in the sections (Figure 1.2, Table 1.1).

RNAi involves sequence-specific recognition of mRNAs and then repressing their expression either by ribonuclease-directed degradation or by blocking their translation. The two main RNAi pathways in plants involve two types of small single-stranded RNAs (~21–24 nucleotides); small interfering RNA (siRNA) and microRNA (miRNA). While miRNAs are encoded by the plant genome and function mainly as regulators of gene expression, siRNAs are a key player in the RNA surveillance system that recognizes and degrades dsRNAs, or highly structured RNAs. As plant viruses replicate their genomes through dsRNA intermediates in the host cell and also use its machinery into translate their mRNAs in the cytoplasm, the siRNA surveillance system is used by the plant to degrade the viral RNA. For the detailed molecular mechanism of RNAi in antiviral defence, we refer the reader to other reviews (Wang et al., 2012; Katoch and Thakur, 2013; Galvez et al., 2014).

Plant viruses with single-stranded RNA as their genetic information replicate using RNA-dependent RNA polymerase, and dsRNA replicative intermediates often form during the process that triggers RNAi in the host (also known as co-suppression; Galvez et al., 2014). Thus, activation of RNAi by overexpression of viral RNA in transgenic plants has proven to be an effective approach for engineering resistance to viruses. This method has been used to create the majority of extant cases of genetically altered crops with viral disease resistance that have been approved for cultivation and consumption (Table 1.1). For example, transgenic squash (*Cucurbita* sp.) and papaya (*Carica papaya*) varieties with resistance to *Watermelon mosaic virus 2* (Tricoli et al., 1995) and *Papaya ringspot virus* (Gonsalves, 1998, 2006), respectively,

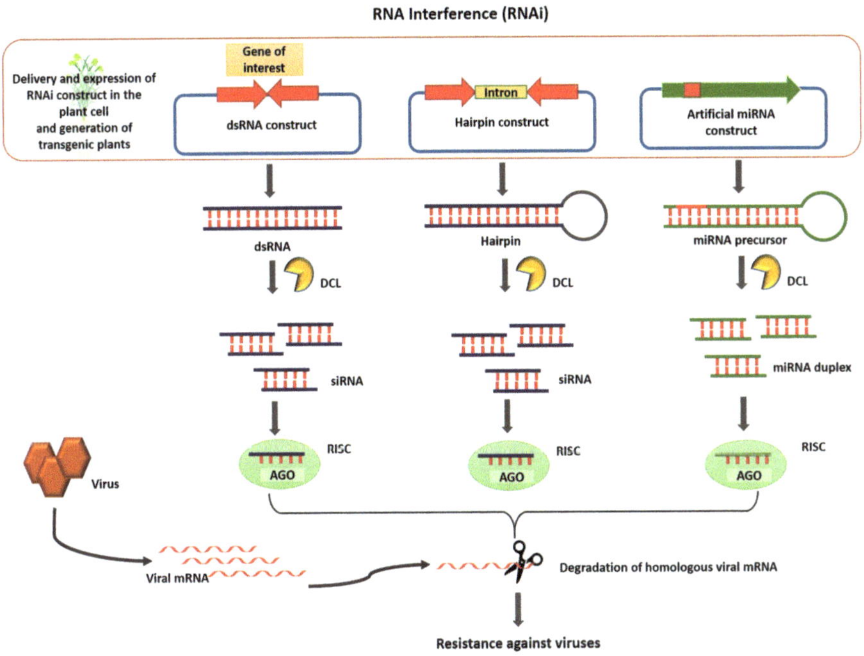

FIGURE 1.2 Overview of RNA interference and the RNAi construct double-stranded (ds) RNA, hairpin construct and artificial miRNA (amiRNA) used for generating viral disease resistance in plants.

created using this approach have been grown commercially in the United States for more than 20 years. According to field data, disease resistance obtained using RNAi-mediated methods is long-lasting and safe. The transgenic squash varieties resistant to multiple viruses have also been created using RNAi constructs (Tricoli et al., 1995; Table 1.1).

Expressing dsRNA using transgene-encoding inverted repeat sequences/hairpin construct is another effective strategy to induce RNAi, particularly in viruses with DNA genomes (Lindbo and Dougherty, 2005; Waterhouse et al., 1998). A commercially successful bean variety resistant to the DNA virus *Bean golden mosaic virus* (BGMV) has been developed using a hairpin construct where a part of the BGMV replication gene AC1 is directionally cloned into an intron-spliced hairpin RNA expression cassette in sense and antisense orientation (Bonfim et al., 2007, Aragão and Faria, 2009). The BGMV-resistant bean variety is now deregulated in Brazil and is grown by farmers in the field. Transgenic expression of an artificially constructed inverted repeat enables the generation of heterogeneous siRNAs that target different genes of the same or multiple viruses, leading to a broad-spectrum resistance. Hairpin RNAi constructs have been used to generate transgenic *N. benthamiana* lines that showed resistance to multiple begomoviruses associated with leaf curl disease in chilli. This study used RNAi constructs targeted at conserved regions of the viral genome and its associated betasatellite. Targeting this conserved region reduced the viral DNA accumulation up to 90% as compared to transgenic

TABLE 1.1
A Summary of the Studies on Genetic Engineering-Mediated Plant Disease Resistance

Genetic Engineering Technique	Disease	Pathogen	Host	Target Genes	References
RNA interference (RNAi)	Viral disease	*Watermelon mosaic virus 2 and zucchini yellow mosaic virus*	Watermelon, zucchini	Viral coat proteins	Tricoli et al. (1995)
	Viral disease	*Cucumber mosaic virus, watermelon mosaic virus 2 and zucchini yellow mosaic virus*	Cucumber, watermelon	Viral coat proteins	Tricoli et al. (1995)
	Viral disease	*Papaya ringspot virus*	Papaya	Viral coat proteins	Gonsalves (1998, 2006)
	Viral disease	*Potato leafroll virus*	Potato	Replicase and helicase	Thomas et al. (1997); Kaniewski and Thomas (2004)
	Viral disease	*Potato virus Y*	Potato	Viral coat proteins	Kaniewski and Thomas (2004)
	Viral disease	*Plum pox virus*	Plum	Coat protein	Ilardi and Tavazza (2015)
	Viral disease	*Bean golden mosaic virus*	Bean	RNA of viral replication protein	Bonfim et al. (2007)
	Viral disease	*Chilli leaf curl virus, Tomato leaf curl New Delhi*	*Nicotiana benthamiana*	Conserved promoter region of the viral genome	Sharma et al. (2015)
	Viral disease	*Rice tungro bacilliform virus, Rice tungro spherical virus*	Rice	Replicase gene of RTBV and RTSV	Sharma et al. (2018)
	Viral disease	*Tomato leaf curl virus, Cucumber mosaic virus*	Tomato, *Nicotiana tabacum*	Exogenous application of double-stranded RNA	Namgial et al. (2019)
	Viral disease	*Tomato leaf curl New Delhi virus*	*Nicotiana tabacum*	Artificial trans-acting small interfering RNA (tasiRNA)	Singh et al. (2015)

(Continued)

TABLE 1.1 (*Continued*)
A Summary of the Studies on Genetic Engineering-Mediated Plant Disease Resistance

Genetic Engineering Technique	Disease	Pathogen	Host	Target Genes	References
Zinc finger nucleases	Viral disease	*Beet severe curly top virus*	*Arabidopsis thaliana*	Blocking the binding of viral replication protein (Rep)	Sera (2005); Takenaka et al. (2007)
	Viral disease	*Tomato yellow leaf curl China virus* and *Tobacco curly shoot virus*		Blocking the binding of viral replication protein (Rep)	Chen et al. (2014)
Transcription activator-like effector nucleases (TALEN)	Rice bacterial blight	*Xanthomonas oryzae* pv. *oryzae*	Rice	Host S genes *OsSWEET14*	Li et al. (2012, 2013); Baufume et al. (2017)
	Citrus canker	*Xanthomonas citri* ssp. *citri*	Citrus	Host S genes *CsLOB1*	Hu et al. (2014); Jia et al. (2017)
	Powdery mildew	*Blumeria graminis* f.sp. *tritici*	Barley	Host S genes *TaMlo-A1*	Wang et al. (2014)
		Tomato yellow leaf curl China virus and *Tobacco curly shoot virus*, *Tomato leaf curl Yunnan virus*	*Nicotiana benthamiana*	Binding site of the replication-associated protein (Rep)	Cheng et al. (2015)
CRISPR/Cas-mediated genome editing		*Turnip mosaic virus*	*A. thaliana*, *N. benthamiana*	Virus RNA genome	Aman et al. (2018)
		Southern rice black-streaked dwarf virus, *Tobacco mosaic virus*	*N. benthamiana*, *Rice*	Virus RNA genome	Zhang et al. (2019)
		Cucumber mosaic virus, *Tobacco mosaic virus*	*A. thaliana*, *N. benthamiana*	Virus RNA genome	Zhang et al. (2018)

(*Continued*)

TABLE 1.1 (*Continued*)
A Summary of the Studies on Genetic Engineering-Mediated Plant Disease Resistance

Genetic Engineering Technique	Disease	Pathogen	Host	Target Genes	References
		Beet curly top virus, Merremia mosaic virus, Tomato yellow leaf curl virus	*N. benthamiana*	Virus DNA Genome Rep, IR and Cp	Ali et al. (2015)
		Bean yellow dwarf virus	*N. benthamiana*	Virus DNA Rep A/Rep and LIR	Baltes et al. (2015)
		Tomato yellow leaf curl virus	*N. benthamiana*	Virus DNA Rep, IR and Cp	Tashkandi et al. (2018)
		Cotton leaf curl Multan virus	*N. benthamiana*	Multiplex editing at Rep and IR	Yin et al. (2019)
		African cassava mosaic virus	*N. benthamiana*	AC2 and AC3	Mehta et al. (2019)
		Chilli leaf curl virus	*N. benthamiana*	Multiplex editing at virus DNA Rep, IR and Cp	Roy et al. (2019)
		Endogenous banana streak virus	Banana	Virus sequences in the host plantain genome	Tripathi et al. (2019)
		Clover yellow vein virus		*Eif4e1*	Bastet et al. (2019)
		Rice tungro spherical virus	Rice	*Eif4g*	Macovei et al. (2018)
		Cassava brown streak disease	Cassava	*nCBP-1, nCBP-2*	Gomez et al. (2019)

(*Continued*)

TABLE 1.1 (*Continued*)
A Summary of the Studies on Genetic Engineering-Mediated Plant Disease Resistance

Genetic Engineering Technique	Disease	Pathogen	Host	Target Genes	References
		Cucumber vein yellowing virus (ipomovirus), Zucchini yellow mosaic virus and Papaya ring spot mosaic virus-W (potyviruses)	Cucumber	eIF4E	Chandrasekaran et al. (2016)
	Powdery mildew	*Oidium neolycopersici*	Tomato	*SlMlo1*	Nekrasov et al. (2017)
	Powdery mildew	*Blumeria graminis f.*sp. *tritici*	Wheat	TaMlo-A1, TaMlo-B1 and TaMlo-D1	Wang et al. (2014)
	Bacterial blight	*Xanthomonas oryzae* pv. *Oryzae*	Rice	OsSWEET13	Kim et al. (2019)
	Bacterial speck disease	*Pseudomonas syringae*	Tomato	*Jaz2*	Ortigosa et al. (2019)
Cas13-independent guide-induced gene silencing (GIGS)	Viral disease	*Turnip mosaic virus (TuMV)*	*Nicotiana benthamiana*	Viral genome	Sharma et al. (2022)

lines (Sharma et al., 2015). RNAi has been successfully used to generate resistance individually against Rice tungro bacilliform virus (RTBV) and Rice tungro spherical virus (RTSV) by expressing different viral ORFs in sense and antisense orientations (Tyagi et al., 2008, Ganesan et al., 2009; Verma et al., 2012). The RTBV ORFIV transgene used for generating resistance against RTBV in tissue culture amenable rice variety Pusa Basmati-1 was subsequently introgressed in ASD 16, a popular rice variety in southern India (Valarmathi et al., 2016). Furthermore, a rice transgenic hairpin RNAi construct targeting both RTBV and RTSV simultaneously caused an approximately 100–500-fold reduction in the levels of both viruses as compared to non-transgenic lines (Sharma et al., 2018). A similar approach has now been used to confer resistance to cassava black streak disease (CBSD) in East and Central Africa. Cassava (*Manihot esculenta*) serves as a staple food crop for this region and is severely affected by the CBSD. CBSD is caused by two related viruses: *Cassava brown streak virus* (CBSV) and *Ugandan cassava brown streak virus* (UCBSV). Beyene et al. (2016) engineered the locally desirable cassava cultivar, TME 204, with transgenes to generate hairpin RNAs corresponding to the coat protein-coding regions of both viruses (Beyene et al., 2016; Wagaba et al., 2016). In May 2020, the Kenya Agricultural Research and Livestock Organization sought permission from the National Biosafety Authority for open field cultivation and marketing of CBSD-resistant cassava (https://cassavaplus.org/news/kalro-seeks-approval-to-introduce-disease-resistant-gm-cassava-2/). The miRNA machinery (Xie et al., 2015) has also been exploited in engineering resistance against viruses, particularly RNA viruses. Here, artificial miRNA with specific nucleotide changes (matching the viral mRNA) in the miRNA-encoding genes to alter targeting specificity is used for engineering resistance to a wide range of plant viruses. Artificial-miRNA-based antiviral strategies have been used against viral pathogens including *Turnip mosaic virus* (TuMV, Niu et al., 2006), *Cucumber mosaic virus* (CMV, Qu et al., 2007; Duan et al., 2008; Zhang et al., 2011), *Potato virus* X (PVX) and *Potato virus* Y (PVY, Ai et al., 2011). Antiviral methods based on artificial miRNA are very promising; however, they remain to be tested in the field. In one of the pioneering works, Singh et al. designed an RNAi vector using the features of trans-acting small interfering RNA (tasiRNA), which is a class of plant-specific endogenous siRNAs that repress gene expression via the cleavage of non-identical mRNA (Yoshikawa, 2013). The tasiRNA-based vector was used to target two genes of *Tomato leaf curl New Delhi virus* (ToLCNDV), and transgenic plants harbouring artificial tasiRNA showed low virus accumulation (Singh et al., 2015).

As RNAi targets mRNA, it can be used to control other plant pathogens, including oomycetes, fungi and nematodes. Unlike viruses, which replicate in the cytoplasmic nucleus depending on their nature, the cell cytoplasm of other pathogens, including oomycetes, fungi and nematodes, remains physically separated from that of the host. The exchange of small regulatory RNAs, including siRNA and miRNA, between plants and their pests (Cai et al., 2018, Hudzik et al., 2020) and their roles as mobile signals (Ham et al., 2017) within the plant have already been proven. Therefore, instead of targeting pathogen RNAs directly within the plant cell, as in the case of viruses, the RNAi inducers, both siRNA and miRNA, are transferred from the plant into the cells of eukaryotic plant pathogens, inducing RNAi within the pathogen as a plant disease control strategy. As the plant host transmits the RNAi inducer molecules to the pathogen, this is also known as host-induced gene silencing (HIGS) (Baulcombe et al., 2015). The transgenic HIGS approach has proven to be effective against the fungus *Blumeria graminis*, the causal agent of powdery mildew disease in many cereal crops. Host barley (*Hordeum vulgare*) plants were engineered to produce hairpin RNAs homologous to the mRNAs for genes encoding pathogenicity factors (effectors Avra10 and Avrk1) in *B. graminis* (Nowara et al., 2010). These transgenic plants showed reduced haustorium and fungal development upon inoculation with *B. graminis* (Nowara et al., 2010). A similar strategy was used against *Meloidogyne incognita*, a nematode that causes root knots in plants. *Arabidopsis thaliana* plants engineered to express a hairpin construct against *M. incognita* mRNA for the 16D10 peptide, a pathogenicity factor secreted into the plant cell, showed reduced gall development and *M. incognita* egg production (Huang et al., 2006). As prokaryotes lack RNAi, this approach has not been successful against bacterial plant pathogens. However, in one such example, RNAi hairpin constructs designed against the *iaa*M and *ipt* mRNAs, the *Agrobacterium tumefaciens* pathogenicity factors encoded by the T-DNA integrated into the plant cell, induce RNAi against the integrated T-DNA and reduce gall development (Escobar et al., 2001).

As discussed above, using transgenic plants to induce RNAi activity against certain pathogens can provide very effective disease control. However, new non-transgenic-based approaches, including bacterial endosymbiont-mediated RNAi (Whitten et al., 2006) and exogenous application of RNAi inducers to plants (Koch et al., 2016; Bennett et al., 2020; Dalakouras et al. 2016; Mitter et al., 2017), are also being explored. Exogenous application of dsRNA conferred resistance against Tomato leaf curl virus (ToLCV). Interestingly, this strategy, termed "RNA-based vaccination," has been used successfully two viruses: ToLCV, a DNA virus and Cucumber mosaic virus (CMV), a RNA virus (Namgial et al., 2019).

1.3 ZINC FINGER NUCLEASES

The first developed genome-editing tool, zinc finger nucleases (ZFNs), are a type of programmable DNA-binding protein that allows for targeted genome editing by causing double-strand breaks in DNA at the targeted locus. Each ZFN consists of two functional domains: a DNA-binding domain and a DNA-cleaving domain (Osakabe et al., 2010; Zhang et al., 2010; Carroll, 2011). The DNA-binding domain comprises a chain of two-finger modules, each recognizing a unique hexamer (6 bp) sequence of DNA. The two-finger modules are stitched together to form a Zinc Finger Protein, each with a specificity of 24 bp. The non-specific DNA-cleaving domain is comprised of the nuclease domain of *Fok* I (Curtin et al., 2011). When the DNA-binding and DNA-cleaving domains are fused, a highly specific pair of "genomic scissors" are created by substituting an alternative recognition hexamer sequence for the natural one (Curtin et al., 2012; Puchta and Fauser, 2013). ZFN-mediated genome editing takes place in the nucleus when a ZFN pair targeting the user's gene of interest is delivered into a parental cell line, either by transfection, electroporation or viral delivery.

The ZFN pair then recognizes and heterodimerizes around the target site, subsequently making double-stranded breaks at the target site. Double-stranded breaks stimulate the cell's natural DNA-repair processes, namely homologous recombination (HR) and non-homologous end joining (NHEJ). NHEJ repairs the genome break by joining the two ends of a DSB, which typically results in random indels of various lengths and can trigger frameshift mutations. HR requires exogenously supplied DNA sequences that are homologous to DSBs to direct precise DSB repair (Doudna and Charpentier, 2014; Abdelrahman et al., 2018; Qi et al., 2013). HR can replace a single nucleotide or extensive genomic areas, depending on the type of exogenously given DNA. As a result, precision gene editing or targeted GE achieved using NHEJ or HR could result in novel plant varieties with either agronomically relevant traits added or undesirable traits removed.

ZFNs have already been utilized for genome editing (Figure 1.3), despite their difficult modular structure in *Arabidopsis*, tobacco (*Nicotiana tabacum*), as well as crops including maize (*Zea mays*), soybean (*Glycine max*) and canola (*Brassica napus*) (Osakabe et al., 2010; Petolino et al., 2010; Townsend et al., 2009; Shukla et al., 2009; Curtin et al., 2011; Gupta et al., 2012; Ainley et al., 2013). However, only examples of ZFNs-engineered crop disease resistance exist. Previously, an artificial zinc finger protein (AZP) without a nuclease domain was designed to target the replication origin of the *Beet severe curly top virus* (BSCTV) by blocking the binding of viral replication protein (Rep). The strategy was effective, and transgenic AZP-expressing *Arabidopsis* plants showed increased resistance to BSCTV, and more than 80% of these transgenic plants exhibited no viral infection symptoms (Sera, 2005; Takenaka et al., 2007). In another example, the same strategy was used to target the Rep binding site of various begomoviruses, including *Tomato yellow leaf curl China virus* (TYLCCNV) and *Tobacco curly shoot virus* (TbCSV, Chen et al., 2014) (Table 1.1).

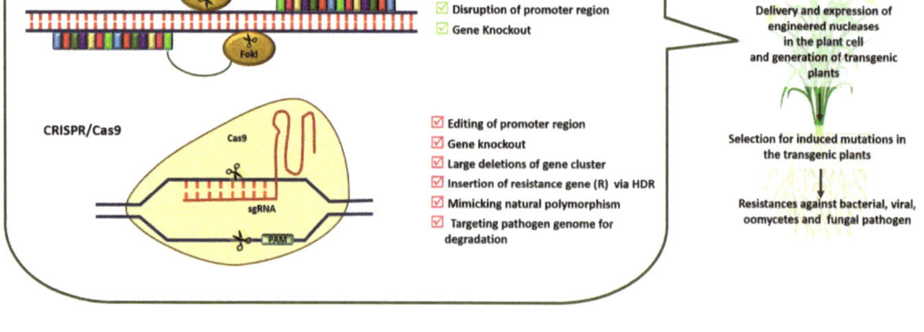

FIGURE 1.3 Overview of genome editing and its applications in generating resistance against various pathogens in plants.

1.4 TRANSCRIPTION ACTIVATOR-LIKE EFFECTOR NUCLEASES

TALENs were developed as an alternative to ZFNs for genome editing and introducing targeted DSBs. Transcription activator-like effectors (TALEs) are proteins secreted by *Xanthomonas* bacteria to alter the transcription of genes in host plant cells by binding to the DNA (Boch and Bonas, 2010). TALE-nucleases (TALEN) were then first conceptualized by fusing the catalytic domain of the restriction endonuclease *Fok*I to the customizable DNA-binding domain of TALEs. Similar to ZFNs, TALEs can be engineered to bind any desirable DNA sequence and introduce DNA breaks at any specific location, which can then be repaired by the cell's HR and NHEJ DNA-repair processes (Miller et al., 2011). The use of TALENs has been demonstrated mainly in engineering disease resistance against bacterial infection in plants (Figure 1.3, Table 1.1).

Rice bacterial blight caused by *Xanthomonas oryzae* pv. *oryzae* (*Xoo*) is a major threat to the production of rice around the world. *Xoo* secretes TALE proteins (AvrXa7, PthXo2) into host cells through the type III secretion system. These TALE proteins bind to the effector-binding element (EBE) in the promoter of target host S genes, *OsSWEET14* (also called *Os11N3*), and upregulate its expression to promote successful infection (Li et al., 2012, 2013). To confer resistance against *Xoo,* TALENs were used to target the EBE and disrupt the binding of AvrXa7 to *OsSWEET14*. As TALENs targeted only the EBE, it did not affect the developmental role of *OsSWEET14* (Li et al., 2012). The promoter of *OsSWEET14* contains multiple EBEs, which interact with different TALEs of geographically distant *Xoo* strains. TALEN-induced mutagenesis was used to study the role and contribution of multiple EBEs in disease development (Baufume et al., 2017). In another study, TALENs were used to target the EBE in the promoter of *CsLOB1*, an S gene for *Xanthomonas citri* ssp. *citri* (*Xcc*), which disrupted the binding of *Xcc* effector PthA4 and conferred resistance to canker disease (Hu et al., 2014; Jia et al., 2017).

The identification of barley *mlo* (*mildew resistance locus o*) mutants is a breakthrough in plant breeding for generating powdery mildew resistance. The common bread wheat has three orthologues of barley *Mlo* (*TaMlo-A1*, *B1* and *D1*). The editing of six copies of *TaMlo-A1* using TALEN conferred resistance against the powdery mildew fungus *Blumeria graminis* f.sp. *tritici* (*Bgt*) (Wang et al., 2014). TALENs have also been used to evaluate broad-spectrum resistance against begomoviruses. TALENs were used to alter the binding of the replication-associated protein (Rep) to the inverted repeat sequence present in the putative promoter region of begomoviruses. The TALEN-engineered *Nicotiana benthamiana* was resistant to TYLCCNV and TbCSV but displayed partial resistance to *Tomato leaf curl Yunnan virus* (TLCYnV, Cheng et al., 2015).

1.5 CRISPR/CAS9-MEDIATED GENOME EDITING

The CRISPR/Cas9 system is the latest entry to the list of genome-editing tools. Precision, robustness, ease of design and delivery both by transient and transgenic means make CRISPR/Cas9 an excellent and highly efficient genome-editing tool. Recent advances in CRISPR technology have enabled the development of a broad range of CRISPR variants with different applications. Here we discuss the CRISPR applications that have been used to successfully engineer disease resistance in plants (Table 1.1).

For details and working of CRISPR/Cas9 system, we refer the readers to other excellent reviews (Chen et al., 2019; Hanna and Doench, 2020). Briefly, CRISPR/Cas9 is a type II adaptive immune system in prokaryotes that provides defence against phages, where it induces site-specific DSBs in the invading phage DNA (Jinek et al., 2012). For genome editing, a short CRISPR-RNA (crRNA) complementary to the target genome is designed and cloned. The CRISPR locus is then delivered to the host cell using transient and transgenic means. The crRNA is transcribed and binds to a trans-activating RNA (tracrRNA) to form the mature guide RNA (gRNA). A Cas9 complex comprising gRNA, endonuclease Cas9 and RNase III is then formed. The gRNA guides this complex to the target genome and binds to it adjacent to the protospacer-associated motif (PAM) site. A DSB is induced three nucleotides upstream of PAM by Cas9, which is then repaired by cellular HR and NHEJ DNA-repair machinery (Wright et al., 2016; O'Connell et al., 2014). Several new CRISPR systems, including Cas12 (Zaidi et al., 2017), Cas13a (Aman et al., 2018; Abudayyeh et al., 2017), Cas13b (Smargon et al., 2017) and fCas9 (Guilinger et al., 2014), deactivated Cas9 (dCas9; Li et al., 2017; Piatek et al., 2015), RNA processing Cas9 (RCas9; Aman et al., 2018; Terns, 2018) and Cas9 fusion proteins such as Cas9-cytidine deaminase fusion (Shimatani et al., 2017), offer a wide range of CRISPR-based applications.

CRISPR-Cas9-induced DSB in DNA followed by NHEJ DNA repair results in the insertion or deletion (indel) of one or more nucleotides at the target site, which introduces a frameshift mutation. This strategy has been used to disrupt the plant susceptibility (S) gene in many cases (Figure 1.3), which alters the plant–pathogen interaction. For example, indel introduced CRISPR-Cas9 in eukaryotic translation initiation factor 4E proteins (eIF4Es), a translational factor that has redundant functions in plants, successfully conferred broad-spectrum resistance against potyviruses (*Zucchini yellow mosaic virus* and *Papaya ringspot mosaic virus*) and Ipomoviruses (*Cucumber vein yellowing virus*) in cucumber (*Cucumis sativus* L.). Potyviruses hijack cellular eIF4Es through virus-encoded protein (VPg), and mutations in this eIF4E gene impair the molecular interaction, leading to reduced viral replication (Chandrasekaran et al., 2016). By mimicking the natural polymorphism of *Pisum sativum* eIF4E alleles by CRISPR-Cas9, Bastet et al. (2019) aimed to broaden the resistance spectrum of the *A. thaliana* eIF4E1 gene, a susceptibility factor to the *Clover yellow vein virus* (ClYVV). The study found that ClYVV resistance in *A. thaliana* can be achieved with just one or two mutations in eIF4E without affecting plant growth and development (Bastet et al., 2019). Rice tungro disease (RTD), a major threat to rice production, is caused by an interaction between RTSV and RTBV. The translation initiation factor 4 gamma (eIF4G) gene controls RTSV resistance as a recessive characteristic. Mutations in eIF4G were induced using CRISPR/Cas9 in the RTSV-susceptible variety IR64, extensively grown in tropical Asia, which conferred resistance to RTSV (Macovei et al., 2018).

Besides targeting the plant susceptibility (S) gene, CRISPR-Cas9 has also been used to target and edit the viral genome, which renders the virus unfit for replication and transcription. The CRISPR/Cas9 system was used to target four locations in *Wheat dwarf virus* (WDV) genome and create WDV-resistant barley transgenic lines (Kis et al. 2019). Using a similar strategy, Ji et al. (2015) were able to reduce the accumulation of *beet severe curly top virus* (BSCTV) by up to 70% in *A. thaliana* and *N. benthamiana* (Ji et al., 2015). In another example, CRISPR/Cas9-induced mutations in the intergenic region (IR) of geminiviruses such as *Merremia mosaic virus* (MeMV), *Tomato yellow leaf curl virus* (TYLCV), *Beet curly top virus* (BCTV), and *Bean yellow dwarf virus* (BeYDV) led to a reduction in viral replication (Ali et al., 2016; Baltes et al., 2015). In another interesting example, CRISPR-Cas9 genome editing was used to prevent correct transcription and translation of integrated endogenous *Banana streak virus* (eBSV) in the B genome of plantain (AAB) (Tripathi et al., 2019). The discovery of Cas13 nucleases that specifically target single-stranded RNA in a CRISPR-RNA (crRNA)-guided manner opened up new avenues for selectively reducing mRNA and manipulating plant traits. CRISPR-Cas13 nucleases have been used to target RNA viruses such as TMV, CMV, TuMV, *Southern rice black-streaked dwarf virus* (SRBSDV) and *Potato virus Y* (PVY) (Zhang et al., 2018; Aman et al., 2018; Zhang et al., 2019; Zhan et al., 2019).

CRISPR/Cas9 genome editing has been an effective strategy to induce mutagenesis in the promoter regions of bacterial blight S genes OsSWEET14 and OsSWEET11 in rice (Jiang et al., 2013). Similar editing in the promoter region of the citrus canker S gene *canker susceptibility lateral organ boundaries* 1 (CsLOB1) helped develop resistance against canker disease (Jia et al., 2016). CRISPR/Cas9 induced targeted mutations in hexaploid wheat, *TaMLO-A1* allele, one of three homoeoalleles that encode Mildew-Resistance Locus (MLO) proteins, resulting in improved disease resistance against downy mildew pathogens (Wang et al., 2014). The editing of *MILDEW RESISTANT LOCUS O* (*Mlo*) in tomato conferred resistance to the powdery mildew fungal pathogen *Oidium neolycopersici* (Nekrasov et al., 2017). The CRISPR/Cas9-induced targeted mutations in the downy mildew resistance 6 (*SlDMR6*-1) gene in tomato conferred broad-spectrum resistance against different pathogens, including *Pseudomonas syringae* pv. *tomato*, *Phytophthora capsici* and *Xanthomonas* spp. (Thomazella et al., 2021). *P. syringae* pv. tomato (*Pto*) DC3000, the causal agent of tomato bacterial speck disease, produces coronatine (COR), which helps bacterial leaf colonization in stomatal guard cells by stimulating stomata opening. CRISPR-Cas9 was used to edit *SlJAZ2*, a major co-receptor of COR, to generate dominant JAZ2 repressors lacking the C-terminal Jas domain, which prevent stomatal opening by COR and confer resistance to *Pto* (Ortigosa et al., 2019).

CRISPR-Cas9 use has moved forward from engineering disease resistance in host plants to dissection of the function of genes in the pathogen involved in plant infection. For example, a CRISPR-Cas9-induced mutation was used to identify the role of RXLR effector gene *Avr4/6* in soybean, an effector of the oomycete pathogen of soybean, *Phytophthora sojae* (*P. sojae*). The point mutations G770V and G839W, as well as a new mutation, 5N837, in the oxysterol-binding protein-related protein-1 (ORP1) gene in *P. capsici* and *P. sojae* that confer resistance to the fungicide oxathiapiprolin, were also validated by GE utilizing CRISPR/Cas9 (Miao et al., 2018). In another example, Zhao et al. (2018) used CRISPR/Cas9 to test the Ptr gene, which is a constitutively expressed atypical resistance gene that offers broad-spectrum blast resistance. The mutant line M2354 was susceptible to *M. oryzae* after a two base pair (bp) deletion within the Ptr protein-coding region, implying that it plays a role in blast resistance (Zhao et al., 2018). These studies add to our understanding of the pathogenomics of other economically important infections by providing new paths for generating resistance.

Despite its enormous potential for plant genome editing, the efficiency of CRISPR-Cas9 system is limited by factors such as off-target activity, PAM specificity and design of sgRNA. Recent advances in technology, including changes in the design of sgRNA and the use of truncated sgRNA, have attempted to overcome these drawbacks. The development of Cas9 variants by introducing mutations into the PAM-interacting domains of wild-type SpCas9 that recognize alternative PAM sequences eliminates the range limitation of NGG-PAM (Kleinstiver et al., 2015). The discovery of Cas12a further expanded the target sites as it recognizes a T-rich Pam site (Zetsche et al., 2015). Whole-genome sequencing of mutants has revealed the off-target frequency of CRISPR-Cas9 is very low in plants (Nekrasov et al., 2017; Feng et al., 2014; Li et al., 2016; Zhang et al., 2016a; Zhang et al., 2014a,b; Zhang et al., 2016b). However, the development of robust methodologies to detect potential off-targets and to design sgRNA in plants, different bioinformatic tools, such as CasOT (Xiao et al., 2014), CRISPRMultiTargeter (Prykhozhij et al., 2015) and Cas-OFFinder (Bae et al., 2014), and new delivery methods, including preassembled complexes of purified Cas9 protein and guide RNA (ribonucleoprotein complexes or RNPs) into protoplasts, reduced off-targeting of CRISPR-Cas9 (Woo et al., 2015). Furthermore, new delivery methods, including the delivery of a mixture of Cas9-encoding mRNA and gRNA (Zhang et al., 2016a) or preassembled RNPs (Yin et al., 2017), would ensure DNA-free genome engineering. The result of a recent study showed the CRISPR guide RNA could induce gene silencing in plants in the absence of Cas nucleases, guide-induced gene silencing (GIGS). GIGS could reduce both viral and endogenous RNA. The results of GIGS could be stably inherited and could be elicited by both crRNAs designed for Cas9 and Cas13. GIGS therefore opens up avenues for generation of *cis*-genics for crop improvement (Sharma et al., 2022).

1.6 FUTURE PERSPECTIVE

Genome-editing techniques have a lot of potential for improving agricultural disease resistance. Our understanding of the plant–pathogen interaction has greatly been augmented with the development of a high-throughput platform for studying sequence polymorphism, expressional changes in proteins, protein modifications and protein–protein interactions.

We have a limited pool of known resistance (R) genes for the development of broad-spectrum and durable resistance. Hence, large-scale identification and functional analysis of new R genes and immune receptors would help enrich the pool of potential candidates for engineering resistance. RenSeq is a comparative genomics approach that can be used to find members of the *NLR* gene family that are involved in disease resistance. RenSeq captures and enriches genomic DNA fragments encoding *NLR*s from resistant germplasms using a library of *NLR* gene-specific DNA fragments as baits (Jupe et al., 2014). The polymorphisms in the *NLR* genes that could account for the observed disease resistance in the resistant germplasm of interest are identified by sequencing the enriched *NLR*-encoding DNA and comparing it to the reference genome (Jupe et al., 2013). For example, *NLR* gene *Rpi-amr3i* and *Rpi-ver1*, which confer resistance against *P. infestans* from wild potato relatives *Solanum americanum* and *Solanum verrucosum*, respectively, were rapidly identified and mapped using RenSeq (Witek et al., 2016; Chen et al., 2018). Similarly, four new rust resistance genes were identified in a wild progenitor species of bread wheat, *Aegilops tauschii* ssp. *Strangulate* using RenSeq (Arora et al., 2019). In addition to the identification of new immune receptors, new synthetic immune receptors with new ligand specificity have been engineered by the fusion of different pathogen recognition sites (PRS). A new chimeric functional *Xa21* for both rice and *Arabidopsis* was engineered by the fusion of Arabidopsis PRR EF-Tu receptor (EFR) to the intracellular domain of the phylogenetically related rice receptor *Xa21* (Holton et al., 2015; Schwessinger et al., 2015). To add to the repertoire of immune receptors, a CRISPR-based mutagenesis platform known as EvolvR was developed recently to introduce nucleotide diversification at specific genomic regions defined by guide RNAs. Variants thus generated can then be screened for resistance (Halperin et al., 2018; Sadanand, 2018).

Besides exploiting genetic variations, increasing evidence suggests that epigenetic variations can also be exploited for crop disease improvement. Increasing evidence has revealed that epigenetic processes such as DNA methylation, histone post-translational modifications, chromatin assembly and remodelling play key roles in the regulation of crop disease resistance against a wide range of phytopathogens, including viruses, bacteria, fungi, oomycetes, nematodes and herbivorous insects (Ding and Wang, 2015; Espinas et al., 2016; Zhu et al., 2016; Ramirez-Prado et al., 2018a, b; Wang et al., 2018; Alonso et al., 2019). Catalytically inactive Cas9 (dCas9) can be exploited to repurpose targeted gene regulation, and it can be used to recruit any major DNA modification domains or chromatin-remodelling complexes, which include histone acetylases and deacetylases, methylases and demethylases, DNA methylases and demethylases, kinases and phosphatases, and others, to enable targeted epigenetic changes to genomic DNA (Doudna and Charpentier, 2014).

There have been remarkable gains both in labs and in the field since the first report of a genetically engineered crop providing disease resistance in 1986, virus-resistant tobacco expressing a viral coat protein gene (Abel et al., 1986). With new technologies such as RenSeq and guided molecular evolution, which allow for the quick discovery of novel immune receptor genes, the pool of deployable genes for increased resistance to other microorganisms has grown significantly. Molecular stacking and targeted gene insertion by genome editing, on the other hand, are predicted to play a key role in building broad-spectrum resistance to viral and nonviral diseases soon. Furthermore, genome-editing methods such as CRISPR-Cas, which allow for a precise change of genome, will fill important knowledge gaps and can be used to develop new strategies for crop improvement. It is also important to develop a mechanism for field trials for promising edited plants to rule out environmental and health concerns.

REFERENCES

Abdelrahman M, Jogaiah S, Burritt DJ, and Tran LP. Legume genetic resources and transcriptome dynamics under abiotic stress conditions. *Plant Cell Environ.* 2018, 41: 1972–1983.

Abel PP, Nelson RS, De B, Hoffmann N, Rogers SG, Fraley RT, et al. Delay of disease development in transgenic plants that express the tobacco mosaic virus coat protein gene. *Science.* 1986, 232: 738–743.

Abudayyeh OO, Gootenberg JS, Essletzbichler P, Han S, Joung J, Belanto JJ, et al. RNA targeting with CRISPR-Cas13. *Nature.* 2017; 550: 280–284.

Ai T, Zhang L, Gao Z, Zhu CX, and Guo X. Highly efficient virus resistance mediated by artificial microRNAs that target the suppressor of PVX and PVY in plants. *Plant Biol (Stuttg).* 2011, 13: 304–316.

Ainley WM, Sastry-Dent L, Welter ME, Murray MG, Zeitler B, and Amora R. Trait stacking via targeted genome editing. *Plant Biotechnol J.* 2013, 11: 1126–1134.

Ali Z, Abulfaraj A, Idris A, Ali S, Tashkandi M, and Mahfouz MM. CRISPR/Cas9-mediated viral interference in plants. *Genome Biol.* 2015, 16(1): 238.

Ali Z, Ali S, Tashkandi M, Zaidi SS, and Mahfouz MM. CRISPR/Cas9-mediated immunity to geminiviruses: differential interference and evasion. Sci Rep. 2016, 6: 26912.

Alonso C, Ramos-Cruz D, and Becker C. The role of plant epigenetics in biotic interactions. *New Phytol.* 2019, 221: 731–737.

Aman R, Ali Z, Butt H, Mahas A, Aljedaani F, Khan MZ, et al. RNA virus interference via CRISPR/Cas13a system in plants. *Genome Biol.* 2018, 19(1): 1.

Aragão FJL, and Faria JC. First transgenic geminivirus-resistant plant in the field. *Nat Biotechnol.* 2009, 27: 1086–1088, author reply 1088–1089.

Arora S, Steuernagel B, Gaurav K, Chandramohan S, Long Y, Matny O, et al. Resistance gene cloning from a wild crop relative by sequence capture and association genetics. *Nat Biotechnol.* 2019, 37: 139–143.

Bae S, Park J, and Kim JS. Cas-OFFinder: a fast and versatile algorithm that searches for potential off-target sites of Cas9 RNA-guided endonucleases. *Bioinformatics* 2014, 30: 1473–1475. doi:10.1093/bioinformatics/btu048

Baltes NJ, Hummel AW, Konecna E, Cegan R, Bruns AN, Bisaro DM, et al. Conferring resistance to geminiviruses with the CRISPR/Cas prokaryotic immune system. *Nat Plants* 2015, 1(10): 15145. doi:10.1038/nplants.2015.145

Barrangou R, and Doudna JA. Applications of CRISPR technologies in research and beyond. *Nat Biotechnol.* 2016, 34: 933–941.

Bastet A, Zafirov D, Giovinazzo N, Guyon-Debast A, Nogué F, Robaglia C, et al. Mimicking natural polymorphism in eIF4E by CRISPR-Cas9 base editing is associated with resistance to potyviruses. *Plant Biotechnol J.* 2019. doi:10.1111/pbi.13096

Baufume SB, Reschke M, Sole M, Auguy F, Doucoure H, Szurek B, et al. Targeted promoter editing for rice resistance to Xanthomonas oryzae pv. oryzae reveals differential activities for SWEET14-inducing TAL effectors. *Plant Biotechnol J.* 2017, 15: 306–317. doi:10.1111/pbi.12613

Baulcombe DC. VIGS, HIGS and FIGS: small RNA silencing in the interactions of viruses or filamentous organisms with their plant hosts. *Curr Opin Plant Biol.* 2015, 26: 141–146.

Bennett M, Deikman J, Hendrix B, and Iandolino A. Barriers to efficient foliar uptake of dsRNA and molecular barriers to dsRNA activity in plant cells. *Front Plant Sci.* 2020, 11: 816.

Beyene G, Chauhan RD, Ilyas M, Wagaba H, Fauquet CM, Miano D, et al. A virus-derived stacked RNAi construct confers robust resistance to cassava brown streak disease. *Front Plant Sci.* 2016, 7, 2052.

Boch J, and Bonas U. Xanthomonas AvrBs3 family-type ø effectors: discovery and function. *Annu Rev Phytopathol.* 2010, 48: 419–436. doi:10.1146/annurev-phyto-080508-081936

Boller T, and Felix G. A renaissance of elicitors: perception of microbe-associated molecular patterns and danger signals by pattern-recognition receptors. *Annu Rev Plant Biol.* 2009, 60: 379–406.

Bonfim K, Faria JC, Nogueira EOPL, Mendes EA, and Aragão FJL. RNAi-mediated resistance to Bean golden mosaic virus in genetically engineered common bean (Phaseolus vulgaris). *Mol Plant Microbe Interact.* 2007, 20: 717–726.

Cai Q, Qiao L, Wang M, He B, Lin F-M, Palmquist J, et al. Plants send small RNAs in extracellular vesicles to fungal pathogen to silence virulence genes. *Science* 2018, 360(6393): 1126–1129.

Carroll D. Genome engineering with zinc-finger nucleases. *Genetics.* 2011, 188: 773–782. doi:10.1534/genetics.111.131433

Chandrasekaran J, Brumin M, Wolf D, Leibman D, Klap C, Pearlsman M, et al. Development of broad virus resistance in non-transgenic cucumber using CRISPR/Cas9 technology. *Mol Plant Pathol.* 2016, 17: 1140–1153. doi:10.1111/mpp.12375

Chen K, Wang Y, Zhang R, Zhang H, and Gao C. CRISPR/Cas genome editing and precision plant breeding in agriculture. *Annu Rev Plant Biol.* 2019, 70: 667–697.

Chen W, Qian Y, Wu X, Sun Y, Wu X, and Cheng X. Inhibiting replication of begomoviruses using artificial zinc finger nucleases that target viral-conserved nucleotide motif. *Virus Genes.* 2014, 48: 494–501. doi:10.1007/ s11262-014-1041-4

Chen X, Lewandowska D, Armstrong MR, Baker K, Lim TY, Bayer M, et al. Identification and rapid mapping of a gene conferring broad-spectrum late blight resistance in the diploid potato species Solanum verrucosum through DNA capture technologies. *Theor Appl Genet.* 2018, 131: 1287–1297.

Cheng X, Li F, Cai J, Chen W, Zhao N, Sun Y, et al. Artificial TALE as a convenient protein platform for engineering broad-spectrum resistance to begomoviruses. *Viruses.* 2015, 7: 4772–4782. doi:10.3390/v7082843

Christou P. Plant genetic engineering and agricultural biotechnology 1983–2013. *Trends Biotechnol.* 2013, 31: 125–127.

Csorba T, Kontra L, and Burgyán J. Viral silencing suppressors: tools forged to fine-tune host-pathogen coexistence. *Virology.* 2015, 479–480: 85–103. doi:10.1016/j.virol.2015.02.028

Curtin SJ, Voytas DF, and Stupar RM. Genome engineering of crops with designer nucleases. *Plant Genome.* 2012, 5: 42–50.

Curtin SJ, Zhang F, Sander JD, Haun WJ, Starker C, Baltes NJ, et al. Targeted mutagenesis of duplicated genes in soybean with zinc-finger nucleases. *Plant Physiol.* 2011, 156: 466–473. doi:10.1104/pp.111.172981

Dalakouras A, Wassenegger M, Mcmillan JN, Cardoza V, Maegele I, Dadami E, et al. Induction of silencing in plants by high-pressure spraying of in vitro-synthesized small RNAs. *Front Plant Sci.* 2016, 7: 1327.

Dangl JL, Horvath DM, and Staskawicz BJ. Pivoting the plant immune system from dissection to deployment. *Science.* 2013, 341: 746–751.

Ding B, and Wang GL. Chromatin versus pathogens: the function of epigenetics in plant immunity. *Front Plant Sci.* 2015, 6: 675. doi:10.3389/fpls.2015.00675

Dodds PN, and Rathjen JP. Plant immunity: towards an integrated view of plant-pathogen interactions. *Nat Rev Genet.* 2010, 11: 539–548.

Doudna JA, and Charpentier E. The new frontier of genome engineering with CRISPR/Cas9. *Science.* 2014, 346: 1258096. doi:10.1126/science.1258096

Duan CG, Wang CH, Fang RX, and Guo HS. Artificial microRNAs highly accessible to targets confer efficient virus resistance in plants. *J Virol.* 2008, 82: 11084–11095.

Escobar MA, Civerolo EL, Summerfelt KR, and Dandekar AM. RNAi-mediated oncogene silencing confers resistance to crown gall tumorigenesis. *Proc Natl Acad Sci USA.* 2008, 98(23): 13437–13442.

Espinas NA, Saze H, and Saijo Y. Epigenetic control of defense signaling and priming in plants. *Front Plant Sci*. 2016, 7: 205. doi:10.3389/fpls.2016.01201

Feng Z, Mao Y, Xu N, Zhang B, Wei P, Yang D-L, et al. Multigeneration analysis reveals the inheritance, specificity, and patterns of CRISPR/Cas-induced gene modifications in Arabidopsis. *Proc Natl Acad Sci USA*. 2014, 111: 4632–4637. doi:10.1073/pnas.1400822111

Fu ZQ, and Dong X. Systemic acquired resistance: turning local infection into global defense. *Annu Rev Plant Biol*. 2013, 64: 839–863.

Galvez LC, Banerjee J, Pinar H, and Mitra A. Engineered plant virus resistance. *Plant Sci*. 2014, 228: 11–25.

Ganesan U, Suri SS, Rajasubramaniam S, Rajam MV, and Dasgupta I. Transgenic expression of coat protein gene of Rice tungro bacilliform virus in rice reduces the accumulation of viral DNA in inoculated plants. *Virus Genes*. 2009, 39: 113–119.

Ghosh S, Watson A, Gonzalez-Navarro OE, Ramirez-Gonzalez RH, Yanes L, Mendoza-Suárez M, et al. Speed breeding in growth chambers and glasshouses for crop breeding and model plant research. *Nat Protoc*. 2018, 13(12): 2944–2963.

Giraldo MC, and Valent B. Filamentous plant pathogen effectors in action. *Nat Rev Microbiol*. 2013, 11: 800–814.

Gomez MA, Daniel Lin Z, Moll T, Deepika Chauhan R, Hayden L, Renninger K, et al. Simultaneous CRISPR/Cas9-mediated editing of cassava eIF4E isoforms nCBP-1 and nCBP-2 reduces cassava brown streak disease symptom severity and incidence. *Plant Biotechnol J*. 2019, 17(2): 421–434.

Gonsalves D. Control of papaya ringspot virus in papaya: a case study. *Annu Rev Phytopathol*. 1998, 36: 415–437.

Gonsalves D. Transgenic papaya: development, release, impact and challenges. *Adv Virus Res*. 2006, 67: 317–354.

Guilinger JP, Thompson DB, and Liu DR. Fusion of catalytically inactive Cas9 to FokI nuclease improves the specificity of genome modification. *Nat Biotechnol*. 2014, 32(6): 577–582.

Gupta M, DeKelver RC, Palta A, Clifford C, Gopalan S, Miller JC, et al. Transcriptional activation of Brassica napus beta-ketoacyl-ACP synthase II with an engineered zinc finger protein transcription factor. *Plant Biotechnol J*. 2012, 10: 783–791. doi:10.1111/j.1467-7652.2012.00695.x

Halperin SO, Tou CJ, Wong EB, Modavi C, Schaffer DV, and Dueber JE. CRISPR-guided DNA polymerases enable diversification of all nucleotides in a tunable window. *Nature*. 2018, 560: 248–252.

Ham BK, and Lucas WJ. Phloem-mobile RNAs as systemic signaling agents. *Annu Rev Plant Biol*. 2017, 68: 173–195.

Hanna RE, and Doench JG. Design and analysis of CRISPR-Cas experiments. *Nat Biotechnol*. 2020, 38(7): 813–823.

He Y, and Zhao Y. Technological breakthroughs in generating transgene-free and genetically stable CRISPR-edited plants. *aBIOTECH*. 2019, 1(1): 88–96.

Holton N, Nekrasov V, Ronald PC, and Zipfel C. The phylogenetically related pattern recognition receptors EFR and XA21 recruit similar immune signaling components in monocots and dicots. *PLoS Pathog*. 2015, 11: e1004602.

Hu Y, Zhang J, Jia H, Sosso D, Li T, Frommer WB, et al. Lateral organ boundaries 1 is a disease susceptibility gene for citrus bacterial canker. *Proc Nat Acad Sci USA*. 2014, 111: E521.

Huang G, Allen R, Davis EL, Baum TJ, and Hussey RS. Engineering broad root-knot resistance in transgenic plants by RNAi silencing of a conserved and essential root-knot nematode parasitism gene. *Proc Natl Acad Sci USA*. 2006, 103(39): 14302–14306.

Hudzik C, Hou Y, Ma W, and Axtell MJ. Exchange of small regulatory RNAs between plants and their pests. *Plant Physiol*. 2020, 182(1): 51–62.

Ji X, Zhang H, Zhang Y, Wang Y, and Gao C. Establishing a CRISPR-Cas-like immune system conferring DNA virus resistance in plants. *Nat Plants*. 2015, 1(10): 15144. doi:10.1038/nplants.2015.144

Jia H, Orbovic V, Jones JB, and Wang N. Modification of the PthA4 effector binding elements in type I CsLOB1 promoter using Cas9/sgRNA to produce transgenic Duncan grapefruit alleviating XccDpthA4: dCsLOB1.3 infection. *Plant Biotechnol J*. 2016, 14: 1291–1301. doi:10.1111/pbi

Jia H, Zhang Y, Orbović V, Xu J, White FF, Jones JB, et al. Genome editing of the disease susceptibility gene CsLOB1 in citrus confers resistance to citrus canker. *Plant Biotechnol J*. 2017, 15: 817–823. doi:10.1111/pbi.12677

Jiang W, Zhou H, Bi H, Fromm M, Yang B, Weeks DP, et al. Demonstration of CRISPR/Cas9/sgRNA-mediated targeted gene modification in Arabidopsis, tobacco, sorghum and rice. *Nucleic Acids Res*. 2013, 41: 188. doi:10.1093/nar/gkt780

Jinek M, Chylinski K, Fonfara I, Hauer M, Doudna JA, and Charpentier R. A programmable dual-RNA-guided DNA endonuclease in adaptive bacterial immunity. *Science*. 2012, 337(6096): 816–821.

Jones JD, and Dangl JL. The plant immune system. *Nature* 2006, 444: 323–329.

Jupe F, Chen X, Verweij W, Witek K, Jones JD, and Hein I. Genomic DNA library preparation for resistance gene enrichment and sequencing (RenSeq) in plants. *Methods Mol Biol*. 2014, 1127: 291–303.

Jupe F, Witek K, Verweij W, Sliwka J, Pritchard L, Etherington GJ, et al. Resistance gene enrichment sequencing (RenSeq) enables reannotation of the NB-LRR gene family from sequenced plant genomes and rapid mapping of resistance loci in segregating populations. *Plant J*. 2013, 76: 530–544.

Katoch R, and Thakur N. Advances in RNA interference technology and its impact on nutritional improvement, disease and insect control in plants. *Appl Biochem Biotechnol*. 2013, 169: 1579–1605.

Kim YA, Moon H, and Park CJ. CRISPR/Cas9-targeted mutagenesis of Os8N3 in rice to confer resistance to Xanthomonas oryzae pv. oryzae. *Rice (N Y)*. 2019, 12(1): 67.

Kis A, Hamar E, Tholt G, Bán R, and Havelda Z. Creating highly efficient resistance against wheat dwarf virus in barley by employing CRISPR/Cas9 system. *Plant Biotechnol J*. 2019, 17(6): 1004–1006. doi:10.1111/pbi.13077

Kleinstiver BP, Prew MS, Tsai SQ, Topkar VV, Nguyen NT, Zheng Z, et al. Engineered CRISPR-Cas9 nucleases with altered PAM specificities. *Nature*. 2015, 523: 481–485. doi:10.1038/nature14592

Koch A, Biedenkopf D, Furch A, Weber L, Rossbach O, Abdellatef E, et al. An RNAi-based control of Fusarium graminearum infections through spraying of long dsRNAs involves a plant passage and is controlled by the fungal silencing machinery. *PLoS Pathog*. 2016, 12(10): e1005901.

Li M, Li X, Zhou Z, Wu P, Fang M, Pan X, et al. Reassessment of the four yield-related genes Gn1a, DEP1, GS3, and IPA1 in rice using a CRISPR/Cas9 system. *Front Plant Sci*. 2016, 7: 377. doi:10.3389/fpls.2016.00377

Li T, Huang S, Zhou J, and Yang B. Designer TAL effectors induce disease susceptibility and resistanceto Xanthomonas oryzae pv. Oryzae in rice. *Mol Plant*. 2013. 6: 781–789. doi:10.1093/mp/sst034

Li T, Liu B, Spalding MH, Weeks DP, and Yang B. High-efficiency TALEN-based gene editing produces disease-resistant rice. *Nat Protocol*. 2012, 30: 5.

Li Z, Zhang D, Xiong X, Yan B, Xie W, Sheen J, et al. A potent Cas9-derived gene activator for plant and mammalian cells. *Nat Plants*. 2017, 3: 930–936.

Lindbo JA, and Dougherty WG. Plant pathology and RNAi: a brief history. *Annu Rev Phytopathol*. 2005, 43: 191–204.

Lomonossoff GP. Pathogen-derived resistance to plant viruses. *Annu Rev Phytopathol.* 1995, 33: 323–343.

Lorence A, and Verpoorte R. Gene transfer and expression in plants. *Methods Mol Biol.* 2004, 267: 329–350.

Macovei A, Sevilla NR, Cantos C, Jonson GB, Slamet-Loedin I, Èermák T, et al. Novel alleles of rice eIF4G generated by CRISPR/Cas9-targeted mutagenesis confer resistance to Rice tungro spherical virus. *Plant Biotechnol J.* 2018, 16: 1918–1927. doi:10.1111/pbi.12927

Mehta D, Stürchler A, Anjanappa RB, Zaidi SS-E-A, Hirsch-Hoffmann M, Gruissem W, et al. Linking CRISPR-Cas9 interference in cassava to the evolution of editing-resistant geminiviruses. *Genome Biol.* 2019, 20(1): 80.

Miao J, Chi Y, Lin D, Tyler BM, and Liu X. Mutations in ORP1 conferring oxathiapiprolin resistance confirmed by genome editing using CRISPR/Cas9 in Phytophthora capsici and P. sojae. *Phytopathology.* 2018, 108: 1412–1419. doi:10.1094/PHYTO-01-18-0010-R

Miller JC, Tan S, Qiao G, Barlow KA, Wang J, Xia DF, et al. A TALE nuclease architecture for efficient genome editing. *Nat Biotechnol.* 2011, 29: 143–148. doi:10.1038/nbt.1755

Mitter N, Worrall EA, Robinson KE, Li P, Jain RG, Taochy C, et al. Clay nanosheets for topical delivery of RNAi for sustained protection against plant viruses. *Nat Plants.* 2017, 3: 16207.

Namgial T, Kaldis A, Chakraborty S, and Voloudakis A. Topical application of double-stranded RNA molecules containing sequences of Tomato leaf curl virus and Cucumber mosaic virus confers protection against the cognate viruses. *Physiol Mol Plant Pathol.* 2019, 108: 101432.

Nekrasov V, Wang C, Win J, Lanz C, Weigel D, and Kamoun S. Rapid generation of a transgene-free powdery mildew resistant tomato by genome deletion. *Sci Rep.* 2017, 7: 482. doi:10.1038/s41598-017-00578-x

Nguyen KL, Grondin A, Courtois B, and Gantet P. Next-generation sequencing accelerates crop gene discovery. *Trends Plant Sci.* 2019, 24(3): 263–274.

Niks R, Li X, and Marcel TC. Quantitative resistance to biotrophic filamentous plant pathogens: concepts, misconceptions, and mechanisms. *Annu Rev Phytopathol.* 2015, 53: 445–470.

Niu QW, Lin SS, Reyes JL, Chen KC, Wu HW, Yeh SD, et al. Expression of artificial microRNAs in transgenic Arabidopsis thaliana confers virus resistance. *Nat Biotechnol.* 2006, 24: 1420–1428.

Nowara D, Gay A, Lacomme C, Shaw J, Ridout C, Douchkov D, et al. HIGS: host-induced gene silencing in the obligate biotrophic fungal pathogen Blumeria graminis. *Plant Cell.* 2010, 22(9): 3130–3141.

O'Connell MR, Oakes BL, Sternberg SH, East-Seletsky A, Kaplan M, and Doudna JA. Programmable RNA recognition and cleavage by CRISPR/Cas9. *Nature.* 2014, 516(7530): 263–266.

Ortigosa A, Gimenez-Ibanez S, Leonhardt N, and Solano R. Design of a bacterial speck resistant tomato by CRISPR/Cas9-mediated editing of SlJAZ2. *Plant Biotechnol J.* 2019, 17: 665–673. doi:10.1111/pbi.13006

Osakabe K, Osakabe Y, and Toki S. Site-directed mutagenesis in Arabidopsis using custom-designed zinc finger nucleases. *Proc Natl Acad Sci USA.* 2010, 107: 12034–12039. doi:10.1073/pnas.1000234107

Petolino JF, Worden A, Curlee K, Connell K, Connell J, Strange Moynahan TL, et al. Zinc finger nuclease-mediated transgene deletion. *Plant Mol Biol.* 2010, 73: 617–628. doi:10.1007/s11103-010-9641-4

Piatek A, Ali Z, Baazim H, Li L, Abulfaraj A, Al-Shareef S, et al. RNA-guided transcriptional regulation in planta via synthetic dCas9-based transcription factors. *Plant Biotechnol J.* 2015, 13(4): 578–589.

Prykhozhij SV, Rajan V, and Gaston Berman JN. CRISPR multitargeter: a web tool to find common and unique CRISPR single guide RNA targets in a set of similar sequences. *PLoS One*. 2015, 10: e0119372. doi:10.1371/journal.pone.0119372

Puchta H, and Fauser F. Gene targeting in plants: 25 years later. *Int J Dev Biol*. 2013, 57: 629–637. doi:10.1387/ijdb.130194hp

Qi Y, Zhang Y, Zhang F, Baller JA, Cleland SC, Ryu Y, et al. Increasing frequencies of site-specific mutagenesis and gene targeting in Arabidopsis by manipulating DNA repair pathways. *Genome Res*. 2013, 23: 547–554. doi:10.1101/gr.145557.112

Qu J, Ye J, and Fang R. Artificial microRNA-mediated virus resistance in plants. *J Virol*. 2007, 81: 6690–6699.

Ramirez-Prado JS, Abulfaraj AA, Rayapuram N, Benhamed M, and Hirt H. Plant immunity: from signaling to epigenetic control of defense. *Trends Plant Sci*. 2018a, 23: 833–844. doi:10.1016/j.tplants.2018.06.004

Ramirez-Prado JS, Piquerez SJM, Bendahmane A, Hirt H, Raynaud C, and Benhamed M. Modify the histone to win the battle: chromatin dynamics in plant-pathogen interactions. *Front Plant Sci*. 2018b, 9: 355. doi:10.3389/fpls.2018.00355

Rosa C, Kuo YW, Wuriyanghan H, and Falk BW RNA interference mechanisms and applications in plant pathology. *Annu Rev Phytopathol*. 2018, 56: 581–610.

Roy A, Zhai Y, Ortiz J, Neff M, Mandal B, Mukherjee SK, et al. Multiplexed editing of a begomovirus genome restricts escape mutant formation and disease development. *PLoS One*. 2019, 14(10): e0223765.

Sadanand S. EvolvR-ing to targeted mutagenesis. *Nat Biotechnol*. 2018, 36: 819.

Schwessinger B, Bahar O, Thomas N, Holton N, Nekrasov V, Ruan D, et al. Transgenic expression of the dicotyledonous pattern recognition receptor EFR in rice leads to ligand-dependent activation of defense responses. *PLoS Pathog*. 2015, 11: e1004809.

Sera T. Inhibition of virusDNA replication by artificial zinc finger proteins. *J Virol*. 2005, 79: 2614–2619. doi:10.1128/jvi.79.4.2614-2619.2005

Sharma S, Kumar G, and Dasgupta I. Simultaneous resistance against the two viruses causing rice tungro disease using RNA interference. *Virus Res*. 2018, 255: 157–164.

Sharma VK, Basu S, and Chakraborty S. RNAi mediated broad-spectrum transgenic resistance in Nicotiana benthamiana to chilli-infecting begomoviruses. *Plant Cell Rep*. 2015, 34(8): 1389–1399.

Sharma VK, Marla S, Zheng W, Mishra D, Huang J, Zhang W, et al. CRISPR guides induce gene silencing in plants in the absence of Cas. *Genome Biol*. 2022, 23: 1–24.

Shimatani Z, Kashojiya S, Takayama M, Terada R, Arazoe T, Ishii H, et al. Targeted base editing in rice and tomato using a CRISPR-Cas9 cytidine deaminase fusion. *Nat Biotechnol*. 2017, 35: 441–443.

Shukla VK, Doyon Y, Miller JC, DeKelver RC, Moehle EA, Worden SE, et al. Precise genome modification in the crop species Zea mays using zinc-finger nucleases. *Nature*. 2009, 459: 437–441. doi:10.1038/nature07992

Singh A, Taneja J, Dasgupta I, and Mukherjee SK. Development of plants resistant to tomato geminiviruses using artificial trans-acting small interfering RNA. *Mol Plant Pathol*. 2015, 16: 724–734.

Smargon AA, Cox DBT, Pyzocha NK, Zheng K, Slaymaker IM, Gootenberg JS, et al. Cas13b is a type VI-B CRISPR-associated RNA-guided RNase differentially regulated by accessory proteins Csx27 and Csx28. *Mol Cell*. 2017, 65(4): 618-30.e7.

Spoel SH, and Dong X. How do plants achieve immunity? Defence without specialized immune cells. *Nat Rev Immunol*. 2012, 12: 89–100.

Takenaka K, Koshimo-kimura Y, Aoyama Y, and Sera T. Inhibition of tomato yellow leaf curl virus replication by artificial zinc-finger proteins. *Nucleic Acids Symp Ser*. 2007, 51: 429–430. doi:10.1093/nass/nrm215

Tashkandi M, Ali Z, Aljedaani F, Shami A, and Mahfouz MM. Engineering resistance against Tomato yellow leaf curl virus via the CRISPR/Cas9 system in tomato. *Plant Signal Behav*. 2018, 13(10): e1525996.

Terns MP. CRISPR-based technologies: impact of RNA-targeting systems. *Mol Cell*. 2018, 72(3): 404–412.

Thomazella DP, Seong K, Mackelprang R, Dahlbeck D, Geng Y, Gill US, Qi T, Pham J, Giuseppe P, Lee CY, Ortega A, Cho MJ, Hutton SF, and Staskawicz B. Loss of function of a DMR6 ortholog in tomato confersbroad-spectrum disease resistance. *PNAS*. 2021, 118:1–11.

Townsend JA, Wright DA, Winfrey RJ, Fu F, Maeder ML, Joung JK, et al. High-frequency modification of plant genes using engineered zinc-finger nucleases. *Nature*. 2009, 459: 442–445. doi:10.1038/nature07845

Tricoli DM, Carney KJ, Russell PF, Mcmaster JR, Groff DW, Hadden KC, et al. Fieldevaluation of transgenic squash containing single or multiple virus coat protein gene constructs for resistance to cucumber mosaic virus, watermelon mosaic virus 2, and zucchini yellow mosaic virus. *Biotechnology* (N Y). 1995, 13: 1458–1465.

Tripathi J, Ntui V, Ron M, Muiruri S, Britt A, and Tripathi L. CRISPR/Cas9 editing of endogenous banana streak virus in the B genome of Musa spp. overcomes a major challenge in banana breeding. *Commun Biol*. 2019, 2: 46. doi:10.1038/s42003-019-0288-7.

Tyagi H, Rajasubramaniam S, Rajam MV, and Dasgupta I. RNA-interference in rice against Rice tungro bacilliform virus results in its decreased accumulation in inoculated rice plants. *Transgenic Res*. 2008, 17: 897–904.

Valarmathi P, Kumar G, Robin S, Manonmani S, Dasgupta I, and Rabindran R. Evaluation of virus resistance and agronomic performance of rice cultivar ASD 16 after transfer of transgene against Rice tungro bacilliform virus by backcross breeding. *Virus Genes*. 2016, 52: 521–529.

Verma V, Sharma S, Devi SV, Rajasubramaniam S, and Dasgupta I. Delay in virus accumulation and low virus transmission from transgenic rice plants expressing rice tungro spherical virus RNA. *Virus Genes*. 2012, 45: 350–359.

Wagaba H, Beyene G, Aleu J, Odipio J, Okao-Okuja G, Chauhan RJ, et al. Field level RNAi-mediated resistance to cassava brown streak disease across multiple cropping cycles and diverse East African agro-ecological locations. *Front Plant Sci*. 2016, 7: 2060.

Wang C, Wang C, Xu W, Zou J, Qiu Y, Kong J, et al. Epigenetic changes in the regulation of Nicotiana tabacum response to cucumber mosaic virus infection and symptom recovery through single-base resolution methylomes. *Viruses*. 2018, 10: 402. doi:10.3390/v10080402

Wang MB, Masuta C, Smith NA, and Shimura H. RNA silencing and plant viral diseases. *Mol Plant Microbe Interact*. 2012, 25: 1275–1285.

Wang Y, Cheng X, Shan Q, Zhang Y, Liu J, Gao C, et al. Simultaneous editing of three homoeoalleles in hexaploid bread wheat confers heritable resistance to powdery mildew. *Nat Biotechnol*. 2014, 32: 947–951.

Waterhouse PM, Graham MW, and Wang MB. Virus resistance and gene silencing in plants can be induced by simultaneous expression of sense and antisense RNA. *Proc Natl Acad Sci USA*. 1998, 95: 13959–13964.

Watson A, Ghosh S, Williams MJ, Cuddy WS, Simmonds J, Rey M-D, et al. Speed breeding is a powerful tool to accelerate crop research and breeding. *Nat Plants*. 2018, 4(1): 23–29.

Watt M, Fiorani F, Usadel B, Rascher U, Muller O, and Schurr U. Phenotyping: new windows into the plant for breeders. *Annu Rev Plant Biol*. 2020, 71: 689–712. doi:10.1146/annurev-arplant-042916-041124

Whitten MM, Facey PD, Del Sol R, Fernández-Martínez LT, Evans MC, Mitchell JJ, et al. Symbiont-mediated RNA interference in insects. *Proc Biol Sci*. 2016, 283(1825): 20160042.

Wirthmueller L, Maqbool A, and Banfield MJ. On the front line: structural insights into plant-pathogen interactions. *Nat Rev Microbiol.* 2013, 11: 761–776.

Witek K, Jupe F, Witek AI, Baker D, Clark MD, and Jones JDG. Accelerated cloning of a potato late blight-resistance gene using RenSeq and SMRT sequencing. *Nat Biotechnol.* 2016, 34: 656–660.

Woo JW, Kim J, Kwon SI, Corvalan C, Cho SW, Kim H, et al. DNA-free genome editing in plants with preassembled CRISPRCas9 ribonucleoproteins. *Nat Biotechnol.* 2015, 33: 1162–1164. doi:10.1038/nbt.3389

Wright AV, Nunez JK, and Doudna JA. Biology and applications of CRISPR systems: harnessing nature's toolbox for genome engineering. *Cell.* 2016, 164(1–2): 29–44.

Xiao A, Cheng Z, Kong L, Zhu Z, Lin S, Gao G, et al. CasOT: a genome-wide Cas9/gRNA off-target searching tool. *Bioinformatics.* 2014, 30: 1180–1182. doi:10.1093/bioinformatics/btt764

Xie M, Zhang S, and Yu B. microRNA biogenesis, degradation and activity in plants. *Cell Mol Life Sci.* 2015, 72: 87–99.

Yin K, Gao C, and Qiu J. Progress and prospects in plant genome editing. *Nat Plants.* 2017, 3: 17107. doi:10.1038/nplants.2017.107

Yin K, Han T, Xie K, Zhao J, Song J, and Liu Y. Engineer complete resistance to Cotton Leaf Curl Multan virus by the CRISPR/Cas9 system in Nicotiana benthamiana. *Phytopathol Res.* 2019, 1: 9.

Yoshikawa M. Biogenesis of trans-acting siRNAs, endogenous secondary siRNAs in plants. *Genes Genetic Syst.* 2013, 88: 77–84.

Zaidi SS, Mahfouz MM, and Mansoor S. CRISPR-Cpf1: a new tool for plant genome editing. *Trends Plant Sci.* 2017, 22(7): 550–553.

Zamore PD. Plant RNAi: how a viral silencing suppressor inactivates siRNA. *Curr Biol.* 2004, 14: R198–R200.

Zetsche B, Gootenberg JS, Abudayyeh OO, Slaymaker IM, Makarova KS, Essletzbichler P, et al. Cpf1 Is a single RNA-guided endonuclease of a Class 2 CRISPR-Cas system. *Cell.* 2015, 163: 759–771. doi:10.1016/j.cell.2015.09.038

Zhan X, Zhang F, Zhong Z, Chen R, Wang Y, Chang L, et al. Generation of virus-resistant potato plants by RNA genome targeting. *Plant Biotechnol J.* 2019, 17(9): 1814–1822.

Zhang XP, Liu DS, Yan T, Fang XD, Dong K, Xu J, et al. Cucumber mosaic virus coat protein modulates the accumulation of 2b protein and antiviral silencing that causes symptom recovery *in planta*. PLoS Pathog 2017, 13(7): e1006522. doi:-org/10.1371/journal.ppat.1006522

Zhang D, Li Z, and Li JF. Targeted gene manipulation in plants using the CRISPR/Cas technology. *J Genet Genom.* 2016a, 43: 251–262. doi:10.1016/j.jgg.2016.03.001

Zhang F, Maeder ML, Unger-Wallace E, Hoshaw JP, Reyon D, Christian M, et al. High frequency targeted mutagenesis in Arabidopsis thaliana using zinc finger nucleases. *Proc Natl Acad Sci USA.* 2010, 107: 12028–12033. doi:10.1073/pnas.0914991107

Zhang F, Wen Y, and Guo X. CRISPR/Cas9 for genome editing: progress, implications and challenges. *Hum Mol Genet.* 2014a, 15: 23. doi:10.1093/hmg/ddu125

Zhang H, Zhang J, Wei P, Zhang B, Gou F, Feng Z, et al. The CRISR/Cas9 system produces specific and homozygous targeted gene editing in rice in one generation. *Plant Biotechnol J.* 2014b, 12: 797–807.

Zhang T, Zhao Y, Ye J, Cao X, Xu C, Chen B, et al. Establishing CRISPR/Cas13a immune system conferring RNA virus resistance in both dicot and monocot plants. *Plant Biotechnol J.* 2019, 17(7): 1185–1187.

Zhang T, Zheng Q, Yi X, An H, Zhao Y, Ma S, et al. Establishing RNA virus resistance in plants by harnessing CRISPR immune system. *Plant Biotechnol J.* 2018, 16(8): 1415–1423.

Zhang X, Li H, Zhang J, Zhang C, Gong P, Ziaf K, et al. Expression of artificial microRNAs in tomato confers efficient and stable virus resistance in a cell-autonomous manner. *Transgenic Res.* 2011, 20: 569–581.

Zhang Y, Liang Z, Zong Y, Wang Y, Liu J, and Chen K. Efficient and transgene-free genome editing in wheat through transient expression of CRISPR/Cas9 DNA or RNA. *Nat Commun.* 2016b, 7: 12617.

Zhao H, Wang X, Jia Y, Minkenberg B, Wheatley M, and Fan J. The rice blast resistance gene Ptr encodes an atypical protein required for broad spectrum disease resistance. *Nat Commun.* 2018, 9: 2039.

Zhu QH, Shan WX, Ayliffe MA, and Wang MB. Epigenetic mechanisms: an emerging player in plant-microbe interactions. *Mol Plant Microbe Interact.* 2016, 29: 187–196.

2 Signal Transduction of *R* Gene-Mediated Plant Resistance against Pathogens

Lessons for Strategic Management of Diseases

Aditi Tailor
Arid Forest Research Institute

Satish C. Bhatla
Ministry of Agriculture and Farmers Welfare

2.1 INTRODUCTION

Biotic stresses, including plant diseases caused by pests and pathogens, continue to cause huge losses to the global economy in terms of a massive reduction in crop harvest (Savary et al., 2019). Major pests and pathogens of plants include members of diverse groups, such as bacteria, fungi, oomycete, insects and nematods. In addition to the diverse pathogens, another major challenge for sustainable crop production is the impact of climate change (Savary et al., 2019; Zayan, 2019; Skendžić et al., 2021). It is projected that with increasing temperatures due to global warming, even larger outbreaks of insects, diseases and weeds are expected worldwide (Zayan, 2019; Skendžić et al., 2021). This will have devastating effects on crop yield, thereby posing a great threat to food security (Zayan, 2019; Skendžić et al., 2021). It is thus of paramount importance to find ways for efficient management of plant diseases by harnessing the natural resistance strategies operative in plants.

Plants have developed versatile strategies to combat biotic stress factors that rely on preformed defense or induced defense responses, which, in turn, entail a complex crosstalk among different signaling pathways (Jones and Dangl, 2006; Dodds and Rathjen, 2010; Zipfel, 2014; Couto and Zipfel, 2016; Jones et al., 2016; Saijo et al., 2018). Defense responses are triggered upon recognition of specific factors of the invading pathogen by the corresponding elements in the host plant. These are encoded by the avirulence (*Avr*) and resistance (*R*) genes, respectively (Jones and Dangl, 2006; Dodds and Rathjen, 2010). *R* genes are crucial for providing resistance

DOI: 10.1201/9781003403142-2

to pathogens in plant hosts by recognizing cognate pathogenic *Avr*-dependent effectors. Following recognition of pathogens, *R* gene-encoded products or R proteins mediate signaling cascade(s) to coordinate specific defense responses in order to disable invading pathogens (Jones and Dangl, 2006; Dodds and Rathjen, 2010; Jones et al., 2016). A plethora of *R* genes have been identified and isolated from different plant systems, with the nucleotide-binding leucine-rich repeat proteins, or NLRs, representing the major class of R proteins (Kourelis and van der Hoorn, 2018; Kourelis et al., 2021). Identification of *R* genes and the functionally-related components of plant defense response and underlying signal transduction has enabled researchers to exploit them for competent introgression of disease resistance in plants (Pandolfi et al., 2017; van Wersch et al., 2020). This has led to the development of varieties with improved resistance to specific, or in some cases, a broader range of pathogens. More recently, specifically, engineering plants with improved disease resistance have been developed using modern biotechnological tools (Dong and Ronald, 2019; van Esse et al., 2020; van Wersch et al., 2020).

The use of resistance genes has been practiced for decades by plant breeders, albeit not fully knowing about their true nature, and this has rendered great benefits to crop production (Pandolfi et al., 2017). Particularly, introgression of *R* genes corresponding to various pathogen effectors has been effective in providing resistance to diseases (Pandolfi et al., 2017; Dong and Ronald, 2019; van Esse et al., 2020; van Wersch et al., 2020). Isolation of more such resistance determinants and understanding the yet elusive molecular mechanisms of *R* gene-mediated resistance is likely to provide acceleration to the development of new regimes for resistance transfer for an extended durability and recognition spectrum. This would promote utilization of *R* genes for imparting resistance in plants. A lot has been deciphered, but a lot remains to be discovered regarding the intricacies of *R* gene signaling. The present review attempts to collate the existing knowledge regarding *R* gene-mediated resistance signaling and underlying signal transduction pathways.

2.2 *R* GENE-MEDIATED SIGNALING FOR THE INDUCTION OF EFFECTOR-TRIGGERED IMMUNITY

Plants have evolved multifaceted immune responses for recognizing pathogens and further activation of defense mechanisms to prevent or minimize pathogen invasion (Jones and Dangl, 2006; Dodds and Rathjen, 2010; Jones et al., 2016). The two major forms of immune responses include Pathogen-Associated Molecular Patterns (PAMPs)-triggered immunity (PTI) and effector-triggered immunity (ETI) (Jones and Dangl, 2006; Zipfel, 2014; Couto and Zipfel, 2016; Saijo et al., 2018). PTI is induced after the recognition of conserved pathogen-associated molecules (or PAMPs) by the host cell surface receptors, i.e., pattern recognition receptors (PRRs) (Jones and Dangl, 2006; Zipfel, 2014; Couto and Zipfel, 2016; Saijo et al., 2018). However, the impact of PTI is more general, and it induces a milder response. For successful invasion, pathogens also deliver some factors, known as effectors, directly into the plant cell to evade PTI. These effectors are encoded by *Avr* genes in pathogen genomes and work to alter the components of plant physiology or suppress plant defense to attenuate PTI in order to increase the success of invasion and virulence (Asai and Shirasu, 2015; Toruño et al., 2016). Plants have also evolved a second tier of immunity, i.e., the ETI, elicited following effector recognition, either directly or indirectly, by the cognate intracellular receptors, the R proteins (Jones and Dangl, 2006; Dodds and Rathjen, 2010; Jones et al., 2016). ETI is a more rapid and stronger immune response and often terminates in the development of a hypersensitive response (HR) (Jones and Dangl, 2006; Dodds and Rathjen, 2010; Jones et al., 2016). HR is accompanied by an oxidative burst due to generation of excess amounts of reactive oxygen species (ROS), accumulation of salicylic acid (SA) and other pathogen-disruptive compounds, and transcriptional reprogramming to activate defense genes for restricting the spread of pathogens in the host plant (Jones and Dangl, 2006; Dodds and Rathjen, 2010; Jones et al., 2016). A successful Avr-R interaction is an absolute necessity for activation of HR, and hence, pathogens tend to rapidly mutate their *Avr* genes to escape recognition (Chrisholm et al., 2006). However, as a countermeasure, plants also coevolve to gain new *R* gene specificities to detect the newly emerged pathogen strains (Chrisholm et al., 2006; Jia et al., 2016; Singh et al., 2018).

2.3 STRUCTURE OF R PROTEINS

R proteins are products of *R* genes and have been isolated from various plant species. A large majority of *R* genes are known to encode R proteins of the nucleotide-binding leucine-rich repeat (NB-LRR) class (Kourelis and van der Hoorn, 2018; Kourelis et al., 2021). NB-LRRs or simply NLRs belong to STAND (Signal Transduction ATPases with Numerous Domains) group of P-loop NTPases and are the most prevalent type of R proteins (Lukasik and Takken, 2009). They possess a multidomain architecture characterized by the presence of conserved domains, i.e., nucleotide-binding domains (that contain nucleotide-binding sites or NBS) and leucine-rich repeats (LRR), crucial for perception, recognition, and activation of downstream signaling cascades to trigger defense responses (Tameling et al., 2006; Lukasik and Takken, 2009). NLRs are distinguished into three phylogenetically distinct classes on the basis of the domain at their N-terminus, i.e., toll/interleukin 1 receptor (TIR), coiled coil (CC) and RPW8 domain. These NLR classes are as follows: (1) the TIR-NB-LRRs (also known as TIR-NLRs or TNLs), (2) the CC-NB-LRRs (CC-NLRs or CNLs) and (3) the RPW8-NB-NLRs (CCR-NBS-LRR or RNLs) (Meyers et al., 2003; Kourelis and van der Hoorn, 2018). The CC and TIR domains facilitate homodimerization and the formation of heterocomplexes essential for activating downstream signal transduction pathways (Maekawa et al., 2011; Nishimura et al., 2017; Zhang et al., 2017a). The central domain of NLRs harbors a nucleotide-binding site (NBS) residing within an evolutionarily conserved domain similar to that present in the apoptosis-related proteins APAF-1 (Apoptotic protease-activating factor) and CED-4 (*Caenorhabditis elegans* death-4) proteins. Hence, they are often called as NB-ARC domains (Tameling et al., 2006; van Ooijen et al., 2008). NBS is responsible for binding the nucleotides ADP/ATP essential for facilitating conformational alterations for NLRs to switch between active and inactive states via replacement of ADP with ATP and vice versa (Tameling et al., 2006; Wang et al., 2019a, b; Ma et al., 2020; Martin et al., 2020). Besides the NBS, the NB-ARC domain also contains ARC1 and ARC2 domains that are involved in interdomain and intermolecular interactions (Wang et al., 2019a; Ma et al., 2020; Martin et al., 2020). The nucleotide-dependent reconfiguration is critical for the operation of NLRs. The C-terminal LRR domain is highly variable and accounts for conferring the ability to comprehend a variety of effector molecules via protein–protein interactions and therefore recognize various pathogens upon these receptors (Parniske et al., 1997; Botella et al., 1998; Ellis et al., 1999). RNLs encompass two subgroups, i.e., the *Nicotiana benthamiana N*-Required Gene 1 (NRG1) protein and the Arabidopsis Activated Disease Resistance Gene 1 (ADR1) protein, that operate as helpers of other CNLs and TNLs, respectively, in transducing signals for induction of defense responses (Collier et al., 2011).

2.4 DEFENSE ACTIVATION BY *R* GENE-MEDIATED SIGNALING

2.4.1 Perception of Pathogen Effector by R Proteins

One of the most important aspects of defense activation is the successful recognition of pathogen effectors, which depends on the perception of the effectors by the corresponding R proteins. In order to detect specific pathogen effectors, R proteins employ direct or indirect modes of recognition. While some R proteins are capable of perceiving pathogens directly through physical interaction with their cognate effectors, many others recognize effectors indirectly (Dangl and Jones, 2001; Jones and Dangl, 2006; van der Hoorn and Kamoun, 2008; Cesari, 2018; Adachi et al., 2019b; Figure 2.1).

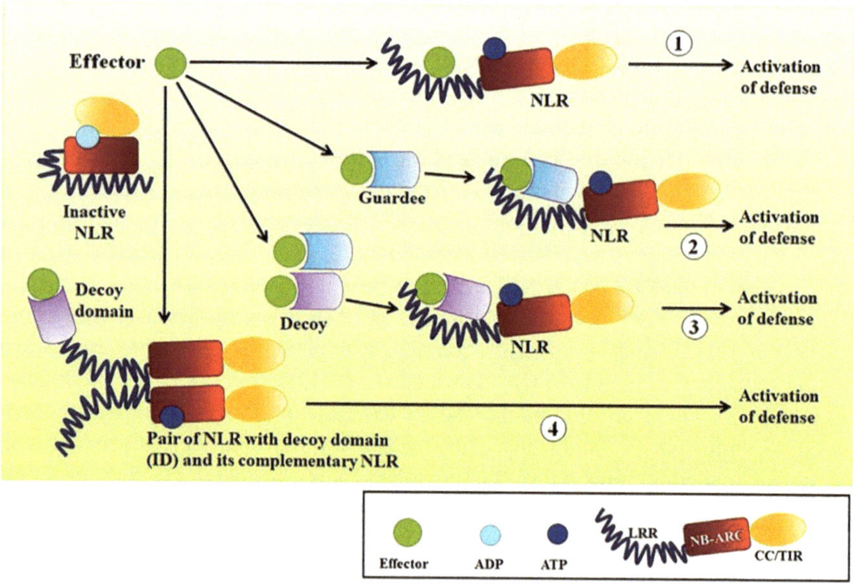

FIGURE 2.1 Different modes of pathogen effector recognition by NLRs – 1. Direct, 2. Guard, 3.

2.4.2 Decoy Integrated Models

The direct pathway of effector detection was described by Flor (1971), and it was proposed that *Avr* and *R* gene products encoded by the pathogen and host genomes, respectively, physically interact and continuously evolve with respect to each other, with the pathogen trying to evade host immunity while the host evolves to gain the ability to recognize this new effector (Flor, 1971). The flax *L567* locus is perhaps the best characterized example of direct recognition, conferring tolerance to *Melampsora lini* carrying *AvrL567* (Ellis et al., 1999; Dodds et al., 2006; Ravensdale et al., 2010). Alternatively, pathogen effectors may be detected indirectly via three modes described by three models, i.e. guard, decoy and integrated decoy models (Cesari, 2018; Grund et al., 2019; Adachi et al., 2019b). In the guard and decoy models, the effectors interact with host targets called 'guardee' proteins, which are guarded by the NLRs or 'decoy' proteins that are structural mimics of effector targets, respectively. The effector-induced modifications of these proteins are monitored by host NLRs (Dangl and Jones, 2001; van der Hoorn and Kamoun, 2008). The best described examples of guard models involve Arabidopsis RIN4 and SUMM2, and *Solanum demissum* BSL1 (Saunders et al., 2012; Zhang et al., 2017b; Redditt et al., 2019). Arabidopsis guardee protein RIN4 (RPM1-Interacting Protein 4) is targeted by several pathogen effectors, such as AvrRpm1, AvrB, AvrRpt2 and HopF2Pto of *Pseudomonas syringae* (Mackey et al., 2002, 2003; Axtell and Staskawicz, 2003; Wilton et al., 2010). Two NLRs, RPM1 (Resistance to *Pseudomonas maculicola* 1) and RPS2 (Resistance to *Pseudomonas syringae* 2), guard RIN4 by monitoring the effector-induced modifications in the protein (Mackey et al., 2002, 2003; Axtell and Staskawicz, 2003). RIPK (Receptor-Interacting Protein Kinase)-induced phosphorylation of RIN4 by AvrB or AvrRpm1 activates RPM1-triggered immunity, while its cleavage by the AvrRpt2 effector induces RPS2-mediated defense responses (Mackey et al., 2002, 2003; Axtell and Staskawicz, 2003). Another example of guardee, BSL1 (BSU-like protein 1), a putative plant phosphatase, is a protein target of *Phytophthora infestans* Avr2, and it is guarded by R2 of *Solanum demissum* (Saunders et al., 2012).

The decoy model was first proposed for recognition of AvrPto from *Pseudomonas syringae* pv. tomato by the Pto/Prf complex in tomato (Zhou and Chai, 2008; Zipfel and Rathjen, 2008). Pto is a kinase that becomes phosphorylated in an AvrPto effector-induced manner, and this alteration is sensed by Prf, an NLR, for triggering ETI (Ntoukakis et al., 2014). Other examples of proteins acting as decoys are PBS1, PBL1 and RCR3 (Shabab et al., 2008; Zhang et al., 2010; Qi et al., 2014; Wang et al., 2015). PBS1 mimics BIK1 (Botrytis-induced Kinase 1) and is also cleaved by *Pseudomonas syringae* effector HopAR1, which triggers RPS5-mediated resistance (Zhang et al., 2010; Qi et al., 2014). In a similar manner, PBL2 (PBS1-like kinase 2), another decoy mimicking BIK1, is targeted and uridylylated by another effector, i.e. AvrAC from *Xanthomonas campestris* pv. *campestris*, and is recruited by the RKS1/ZAR1 immune receptor complex to induce cell death (Wang et al., 2015). RCR3 is a decoy of PIP1 and traps Avr2 to activate Cf-2-dependent resistance (Shabab et al., 2008).

The integrated decoy model requires physically interacting pairs of NLRs, with one acting as the 'sensor' for effector detection while the other serves as an 'executor' NLR for defense activation (Cesari et al., 2014a; Le Roux et al., 2015; Maqbool et al., 2015). Many such NLRs have atypical integrated domains (IDs), which mimic the targets of effector targets into their structure (Cesari et al., 2014a; Le Roux et al., 2015; Maqbool et al., 2015; Ortiz et al., 2017). The genes of these complementary NLR pairs are co-regulated as they are tightly linked (Cesari et al., 2014b; Williams et al., 2014; Saucet et al., 2015). Important examples of these complementary NLR pairs include both TNL/TNL and CNL/CNL pairs, such as the Arabidopsis TNL pairs RPS4/RRS1-R, RPP2A/RPP2B and the rice CNL pairs RGA4/RGA5 and Pik-1/Pik-2 (Sinapidou et al., 2004; Cesari et al., 2013; Le Roux et al., 2015; Maqbool et al., 2015). RPP2A/RPP2B mediates resistance to the oomycete *Hyaloperonospora arabidopsidis* by detecting the effector protein ATR2 (Sinapidou et al., 2004). In the RRS1-R/RPS4 pair, RRS1-R (Resistance to *Ralstonia solanacearum* 1) is an atypical TNL containing an integrated WRKY domain (a characteristic DNA-binding domain of the WRKY family of TFs) that serves as a sensor for AvrRps4 and PopP2 effectors targeting other WRKY proteins, while RPS4 is the executor NLR that induces cell death (Le Roux et al., 2015; Guo et al., 2020). Conformational rearrangements induced in RRS1-R by AvrRps4 or PopP2 result in the activation of RPS4-mediated resistance (Guo et al., 2020). RGA5 and Pik-1 contain HMA (Heavy Metal-Associated) IDs, and they are involved in RGA4- and Pik-2-triggered immune responses, respectively, for providing resistance to the rice blast fungus, *Magnaporthe oryzae*, through direct binding with a number of effectors at their respective IDs (Cesari et al., 2013; Maqbool et al., 2015).

2.5 TRANSDUCTION OF DOWNSTREAM SIGNALING

Following successful R-Avr recognition, activation of defense responses requires the participation of several downstream components to initiate a conserved signaling cascade to accomplish the targeted activation of cell death and trigger systemic acquired resistance (SAR) (Dodds and Rathjen, 2010; Cui et al., 2015; Figure 2.2). *R* gene-mediated signaling requires NLRs to first attain an active conformation and then transduce the signal to downstream components, such as mitogen-activated protein kinases (MAPKs), phytohormones and transcription factors (TFs). It further involves Ca^{2+} influx, ROS burst, MAPK (Mitogen-Activated Protein Kinases) activation and alteration of defense gene expression (Dodds and Rathjen, 2010; Cui et al., 2015; Jones et al., 2016). *R* gene-mediated signaling is also associated with SA-mediated signaling, which results in amplification of the immunity response (Zhang and Li, 2019; Sun et al., 2021).

2.6 ACTIVATION OF NLR FOR DEFENSE SIGNALING

In unchallenged conditions, defense responses are kept at the basal level by maintaining a low expression level and an inactive state of NLRs (Lukasik and Takken, 2009; Takken and Goverse, 2012; van Wersch et al., 2016). It is important to preserve this inactive state of NLRs since constitutive NLR signaling often results in growth and fitness penalties (McHale et al., 2006; van Wersch et al., 2016). In the absence of pathogen interference, NLRs are maintained in an inactive state through auto-inhibitory mechanisms that involve intramolecular interactions between particular

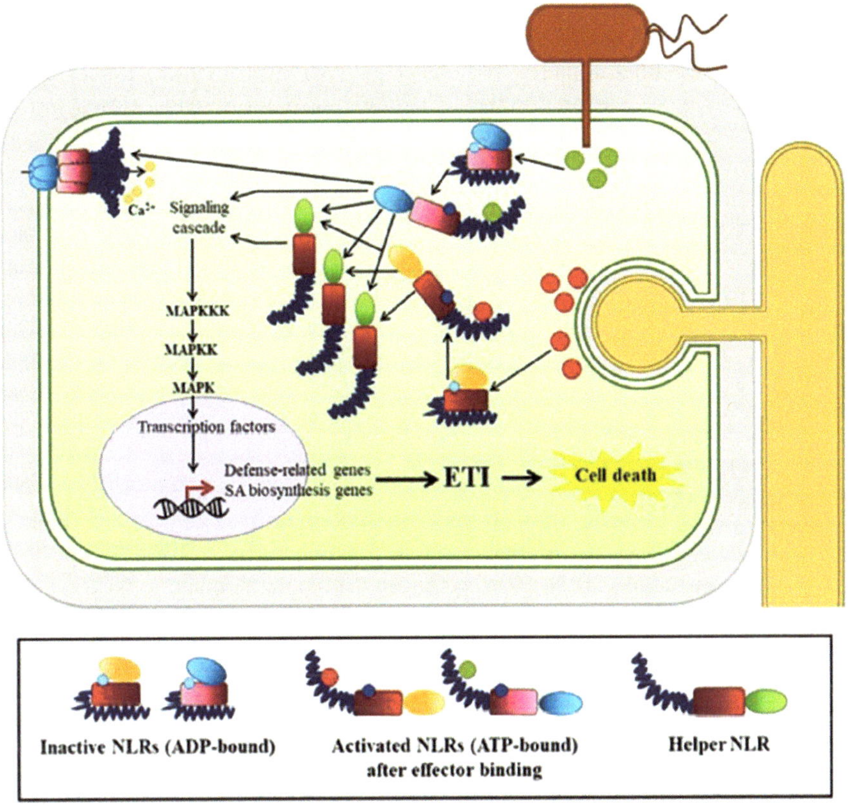

FIGURE 2.2 NLT-mediated signaling for activation of ETI and cell death. After binding of the effector, NLR becomes active via ADP/ATP exchange. Some NLRs, like ZAR1, are capable of oligomerizing into resistosomes for activating cell death, while others depend on helper NLRs for signaling. The association between NLRs and helper NLRs may be mediated by EDS1, which is a downstream component essential for signaling. EDS1 functions in two heteromeric complexes, EDS1/SAG101 and EDS1/PAD4, forming two exclusive signaling branches by associating with NRG1 and ADR1, respectively.

domains in their structure or through intermolecular interactions with other proteins in order to keep the specific motifs in the NB-ARC domain, such as ARC1 (or HD1) and ARC2 (or WHD), in a closed configuration. This has been shown for NLRs such as Rx, RPS5 (Resistance to *Pseudomonas syringae* 5) and ZAR1 (HopZ-Activated Resistance) (Takken and Goverse, 2012; Qi and Innes, 2013; Slootweg et al., 2013; Wang et al., 2019a, b). For example, the LRR domain takes up a horseshoe conformation to interact with the NB-ARC domain to maintain the resting or 'OFF' state of NLR (Lukasik and Takken, 2009; Takken and Goverse, 2012; Qi and Innes, 2013; Slootweg et al., 2013). The NB-ARC domain also assists in NLR autoinhibition in the absence of a pathogen (e.g. in flax L6) by preventing its self-association via the TIR domain (Bernoux et al., 2011). Furthermore, stabilization of NLR in the resting state may involve accessory proteins. For example, RIN4 interacts with and negatively regulates its partner NLR RPS2 to maintain its autoinhibitory state (Day et al., 2005). When inactive, NLRs exist in their ADP-bound state which could be attributed to either a binding preference for ADP over ATP to a higher intrinsic ATPase activity as compared to the active state, as shown by RPP1 (Recognition of *Peronospora parasitica* 1) (Tameling et al., 2006; Ma et al., 2020). This ADP-bound state facilitates the LRR domain to interact with NB-ARC to maintain an NLR in its autoinhibited state (Wang et al., 2019b). In the inactive ADP-bound state, LRR domains keep ZAR1 in an autoinhibited form (Wang et al., 2019b). Transformation into an active configuration requires the exchange of ADP for ATP by way of attaining a more open configuration to allow the nucleotide exchange (Williams et al., 2011). Reconfiguration of ZAR1 after effector perception causes the WHD (or ARC2) motif to be repositioned away from the ADP binding pocket, thus enabling ADP-ATP exchange to occur (Wang et al., 2019b).

2.7 OLIGOMERIZATION OF NLRs

Oligomerization of NLRs through intra- and intermolecular interactions is another requirement for NLRs to exert their immune function (Shirasu, 2009; Bernoux et al., 2011; Maekawa et al., 2011). Upon pathogen detection, monomeric NLRs undergo homodimerization by self-interaction and heterodimerization with other proteins (Bernoux et al., 2011; Maekawa et al., 2011; Williams et al., 2014). Both TIR and CC domains have been shown to promote this oligomerization (Williams et al., 2014; Casey et al., 2016; El Kasmi and Nishimura, 2016; Nishimura et al., 2017; Zhang et al., 2017a; Zhao and Song, 2021). Their role in oligomerization was first brought into light when crystal structures of the NLRs—barley MLA and flax L6—were revealed to self-associate through their CC and TIR domains, respectively (Bernoux et al., 2011; Maekawa et al., 2011). This self-association is essential for their ability to trigger downstream defense signaling (Bernoux et al., 2011; Maekawa et al., 2011). Recently, it has also been shown that this self-interaction is also required for the NAD^+-cleaving ability of some TNLs, which, in turn, is important for stimulation of cell death (Horsefield et al., 2019; Wan et al., 2019). Dimerization also seems to be important for functioning of truncated NLRs, such as the TIR-only RBA1, which self-associates into a dimer at the TIR domain as a consequence of its interaction with the bacterial effector HopBA1 (Nishimura et al., 2017).

Heteromeric cooperation is observed between complementary NLR pairs and between NLRs and the 'helper' RNLs or NLRs of the NRC clade (NLRs Required for Cell Death) in asterid groups (for example, the Solanaceae family), which work in association with the 'sensor' NLRs to mediate effector recognition (Cesari et al., 2014b; Williams et al., 2014; Adachi et al., 2019b; Jubic et al., 2019; Wu et al., 2019). Helper NLRs belong to the RNL (CCR-NLRs) class of highly conserved NLRs with two distinct clades, designated as the NRG1 (*Nicotiana benthamiana* N Requirement Gene 1) and ADR1 (Activated Disease Resistance 1), which are found in most angiosperm families, only with a few exceptions, including monocots and dicot *Aquilegia coerulea* (Bonardi et al., 2011; Collier et al., 2011; Dong et al., 2016; Shao et al., 2016; Castel et al., 2019; Wu et al., 2019). An additional conserved class of helper NLRs distinct from the RNL clade has been described in the Solanaceae family. It is called the NLR-REQUIRED FOR CELL DEATH (NRCs), representing a slow-evolving CNL class required downstream of a subset of sensor CNLs (Gabriëls et al., 2007; Wu et al., 2019). Many NLRs function via these 'helper' NLRs, NRG1 and ADR1, to transduce immune signals depending on their structural preference (Castel et al., 2019; Saile et al., 2020). Largely, NRG1 is involved downstream of TNL-mediated immunity, such as that mediated by N, ROQ1, RPS4/RRS1 and RPP1, while ADR1 family mediates responses of both TNLs and CNLs, including RPS4/RRS1, RPP2 and RPS2 (Bonardi et al., 2011; Collier et al., 2011; Castel et al., 2019; Wu et al., 2019; Saile et al., 2020). The NRC clade members, such as the NRC1 and its paralogs (NRC2, NRC3 and NRC4), are required during resistance responses mediated by several NLRs, such as the tomato Cf-4/9, Pto/Prf, Mi-1 and Ve1 and potato R1, Rx and Rpi-blb2 (Gabriëls et al., 2007; Fradin et al., 2009; Wu et al., 2016, 2019).

2.8 FORMATION OF 'RESISTOSOME'

An even higher order of NLR assembly has been revealed through cryo-electron microscopic structural investigations on the Arabidopsis NLRs, ZAR1 and RPP1, and tobacco ROQ1 (Recognition of XopQ 1) (Wang et al., 2019a, b; Ma et al., 2020; Martin et al., 2020). Arabidopsis ZAR1 (a CNL) triggers defense in response to effector AvrAC form *Xanthomonas campestris* and forms a pentameric structure upon activation, designated as 'resistosome' (Wang et al., 2019a, b; Bi et al., 2021). In its ADP-bound inactive state, the LRR domain of the NLR interacts with its ARC2 domain to keep it in resting state (Wang et al., 2019a, b). However, it also interacts with RKS1 (Resistance-Related Kinase 1) pseudokinase, a member of the RLCK subfamily XII-2, such that ZAR1 forms heterodimeric complexes (Wang et al., 2019a, b). Together with RKS1 in the ZAR1:RKS1 complex, ZAR1 facilitates detection of AvrAC-induced uridylylation of PBL2 (PBS1-Like Kinase 2), a decoy protein of BIK1 (Botrytis-Induced Kinase 1), which is the authentic target of effector (Zhang et al., 2010; Liu et al., 2013). The uridylylated PBL2 ($PBL2^{UMP}$) is recruited by RKS1 in the ZAR1:RKS1 complex, which, in turn, activates its interactive partner ZAR1 by inducing conformational changes in the latter to favor replacement of ADP by ATP and eventual organization of a wheel-like, pentameric 'resistosome' structure by such heterotrimeric ZAR1-RKS1-$PBL2^{UMP}$ protomers (Wang et al., 2019a, b). This activated ZAR1 resistosome is translocated to the plasma membrane, where it is inserted through its CC domains and forms a Ca^{2+}-permeable channel, thereby stimulating Ca^{2+} influx, production of ROS and effectuation of HR-associated cell death (Bi et al., 2021). The conserved MADA motif present in the pore-forming $\alpha 1$ helix of ZAR1 is important for inducing cell death (Adachi et al., 2019a).

Similar arrangements have been unraveled from structural studies on TNLs ROQ1 and RPP1, which directly bind their cognate effectors via the LRR domains (Ma et al., 2020; Martin et al., 2020). ROQ mediates direct recognition of HopQ1 (from *P. syringae*) and its close homologs XopQ (from *Xanthomonas* spp.) and RipB (from *Ralstonia* spp.), whereas the polymorphic RPP1 variants detect specific ATR1 variants from *Hyaloperonospora arabidopsidis* (Steinbrenner et al., 2015; Schultink et al., 2017; Thomas et al., 2020). However, in contrast to ZAR1, ROQ1 and RPP1 oligomerize into tetrameric resistosomes resembling a four-leaf clover in an effector-induced manner through their NB-ARC domains (Ma et al., 2020; Martin et al., 2020). This arrangement also helps bring together their TIR domains to enable the formation of the NAD^+ hydrolase active sites, with each tetrameric resistosome having two of such active sites for NAD^+ cleavage (Horsefield et al., 2019; Wan et al., 2019). Products formed as a result of this NAD^+ hydrolase activity give rise to a plant-specific variant of cyclic ADPR (v-cADPR), along with nicotinamide (Wan et al., 2019), which is suggested to potentially facilitate downstream immune signaling involving helper NLR EDS1 to induce cell death (Wan et al., 2019). The TIR domains have also been shown to possess the ability to hydrolyze RNA or DNA to 2′,3′-cAMP/cGMP, which is proposed to be a signaling molecule (Yu et al., 2022).

2.9 DOWNSTREAM SIGNALING: ITS COMPONENTS AND REGULATION

The two components of central importance to *R* gene-mediated signaling are the two non-NLR proteins, namely EDS1 (Enhanced Disease Sensitivity 1) and NDR1 (Non-race-specific Disease Resistance 1) (Century et al., 1995; Knepper et al., 2011a, b; Dongus and Parker, 2021). EDS1 is a lipase-like protein that is engaged in the signaling pathway of TNL-type R proteins and is also a positive regulator of SA signaling (Aarts et al., 1998; Feys et al., 2001, 2005; Wagner et al., 2013). NDR1, on the other hand, is a membrane-associated protein and functions as a downstream component in defense responses largely mediated by CNL-type R proteins, such as RPS2, RPM1 and RPS5 (Century et al., 1995, 1997; Aarts et al., 1998; Knepper et al., 2011a, b). The EDS1-dependent pathway operates in association with two other EDS1 family members—PAD4 (PHYTOALEXIN DEFICIENT 4) and SAG101 (SENESCENCE ASSOCIATED GENE 101) (Feys et al., 2001, 2005; Wagner et al., 2013). EDS1 interacts with SAG101 and PAD4 in heterodimeric complexes, mediated by their N-terminal lipase-like α/β-hydrolase domain and the C-terminal EP (termed based on EDS1-PAD4) domain consisting of α-helical bundles (Wagner et al., 2013). The EDS1–SAG101 and EDS1–PAD4 heterodimeric complexes are utilized differentially in various plant species, with the former contributing to defense responses of TNLs and the latter to both CNL and TNL-mediated responses (Dong et al., 2016; Bhandari et al., 2019; Gantner et al., 2019; Lapin et al., 2019; Wu et al., 2019; Sun et al., 2021). Furthermore, it has been suggested that EDS1–SAG101 and EDS1–PAD4 modules have co-evolved with helper RNL clades NRG1 and ADR1, respectively (Saile et al., 2020). They form two distinct signaling complexes—EDS1/SAG101/NRG1 and EDS1/PAD4/ADR1, which may act parallelly to contribute differentially or in a compensatory manner with respect to each other (Qi et al., 2018; Lapin et al., 2019; Wu et al., 2019; Sun et al., 2021). EDS1/SAG101/NRG1 is utilized in TNL-mediated immunity, while EDS1/PAD4/ADR1 is employed both by TNL and CNL receptors (Castel et al., 2019; Lapin et al., 2019; Saile et al., 2020). A differential contribution from the two signaling branches is evident in Arabidopsis and *Nicotiana benthamiana*. In the former, both EDS1/SAG101/NRG1 and EDS1/PAD4/ADR1 signaling complexes play a major role in immunity, and in the latter case, EDS1/SAG101/NRG1 is majorly employed for TNL-mediated HR (Gantner et al.,

2019; Lapin et al., 2019; Wu et al., 2019, 2021; Sun et al., 2021). In Arabidopsis, it is only the EDS1/SAG101/NRG1 complex that is required for TNL-induced response, whereas EDS1/PAD4/ADR1 is involved in basal immunity with a very limited role in cell death (Gantner et al., 2019; Lapin et al., 2020; Sun et al., 2021).

Involvement of the MAPK cascade in effector-triggered immunity is evident during the induction of HR-induced cell death by NLR; for example, N and Cf-9 rely on activation of WIPK and SIPK (Zhang and Klessig 1998; Romeis et al., 1999). EDS1 and NDR1 have recently been implicated in promoting the activities of MPK3/6 for enhanced expression of NLR genes AT3G04220 and AT4G11170 (Lang et al., 2022). The Pto/Prf-induced recognition of AvrPto is also dependent on phosphorylation and subsequent activation of Pto kinase after interaction with the effector, thereby inducing a defense response through Pto-induced phosphorylation of downstream signaling components (Bogdanove and Martin, 2000). On the other hand, SUMM2-induced cell death involves HopAI1-induced disturbances in the phosphorylation status of CRCK3 (Zhang et al., 2017b). Additionally, perturbation of the MAPK cascade negatively affects *R* gene-mediated cell death, such as that observed in the case of N-mediated signaling in tobacco in response to silencing of NTF6, MEK1 and NPK1 and in the case of Pto-mediated signaling due to silencing of NbMAPKKKα (Jin et al., 2002; Ekengren et al., 2003; del Pozo et al., 2004; Liu et al., 2004). This evidence suggests that MAPK cascade components are also essential elements of *R* gene-mediated signaling. An increase in cytoplasmic Ca^{2+} concentration is another important factor for the activation of HR, and it lies upstream of the oxidative burst, as indicated by the Ca^{2+} levels induced by RPM1 in Arabidopsis in response to the effectors AvrRpm1 and AvrB of *Pseudomonas syringae* pv. tomato (Grant et al., 2000; Bi et al., 2021). More recently, the ZAR1 resistosome has also been shown to get inserted in the plasma membrane, where it acts as a Ca^{2+} channel to stimulate Ca^{2+} influx (Wang et al., 2019a, b; Bi et al., 2021). In a similar manner as the ZAR1 resistosome, NRG1.1 DV and ADR1, the two helper NLRs, have also been highlighted to induce Ca^{2+} influx (Jacob et al., 2021).

For the activation of defense-related genes, NLRs facilitate transcriptional reprogramming by regulating the activity of related transcription factors (TFs), which may follow downstream of the MAPK cascade. Many activated NLRs are capable of migrating to the nucleus to directly influence gene expression by interacting with TFs to activate them or otherwise relieve their suppression (Zhu et al., 2010; Padmanabhan et al., 2013; Chang et al., 2013a; Xu et al., 2014). For example, RPP4-mediated resistance and basal defense against *Hyaloperonospora arabidopsidis* require a WRKY TF AtWRKY70 (Knoth et al., 2007). In a recent report, a putative resistance gene analog (*CaRGA*) from chickpea, encoding a CNL, has been shown to interact with WRKY64 to promote EDS1-dependent resistance to *Fusarium oxysporum* f.sp. *ciceri* race1 (Chakraborty et al., 2018). This putative CNL possibly works by phosphorylating the TF, which protects the latter from degradation by ubiquitination and proteasome-mediated degradation. The TNL SNC1 (Suppressor of NPR1-1, Constitutive 1) interacts with transcriptional co-repressor TPR1 (Topless-Related 1), which further targets negative regulators (DND1 and DND2, Defense No Death 1 and 2), thereby relieving the repression of immune responses (Zhu et al., 2010). NLR-mediated transcriptional reprogramming also triggers accumulation of salicylic acid (SA) and promotes systemic acquired resistance via activation of *ICS1* (*Isochorismate synthase* 1) gene and other SA-responsive genes, and by relieving the antagonizing effect of jasmonic acid (JA) through the TF MYC2, which is a positive regulator of JA signaling and is suggested to involve the EDS1-PAD4 complex (Cui et al., 2018).

Although some diverse signaling components have been recognized, a significant proportion of the components of *R* gene-mediated pathways still remain obscure. Thus, more investigations are required to understand *R* gene-mediated signaling, unraveling its components and connecting links among the existing players to decipher how cell death occurs in response to pathogens in plants.

2.10 REGULATION OF NLRs

As mentioned previously, tight regulation of NLR expression (and abundance) as well as their functions is vital for timely activation of defense and to check improper immune signaling to avoid growth penalties (van Wersch et al., 2020). Post-translational modifications (PTMs) are important regulatory mechanisms that determine the functional state of NLRs or may contribute to their stability (Bhattacharjee et al., 2011). Phosphorylation, for example, has been shown to enable reconfiguration of the RRS1-R/RPS4 TNL pair, and it also promotes its activation (Guo et al., 2020). Recognition of effector PopP2 by RRS1-R relies on a phosphorylation event in the C-terminus (Guo et al., 2020). Ubiquitination, another PTM, is significant for the turnover of NLRs mediated by SKP1-CULLIN1-F-box (SCF) E3 ubiquitin ligase complex for degradation by the 26S proteasome (Cheng and Li, 2012). The F-box protein CPR1 (Constitutive Expresser of *PR-1*, also known as CPR30) governs the abundance of SNC1 and RPS2 (Cheng et al., 2011). This is evident in *cpr1* mutants, which show higher accumulation of these NLR proteins, while *CPR1* overexpression mitigates the autoimmune phenotype of *snc1* mutants by reducing SNC1 accumulation (Cheng et al., 2011). Another regulatory protein, SRFR1, negatively regulates the accumulation of NLRs such as SNC1, RPS2 and RPS4 (Li et al., 2010). SGT1 (Suppressor of G2 allele of SKP1), which is a part of the chaperone complex HSP90-RAR1-SGT1, is a co-chaperone of HSP90 and a component of Skp1-Cullin F-box protein (SCF) E3 ubiquitin ligase complex, which negatively regulates R protein accumulation (Holt et al., 2005; Mayor et al., 2007; Seo et al., 2008; Hoser et al., 2014). The chaperone complex HSP90-RAR1-SGT1 regulates the folding and accumulation of NLRs (Azevedo et al., 2002; Muskett and Parker, 2003; Shirasu, 2009). HSP90 (heat shock protein 90) is a chaperone from the heat shock protein family, and RAR1 (Required for *Mla*-dependent Resistance 1) is a small CHORD-containing protein that antagonistically interacts with SGT1 to regulate R protein accumulation (Holt et al., 2005).

For exerting their function in immune responses, changes in the subcellular localization of NLRs as well as their associated proteins are important (García et al., 2010; Chang et al., 2013b). Some NLRs, like RPM1, are localized on the plasma membrane (Gao et al., 2011). Their downstream components, such as NDR1, are also associated with the plasma membrane (Knepper et al., 2011a, b). Several NLRs are known to shuttle between cytoplasm and nucleus and show nuclear accumulation in response to pathogen attack, including both CNLs and TNLs (Shen et al., 2007; Wirthmueller et al., 2007; Slootweg et al., 2010; Tameling et al., 2010; Hoser et al., 2013; Inoue et al., 2013; Ma et al., 2013; Padmanabhan et al., 2013). The barley CNL MLA10, after getting translocated into the nucleus, regulates WRKY TFs to relieve repression of MAMP-triggered immunity in response to *Blumeria graminis* f.sp. *hordei* (Shen et al., 2007). Maintenance of both the nuclear and cytoplasmic pools of Rx1, a Solanaceae CNL, is essential for resistance to Potato virus X (PVX), which suppresses host immunity by promoting Rx activation in the cytoplasm and forcing its hyperaccumulation in the nucleus (Slootweg et al., 2010). The compromised resistance due to nuclear hyperaccumulation of Rx could be alleviated by stabilizing the cytoplasmic pool of Rx by overexpressing RanGAP2, a factor that prevents nuclear localization of Rx and promotes its cytoplasmic retention, thereby suggesting the predominant role of cytoplasmic Rx in limiting PVX replication (Tameling et al., 2010). Similarly, RPS4 forms a signaling complex with EDS1, a downstream helper NLR, after AvrRps4 detection and becomes partitioned into both nucleus and cytoplasm pools, both subcellular pools being important for coordinating complete immunity (Heidrich et al., 2011). Additionally, the localization of helper NLRs and other downstream components is also important for NLR signaling and activation of defense. EDS1/PAD4 displays nucleocytoplasmic distribution, and a balance between the cytoplasmic and nuclear pools of EDS1 is vital for complete immunity (García et al., 2010; Bhattacharjee et al., 2011; Heidrich et al., 2011). Previously, it was believed that under unchallenged conditions, EDS1 exists either as homodimers or in interaction with PAD4/SAG101. However, more recently, it has been shown that EDS1 is able to exist as a stable monomer in an inactive or pre-activated state, wherein its active sites are masked, and does not form homodimers (Voss et al., 2019). EDS1/SAG101, on the contrary, is exclusive to the nucleus and is responsible for the regulation of defense (Feys et al., 2005).

NLR-mediated immunity may also be negatively regulated by the elements of NLR signaling and additional proteins. RIN4 negatively regulates the two NLRs—RPM1 and RPS2—in the absence of pathogens (Belkhadir et al., 2004). Their repression is relieved by post-effector-induced modification of RIN4 (Belkhadir et al., 2004; Day et al., 2005). The activation of resistance signaling also requires association of RIN4 with NDR1 (Day et al., 2006). NRG1C, an Arabidopsis RNL, also acts as a negative regulator of NLR-triggered defense responses, as suggested by compromised resistance and cell death in plants overexpressing NRG1C (Wu et al., 2021). An NRC member (NRCX) has also been recently implicated in the negative regulation of immunity mediated by other NRC members (Adachi et al., 2023).

2.11 WORKING MODELS FOR CNL- AND TNL-MEDIATED IMMUNE SIGNALING

In the absence of biotic challenge, NLRs are maintained in their inactive or resting state. Following recognition of pathogen, CNLs and TNLs become activated by virtue of ADP/ATP exchange and undergo oligomerization to form resistosomes (Figure 2.3). While some CNLs are capable of inducing signaling by themselves, others, along with many other TNLs, work in complementary pairs with other NLRs

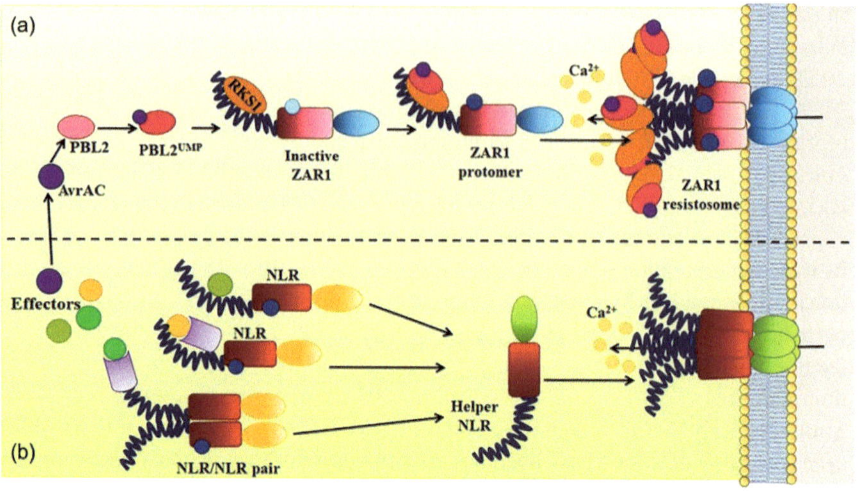

FIGURE 2.3 Formation of resistosomes by ZAR1 (a CNL) and Helper NLRs (such as NRG1). (a) ZAR1 is capable of signaling independently of helper NLRs; (b) others function by employing the aid of helper NLRs, which may form a resistosome structure.

(NLR–NLR pair), and others rely on 'helper' NLRs (Cesari et al., 2014b; Williams et al., 2014; Adachi et al., 2019b; Wu et al., 2019; Jubic et al., 2019). Arabidopsis ZAR1 is one such CNL capable of triggering cell death autonomously without the involvement of downstream helper NLRs by acting as Ca^{2+}-channels upon oligomerization (Bi et al., 2021). Pore formation triggers Ca^{2+} influx, which, in turn, induces an oxidative burst by ROS production required for cell death (Bi et al., 2021). The α1 helix (containing the MADA motif) of ZAR1 is responsible for pore formation in the plasma membrane once the ZAR1 resistosome gets inserted into the plasma membrane (Bi et al., 2021). This may be a conserved strategy among CNLs since many of them (~ 20% of the total analyzed CNLs) contain the characteristic MADA motif required for pore formation and cell death (Adachi et al., 2019a). TNL resistosomes possess NAD^+-cleaving activity and give rise to NAD^+-cleavage products that are crucial for defense signaling, probably through helper NLRs (or RNLs) and downstream components of the EDS1 family (Wan et al., 2019; Ma et al., 2020; Martin et al., 2020). The CNL- or TNL-mediated signaling cascades eventually trigger HR-associated cell death. It has been proposed that oligomerized TNLs directly associate with downstream EDS1 heterodimers (EDS1/SAG101 and EDS1/PAD4) (Sun et al., 2021). Engagement of helper RNLs, NRG1 and ADR1, for the execution of defense responses may come by way of their specific interaction with the two heterodimers EDS1/SAG101 and EDS1/PAD4, to form exclusive signaling complexes, EDS1/SAG101/NRG1 and EDS1/PAD4/ADR1, since no direct protein–protein interaction has so far been reported between TNLs and RNLs (Sun et al., 2021). The formation of NAD^+ hydrolysis products, including v-cADPR, catalyzed by the activated TNL resistosomes may be involved in the promotion of the two EDS1-RNL protein complexes through the perception of these products either by EDS1 or RNL (Horsefield et al., 2019; Wan et al., 2019). These RNLs may, in turn, form Ca^{2+} permeable channels in a mechanism that may be similar to the ZAR1 resistosome, such as that demonstrated for the two helper NLRs, NRG1.1 and ADR1 (Jacob et al., 2021).

2.12 PERSPECTIVES

Sustainable crop production is a challenging target owing to the wide array of prevailing and rapidly evolving biotic stress factors. For efficient disease management, therefore, a complete understanding of the *R* gene-mediated pathogen resistance is a prerequisite. Along with the identification of broad-spectrum *R* genes, there is also a great need to decipher the yet unknown components of *R* gene-induced signaling to fill the missing links. Revelation of resistosome formation by CNLs and TNLs has been a major breakthrough that has significantly contributed to the existing knowledge about how NLR signaling is activated. Although significant progress has been made toward understanding *R* gene signaling, there is still a lot more that remains elusive and requires further extensive research. More detailed investigations are required to elucidate how NLRs are regulated through post-translational modifications and the activated TNLs drive specific associations between RNLs, ADR1 and NRG1, and EDS1 complexes (EDS1/PAD4 and EDS1/SAG101), respectively. Research is also required to identify the NAD^+ cleavage products formed by TNLs and to decipher their roles in assisting EDS1-RNL interactions. It is also to be determined whether the formation of resistosomes is a universal feature and, if not, what are the other means by which NLRs trigger cell death. Drawing lessons from the present knowledge and structuring future research around deciphering the unknown facets of *R* gene-mediated signaling will help us gain more insights into the regulation of NLRs and how the knowledge so generated can be used to improve resistance and achieve effective disease management.

REFERENCES

Aarts N, Metz M, Holub E, Staskawicz BJ, Daniels MJ and Parker JE (1998) Different requirements for EDS1 and NDR1 by disease resistance genes define at least two R gene-mediated signaling pathways in Arabidopsis. *Proc Natl Acad Sci USA* 95: 10306–10311. doi:10.1073/pnas.95.17.10306

Adachi H, Contreras MP, Harant A, Wu CH, Derevnina L, Sakai T, Duggan C, Moratto E, Bozkurt TO, Maqbool A, Win J and Kamoun S (2019a) An N-terminal motif in NLR immune receptors is functionally conserved across distantly related plant species. *Elife* 8: e49956. doi:10.7554/eLife.49956

Adachi H, Derevnina L and Kamoun S (2019b) NLR singletons, pairs, and networks: evolution, assembly, and regulation of the intracellular immunoreceptor circuitry of plants. *Curr Opin Plant Biol* 50: 121–131. doi:10.1016/j.pbi.2019.04.007

Adachi H, Sakai T, Harant A, Pai H, Honda K, Toghani A, et al. (2023) An atypical NLR protein modulates the NRC immune receptor network in *Nicotiana benthamiana*. PLoS Genet 19(1): e1010500. https://doi.org/10.1371/journal.pgen.1010500

Asai S and Shirasu K (2015) Plant cells under siege: plant immune system versus pathogen effectors. *Curr Opin Plant Biol* 28: 1–8. doi:10.1016/j.pbi.2015.08.008

Axtell MJ and Staskawicz BJ (2003) Initiation of RPS2-specified disease resistance in Arabidopsis is coupled to the AvrRpt2-directed elimination of RIN4. *Cell* 112: 369–377. doi:10.1016/s0092-8674(03)00036-9

Azevedo C, Sadanandom A, Kitagawa K, Freialdenhoven A, Shirasu K and Schulze-Lefert P (2002) The RAR1 interactor SGT1, an essential component of *R* gene-triggered disease resistance. *Science* 295: 2073–2076. doi:10.1126/science.1067554

Belkhadir Y, Nimchuk Z, Hubert DA, Mackey D and Dangl JL (2004) Arabidopsis RIN4 negatively regulates disease resistance mediated by RPS2 and RPM1 downstream or independent of the NDR1 signal modulator and is not required for the virulence functions of bacterial type III effectors of AvrRpt2 or AvrRpm1. *Plant Cell* 16: 2822–2835. doi:10.1105/tpc.104.24117

Bernoux M, Ve T, Williams S, Warren C, Hatters D, Valkov E, Zhang X, Ellis JG, Kobe B and Dodds PN (2011) Structural and functional analysis of a plant resistance protein TIR domain reveals interfaces for self-association, signaling, and autoregulation. *Cell Host Microbe* 9: 200–211. doi:10.1016/j.chom.2011.02.009

Bhandari DD, Lapin D, Kracher B, von Born P, Bautor J, Niefind K and Parker JE (2019) An EDS1 heterodimer signalling surface enforces timely reprogramming of immunity genes in Arabidopsis. *Nat Commun* 10: 772. doi:10.1038/s41467-019-08783-0

Bhattacharjee S, Halane MK, Kim SH and Gassmann W (2011) Pathogen effectors target Arabidopsis EDS1 and alter its interactions with immune regulators. *Science* 334: 1405–1408. doi:10.1126/science.1211592

Bi G, Su M, Li N, Liang Y, Dang S, Xu J, Hu M, Wang J, Zou M, Deng Y, Li Q, Huang S, Li J, Chai J, He K, Chen YH and Zhou JM (2021) The ZAR1 resistosome is a calcium-permeable channel triggering plant immune signaling. *Cell* 184: 3528–3541.e12. doi:10.1016/j.cell.2021.05.003

Bogdanove AJ and Martin GB (2000) AvrPto-dependent Pto-interacting proteins and AvrPto-interacting proteins in tomato. *Proc Natl Acad Sci USA* 97: 8836–8840. doi:10.1073/pnas.97.16.8836

Bonardi V, Tang S, Stallmann A, Roberts M, Cherkis K and Dangl JL (2011) Expanded functions for a family of plant intracellular immune receptors beyond specific recognition of pathogen effectors. *Proc Natl Acad Sci USA* 108: 16463–16468. doi:10.1073/pnas.1113726108

Botella MA, Parker JE, Frost LN, Bittner-Eddy PD, Beynon JL, Daniels MJ, Holub EB and Jones JD (1998) Three genes of the Arabidopsis RPP1 complex resistance locus recognize distinct *Peronospora parasitica* avirulence determinants. *Plant Cell* 10: 1847–1860. doi:10.1105/tpc.10.11.1847

Casey LW, Lavrencic P, Bentham AR, Cesari S, Ericsson DJ, Croll T, Turk D, Anderson PA, Mark AE, Dodds PN, Mobli M, Kobe B and Williams SJ (2016) The CC domain structure from the wheat stem rust resistance protein Sr33 challenges paradigms for dimerization in plant NLR proteins. *Proc Natl Acad Sci USA* 113: 12856–12861. doi:10.1073/pnas.1609922113

Castel B, Ngou PM, Cevik V, Redkar A, Kim DS, Yang Y, Ding P and Jones JDG (2019) Diverse NLR immune receptors activate defence via the RPW8-NLR NRG1. *New Phytol* 222: 966–980. doi:10.1111/nph.15659

Century KS, Holub EB and Staskawicz BJ (1995) NDR1, a locus of *Arabidopsis thaliana* that is required for disease resistance to both a bacterial and a fungal pathogen. *Proc Natl Acad Sci USA* 92: 6597–6601. doi:10.1073/pnas.92.14.6597

Century KS, Shapiro AD, Repetti PP, Dahlbeck D, Holub E and Staskawicz BJ (1997) NDR1, a pathogen-induced component required for *Arabidopsis* disease resistance. *Science* 278: 1963–1965. doi:10.1126/science.278.5345.1963

Cesari S (2018) Multiple strategies for pathogen perception by plant immune receptors. New Phytol 219: 17–24. doi:10.1111/nph.14877

Cesari S, Bernoux M, Moncuquet P, Kroj T and Dodds PN (2014a) A novel conserved mechanism for plant NLR protein pairs: the "integrated decoy" hypothesis. *Front Plant Sci* 5: 606. doi:10.3389/fpls.2014.00606

Cesari S, Kanzaki H, Fujiwara T, Bernoux M, Chalvon V, Kawano Y, Shimamoto K, Dodds P, Terauchi R and Kroj T (2014b) The NB-LRR proteins RGA4 and RGA5 interact functionally and physically to confer disease resistance. *EMBO J* 33: 1941–1959. doi:10.15252/embj.201487923

Cesari S, Thilliez G, Ribot C, Chalvon V, Michel C, Jauneau A, Rivas S, Alaux L, Kanzaki H, Okuyama Y, Morel JB, Fournier E, Tharreau D, Terauchi R and Kroj T (2013) The rice resistance protein pair RGA4/RGA5 recognizes the *Magnaporthe oryzae* effectors AVR-Pia and AVR1-CO39 by direct binding. *Plant Cell* 25: 1463–1481. doi:10.1105/tpc.112.107201

Chakraborty J, Priya P, Dastidar SG and Das S (2018) Physical interaction between nuclear accumulated CC-NB-ARC-LRR protein and WRKY64 promotes EDS1 dependent Fusarium wilt resistance in chickpea. *Plant Sci* 276: 111–133. doi:10.1016/j.plantsci.2018.08.008

Chang C, Yu D, Jiao J, Jing S, Schulze-Lefert P and Shen QH (2013a) Barley MLA immune receptors directly interfere with antagonistically acting transcription factors to initiate disease resistance signaling. *Plant Cell* 25: 1158–1173. doi:10.1105/tpc.113.109942

Chang C, Zhang L and Shen Q-H (2013b) Partitioning, repressing and derepressing: dynamic regulations in MLA immune receptor triggered defense signaling. *Front Plant Sci* 4: 396. doi:10.3389/fpls.2013.00396

Cheng YT and Li X (2012) Ubiquitination in NB-LRR-mediated immunity. *Curr Opin Plant Biol* 15: 392–399. doi:10.1016/j.pbi.2012.03.014

Cheng YT, Li Y, Huang S, Huang Y, Dong X, Zhang Y and Li X (2011) Stability of plant immune-receptor resistance proteins is controlled by SKP1-CULLIN1-F-box (SCF)-mediated protein degradation. *Proc Natl Acad Sci USA* 108: 14694–14699. doi:10.1073/pnas.1105685108

Chisholm ST, Coaker G, Day B and Staskawicz BJ (2006) Host-microbe interactions: shaping the evolution of the plant immune response. *Cell* 124: 803–814. doi:10.1016/j.cell.2006.02.008

Collier SM, Hamel LP and Moffett P (2011) Cell death mediated by the N-terminal domains of a unique and highly conserved class of NB-LRR protein. *Mol Plant Microbe Interact* 24: 918–931. doi:10.1094/MPMI-03-11-0050

Couto D and Zipfel C (2016) Regulation of pattern recognition receptor signalling in plants. *Nat Rev Immunol* 16: 537–552. doi:10.1038/nri.2016.77

Cui H, Qiu J, Zhou Y, Bhandari DD, Zhao C, Bautor J and Parker JE (2018) Antagonism of transcription factor MYC2 by EDS1/PAD4 complexes bolsters salicylic acid defense in Arabidopsis effector-triggered immunity. *Mol Plant* 11: 1053–1066. doi:10.1016/j.molp.2018.05.007

Cui H, Tsuda K and Parker JE (2015) Effector-triggered immunity: from pathogen perception to robust defense. *Annu Rev Plant Biol* 66: 487–511. doi:10.1146/annurev-arplant-050213-040012

Dangl JL and Jones JD (2001) Plant pathogens and integrated defence responses to infection. *Nature* 411: 826–833. doi:10.1038/35081161

Day B, Dahlbeck D, Huang J, Chisholm ST, Li D and Staskawicz BJ (2005) Molecular basis for the RIN4 negative regulation of RPS2 disease resistance. *Plant Cell* 17: 1292–1305. doi:10.1105/tpc.104.030163

Day B, Dahlbeck D and Staskawicz BJ (2006) NDR1 interaction with RIN4 mediates the differential activation of multiple disease resistance pathways in Arabidopsis. *Plant Cell* 18: 2782–2791. doi:10.1105/tpc.106.044693

del Pozo O, Pedley KF and Martin GB (2004) MAPKKKα is a positive regulator of cell death associated with both plant immunity and disease. *EMBO J* 23: 3072–3082. doi:10.1038/sj.emboj.7600283

Dodds PN, Lawrence GJ, Catanzariti A-M, The T, Wang C-IA, Ayliffe MA, Kobe B and Ellis JG (2006) Direct protein interaction underlies gene-for-gene specificity and coevolution of the flax resistance genes and flax rust avirulence genes. *Proc Natl Acad Sci USA* 103: 8888–8893. doi:10.1073/pnas.0602577103

Dodds PN and Rathjen JP (2010) Plant immunity: towards an integrated view of plant-pathogen interactions. *Nat Rev Genet* 11: 539–548. doi:10.1038/nrg2812

Dong OX and Ronald PC (2019) Genetic engineering for disease resistance in plants: recent progress and future perspectives. *Plant Physiol* 180: 26–38. doi:10.1104/pp.18.01224

Dong OX, Tong M, Bonardi V, El Kasmi F, Woloshen V, Wünsch LK, Dangl JL and Li X (2016) TNL-mediated immunity in Arabidopsis requires complex regulation of the redundant ADR1 gene family. *New Phytol* 210: 960–973. doi:10.1111/nph.13821

Dongus JA and Parker JE (2021) EDS1 signalling: at the nexus of intracellular and surface receptor immunity. *Curr Opin Plant Biol* 62: 102039. doi:10.1016/j.pbi.2021.102039

Ekengren SK, Liu Y, Schiff M, Dinesh-Kumar S and Martin GB (2003) Two MAPK cascades, NPR1, and TGA transcription factors play a role in Pto-mediated disease resistance in tomato. *Plant J* 36: 905–917. doi:10.1046/j.1365-313x.2003.01944.x

El Kasmi F and Nishimura MT (2016) Structural insights into plant NLR immune receptor function. *Proc Natl Acad Sci USA* 113: 12619–12621. doi:10.1073/pnas.1615933113

Ellis JG, Lawrence GJ, Luck JE and Dodds PN (1999) Identification of regions in alleles of the flax rust resistance gene L that determine differences in gene-for-gene specificity. *Plant Cell* 11: 495–506. doi:10.1105/tpc.11.3.495

Feys BJ, Moisan LJ, Newman MA and Parker JE (2001) Direct interaction between the Arabidopsis disease resistance signaling proteins, EDS1 and PAD4. *EMBO J* 20: 5400–5411. doi:10.1093/emboj/20.19.5400

Feys BJ, Wiermer M, Bhat RA, Moisan LJ, Medina-Escobar N, Neu C, Cabral A and Parker JE (2005) Arabidopsis SENESCENCE-ASSOCIATED GENE101 stabilizes and signals within an ENHANCED DISEASE SUSCEPTIBILITY1 complex in plant innate immunity. *Plant Cell* 17: 2601–2613. doi:10.1105/tpc.105.033910

Flor HH (1971) Current status of the gene-for-gene concept. *Annu Rev Phytopathol* 9: 275–296. doi:10.1146/annurev.py.09.090171.001423

Fradin EF, Zhang Z, Ayala JCJ, Castroverde CD, Nazar RN, Robb J, Liu C-M and Thomma BPHJ (2009) Genetic dissection of *Verticillium* wilt resistance mediated by tomato Ve1. *Plant Physiol* 150: 320–332. doi:10.1104/pp.109.136762

Gabriëls SH, Vossen JH, Ekengren SK, van Ooijen G, Abd-El-Haliem AM, van den Berg GC, Rainey DY, Martin GB, Takken FL, de Wit PJ and Joosten MHAJ (2007) An NB-LRR protein required for HR signalling mediated by both extra- and intracellular resistance proteins. *Plant J* 50: 14–28. doi:10.1111/j.1365-313X.2007.03027.x

Gantner J, Ordon J, Kretschmer C, Guerois R and Stuttmann J (2019) An EDS1-SAG101 complex is essential for TNL-mediated immunity in *Nicotiana benthamiana*. *Plant Cell* 31: 2456–2474. doi:10.1105/tpc.19.00099

Gao Z, Chung EH, Eitas TK and Dangl JL (2011) Plant intracellular innate immune receptor Resistance to *Pseudomonas syringae* pv. *maculicola* 1 (RPM1) is activated at, and functions on, the plasma membrane. *Proc Natl Acad Sci USA* 108: 7619–7624. doi:10.1073/pnas.1104410108

García AV, Blanvillain-Baufumé S, Huibers RP, Wiermer M, Li G, Gobbato E, Rietz S and Parker JE (2010) Balanced nuclear and cytoplasmic activities of EDS1 are required for a complete plant innate immune response. *PLoS Pathog* 6: e1000970. doi:10.1371/journal.ppat.1000970.

Grant M, Brown I, Adams S, Knight M, Ainslie A and Mansfield J (2000) The RPM1 plant disease resistance gene facilitates a rapid and sustained increase in cytosolic calcium that is necessary for the oxidative burst and hypersensitive cell death. *Plant J* 23: 441–450. doi:10.1046/j.1365-313x.2000.00804.x

Grund E, Tremousaygue D and Deslandes L (2019) Plant NLRs with integrated domains: unity makes strength. *Plant Physiol* 179: 1227–1235. doi:10.1104/pp.18.01134

Guo H, Ahn HK, Sklenar J, Huang J, Ma Y, Ding P, Menke FLH and Jones JDG (2020) Phosphorylation-regulated activation of the Arabidopsis RRS1-R/RPS4 immune receptor complex reveals two distinct effector recognition mechanisms. *Cell Host Microbe* 27: e6. doi:10.1016/j.chom.2020.03.008

Heidrich K, Wirthmueller L, Tasset C, Pouzet C, Deslandes L and Parker JE (2011) Arabidopsis EDS1 connects pathogen effector recognition to cell compartment-specific immune responses. *Science* 334: 1401–1404. doi:10.1126/science.1211641

Holt BF 3rd, Belkhadir Y and Dangl JL (2005) Antagonistic control of disease resistance protein liability in the plant immune system. *Science* 309: 929–932. doi:10.1126/science.1109977

Horsefield S, Burdett H, Zhang X, Manik MK, Shi Y, Chen J, Qi T, Gilley J, Lai JS, Rank MX, Casey LW, Gu W, Ericsson DJ, Foley G, Hughes RO, Bosanac T, von Itzstein M, Rathjen JP, Nanson JD, Boden M, Dry IB, Williams SJ, Staskawicz BJ, Coleman MP, Vc T, Dodds PN and Kobe B (2019) NAD(+) cleavage activity by animal and plant TIR domains in cell death pathways. *Science* 365: 793–799. doi:10.1126/science.aax1911

Hoser R, Lichocka M, Żurczak M, Hennig J and Krzymowska M (2014) Emerging role of SGT1 as a regulator of NB-LRR-receptor nucleocytoplasmic partitioning. *Plant Signal Behav* 9: e28724. doi:10.4161/psb.28724

Hoser R, Żurczak M, Lichocka M, Zuzga S, Dadlez M, Samuel MA, Ellis BE, Stuttmann J, Parker JE, Hennig J and Krzymowska M (2013) Nucleocytoplasmic partitioning of tobacco N receptor is modulated by SGT1. *New Phytol* 200: 158–171. doi:10.1111/nph.12347

Inoue H, Hayashi N, Matsushita A, Xinqiong L, Nakayama A, Sugano S, Jiang CJ and Takatsuji H (2013) Blast resistance of CC-NB-LRR protein Pb1 is mediated by WRKY45 through protein-protein interaction. *Proc Natl Acad Sci USA* 110: 9577–9582. doi:10.1073/pnas.1222155110

Jacob P, Kim NH, Wu F, El-Kasmi F, Chi Y, Walton WG, Furzer OJ, Lietzan AD, Sunil S, Kempthorn K, Redinbo MR, Pei ZM, Wan L and Dangl JL (2021) Plant "helper" immune receptors are Ca^{2+}-permeable nonselective cation channels. *Science* 373: 420–425. doi:10.1126/science.abg7917

Jia Y, Zhou E, Lee S and Bianco T (2016) Coevolutionary dynamics of rice blast resistance gene *Pi-ta* and *Magnaporthe oryzae* avirulence gene *AVR-Pita 1*. *Phytopathology* 106: 676–683. doi:10.1094/PHYTO-02-16-0057-RVW

Jin H, Axtell MJ, Dahlbeck D, Ekwenna O, Zhang S, Staskawicz B and Baker B (2002) NPK1, an MEKK1-like mitogen-activated protein kinase kinase kinase, regulates innate immunity and development in plants. *Dev Cell* 3: 291–297. doi:10.1016/s1534-5807(02)00205-8

Jones JD and Dangl JL (2006) The plant immune system. *Nature* 444: 323–329. doi:10.1038/nature05286

Jones JDG, Vance RE and Dangl JL (2016) Intracellular innate immune surveillance devices in plants and animal. *Science* 354: aaf6395. doi:10.1126/science.aaf6395

Jubic LM, Saile S, Furzer OJ, El Kasmi F and Dangl JL (2019) Help wanted: helper NLRs and plant immune responses. *Curr Opin Plant Biol* 50: 82–94. doi:10.1016/j.pbi.2019.03.013

Knepper C, Savory EA and Day B (2011a) Arabidopsis NDR1 is an integrin-like protein with a role in fluid loss and plasma membrane-cell wall adhesion. *Plant Physiol* 156: 286–300. doi:10.1104/pp.110.169656

Knepper C, Savory EA and Day B (2011b) The role of NDR1 in pathogen perception and plant defense signaling. *Plant Signal Behav* 6: 1114–1116. doi:10.4161/psb.6.8.15843

Knoth C, Ringler J, Dangl JL and Eulgem T (2007) Arabidopsis WRKY70 is required for full RPP4-mediated disease resistance and basal defense against *Hyaloperonospora parasitica*, *Mol Plant-Microbe Interact* 20: 120–128. doi:10.1094/MPMI-20-2-0120

Kourelis J and van der Hoorn RAL (2018) Defended to the nines: 25 years of resistance gene cloning identifies nine mechanisms for R protein function. *Plant Cell* 30: 285–299. doi:10.1105/tpc.17.00579

Kourelis J, Sakai T, Adachi H and Kamoun S (2021) RefPlantNLR is a comprehensive collection of experimentally validated disease proteins from the NLR family. *PLoS Biol* 19: e3001124. doi:10.1371/journal.pbio.3001124resistance

Lang J, Genot B, Bigeard J and Colcombet J (2022) MAP kinases 3 and 6 control salicylic acid signaling by upregulating NLR receptors during pattern- and effector-triggered immunity. *J Exp Bot* 73: 2190–2205. doi:10.1093/jxb/erab544 [Corrected Proof]

Lapin D, Kovacova V, Sun X, Dongus JA, Bhandari D, von Born P, Bautor J, Guarneri N, Rzemieniewski J and Stuttmann J (2019) A coevolved EDS1-SAG101-NRG1 module mediates cell death signaling by TIR-domain immune receptors. *Plant Cell* 31: 2430–2455. doi:10.1105/tpc.19.00118

Le Roux C, Huet G, Jauneau A, Camborde L, Tremousaygue D, Kraut A, Zhou B, Levaillant M, Adachi H, Yoshioka H, Raffaele S, Berthome R, Coute Y, Parker JE and Deslandes L (2015) A receptor pair with an integrated decoy converts pathogen disabling of transcription factors to immunity. *Cell* 161: 1074–1088. doi:10.1016/j.cell.2015.04.025

Li Y, Li S, Bi D, Cheng YT, Li X and Zhang Y (2010) SRFR1 negatively regulates plant NB-LRR resistance protein accumulation to prevent autoimmunity. *PLoS Pathog* 6: e1001111. doi:10.1371/journal.ppat.1001111

Liu Y, Schiff M and Dinesh-Kumar S (2004) Involvement of MEK1 MAPKK, NTF6 MAPK, WRKY/MYB transcription factors, COI1 and CTR1 in N-mediated resistance to tobacco mosaic virus. *Plant J* 38: 800–809. doi:10.1111/j.1365-313X.2004.02085.x

Liu Z, Wu Y, Yang F, Zhang Y, Chen S, Xie Q, Tian X and Zhou JM (2013) BIK1 interacts with PEPRs to mediate ethylene-induced immunity. *Proc Natl Acad Sci USA* 110: 6205–6210. doi:10.1073/pnas.1215543110

Lukasik E and Takken FL (2009) STANDing strong, resistance proteins instigators of plant defence. *Curr Opin Plant Biol* 12: 427–36. doi:10.1016/j.pbi.2009.03.001

Ma L, Cornelissen BJ and Takken FL (2013) A nuclear localization for Avr2 from *Fusarium oxysporum* is required to activate the tomato resistance protein I-2. *Front Plant Sci* 4: 94. doi:10.3389/fpls.2013.00094

Ma S, Lapin D, Liu L, Sun Y, Song W, Zhang X, Logemann E, Yu D, Wang J, Jirschitzka J, Han Z, Schulze-Lefert P, Parker JE and Chai J (2020) Direct pathogen-induced assembly of an NLR immune receptor complex to form a holoenzyme. *Science* 370: eabe3069. doi:10.1126/science.abe3069

Mackey D, Belkhadir Y, Alonso JM, Ecker JR and Dangl JL (2003) Arabidopsis RIN4 is a target of the type III virulence effector AvrRpt2 and modulates RPS2-mediated resistance. *Cell* 112: 379–389. doi:10.1016/s0092-8674(03)00040-0

Mackey D, Holt BF 3rd, Wiig A and Dangl JL (2002) RIN4 interacts with *Pseudomonas syringae* type III effector molecules and is required for RPM1-mediated resistance in Arabidopsis. *Cell* 108: 743–754. doi:10.1016/S0092-8674(02)00661-X

Maekawa T, Cheng W, Spiridon LN, Toller A, Lukasik E, Saijo Y, Liu P, Shen QH, Micluta MA, Somssich IE, Takken FLW, Petrescu AJ, Chai J and Schulze-Lefert P (2011) Coiled-coil domain-dependent homodimerization of intracellular barley immune receptors defines a minimal functional module for triggering cell death. *Cell Host Microbe* 9: 187–199. doi:10.1016/j.chom.2011.02.008

Maqbool A, Saitoh H, Franceschetti M, Stevenson CEM, Uemura A, Kanzaki H, Kamoun S, Terauchi R and Banfield MJ (2015) Structural basis of pathogen recognition by an integrated HMA domain in a plant NLR immune receptor. *eLife* 4: e08709. doi:10.7554/eLife.08709

Martin R, Qi T, Zhang H, Liu F, King M, Toth C, Nogales E and Staskawicz BJ (2020) Structure of the activated ROQ1 resistosome directly recognizing the pathogen effector XopQ. *Science* 370: eabd9993. doi:10.1126/science.abd9993

Mayor A, Martinon F, De Smedt T, Petrilli V and Tschopp J (2007) A crucial function of SGT1 and HSP90 in inflammasome activity links mammalian and plant innate immune responses. *Nat Immunol* 8: 497–503. doi:10.1038/ni1459

McHale L, Tan X, Koehl P and Michelmore RW (2006) Plant NBS-LRR proteins: adaptable guards. *Genome Biol* 7: 212. doi:10.1186/gb-2006-7-4-212

Meyers BC, Kozik A, Griego A, Kuang H and Michelmore RW (2003) Genome-wide analysis of NBS-LRR-encoding genes in Arabidopsis. *Plant Cell* 15: 809–834. doi:10.1105/tpc.009308

Muskett P and Parker J (2003) Role of SGT1 in the regulation of plant *R* gene signalling. *Microbes Infect* 5: 969–976. doi:10.1016/s1286-4579(03)00183-7

Nishimura MT, Anderson RG, Cherkis KA, Law TF, Liu QL, Machius M, Nimchuk ZL, Yang L, Chung EH, El Kasmi F, Hyunh M, Osborne Nishimura E, Sondek JE and Dangl JL (2017) TIR-only protein RBA1 recognizes a pathogen effector to regulate cell death in Arabidopsis. *Proc Natl Acad Sci USA* 114: E2053–E2062. doi:10.1073/pnas.1620973114

Ntoukakis V, Saur IML, Conlan B and Rathjen JP (2014) The changing of the guard: the Pto/Prf receptor complex of tomato and pathogen recognition. *Curr Opin Plant Biol* 20: 69–74. doi:10.1016/j.pbi.2014.04.002

Ortiz D, de Guillen K, Cesari S, Chalvon V, Gracy J, Padilla A and Kroj T (2017) Recognition of the *Magnaporthe oryzae* effector AVR-Pia by the decoy domain of the rice NLR immune receptor RGA5. *Plant Cell* 29: 156–168. doi:10.1105/tpc.16.00435

Padmanabhan MS, Ma S, Burch-Smith TM, Czymmek K, Huijser P and Dinesh-Kumar SP (2013) Novel positive regulatory role for the SPL6 transcription factor in the N TIR-NB-LRR receptor-mediated plant innate immunity. *PLoS Pathog* 9: e1003235. doi:10.1371/journal.ppat.1003235

Pandolfi V, Neto JRCF, da Silva MD, Amorim LLB, Wanderley-Nogueira AC, de Oliveira Silva RL, Kido EA, Crovella S and Iseppon AMB (2017) Resistance (R) genes: applications and prospects for plant biotechnology and breeding. *Curr Protein Pept Sci* 18: 323–334. doi:10.2174/1389203717666160724195248

Parniske M, Hammond-Kosack KE, Golstein C, Thomas CM, Jones DA, Harrison K, Wulff BB and Jones JD (1997) Novel disease resistance specificities result from sequence exchange between tandemly repeated genes at the Cf-4/9 locus of tomato. *Cell* 91: 821–832. doi:10.1016/s0092-8674(00)80470-5

Qi D, Dubiella U, Kim SH, Sloss DI, Dowen RH, Dixon JE and Innes RW (2014) Recognition of the protein kinase AVRPPHB SUSCEPTIBLE1 by the disease resistance protein RESISTANCE TO PSEUDOMONAS SYRINGAE5 is dependent on S-acylation and an exposed loop in AVRPPHB SUSCEPTIBLE1. *Plant Physiol* 164: 340–351. doi:10.1104/pp.113.227686

Qi D and Innes RW (2013) Recent advances in plant NLR structure, function, localization, and signaling. *Front Immunol* 4: 348. doi:10.3389/fimmu.2013.00348

Qi T, Seong K, Thomazella DPT, Kim JR, Pham J, Seo E, Cho MJ, Schultink A and Staskawicz BJ (2018) NRG1 functions downstream of EDS1 to regulate TIR-NLR-mediated plant immunity in *Nicotiana benthamiana*. *Proc Natl Acad Sci USA* 115: E10979–E10987. doi:10.1073/pnas.1814856115

Ravensdale M, Nemri A, Thrall PH, Ellis JG and Dodds PN (2010) Co-evolutionary interactions between host resistance and pathogen effector genes in flax rust disease. *Mol Plant Pathol* 12: 93–102. doi:10.1111/j.1364-3703.2010.00657.x

Redditt TJ, Chung E-H, Karimi HZ, Rodibaugh N, Zhang Y, Trinidad JC, Kim JH, Zhou Q, Shen M, Dangl JL, Mackey D and Innes RW (2019) AvrRpm1 functions as an ADP-ribosyl transferase to modify NOI-domain containing proteins, including Arabidopsis and soybean RPM1-interacting protein 4. *Plant Cell* 31: 2664–2681. doi:10.1105/tpc.19.00020R2

Romeis T, Piedras P, Zhang S, Klessig DF, Hirt H and Jones JD (1999) Rapid Avr9-and Cf-9-dependent activation of MAP kinases in tobacco cell cultures and leaves: convergence of resistance gene, elicitor, wound, and salicylate responses. *Plant Cell* 11: 273–287. doi:10.1105/tpc.11.2.273

Saijo Y, Loo EP-L and Yasuda S (2018) Pattern recognition receptors and signaling in plant-microbe interactions. *Plant J* 93: 592–613. doi:10.1111/tpj.13808

Saile SC, Jacob P, Castel B, Jubic LM, Salas-Gonzáles I, Bäcker M, Jones JDG, Dangl JL and El Kasmi F (2020) Two unequally redundant "helper" immune receptor families mediate *Arabidopsis thaliana* intracellular "sensor" immune receptor functions. *PLoS Biol* 18: e3000783. doi:10.1371/journal.pbio.300078

Saucet SB, Ma Y, Sarris PF, Furzer OJ, Sohn KH and Jones JD (2015) Two linked pairs of Arabidopsis TNL resistance genes independently confer recognition of bacterial effector AvrRps4. *Nat Commun* 6: 6338. doi:10.1038/ncomms7338

Saunders DGO, Breen S, Win J, Schornack S, Hein I, Bozkurt TO, Champouret N, Vleeshouwers VGAA, Birch PRJ, Gilroy EM and Kamoun S (2012) Host protein BSL1 associates with *Phytophthora infestans* RXLR effector AVR2 and the *Solanum demissum* immune receptor R2 to mediate disease resistance. *Plant Cell* 24: 3420–3434. doi:10.1105/tpc.112.099861

Savary S, Willocquet L, Pethybridge SJ, Esker P, McRoberts N and Nelson A (2019) The global burden of pathogens and pests on major food crops. *Nat Ecol E*vol 3: 430–439. doi:10.1038/s41559-018-0793-y

Schultink A, Qi T, Lee A, Steinbrenner AD and Staskawicz B (2017) Roq1 mediates recognition of the *Xanthomonas* and *Pseudomonas* effector proteins XopQ and HopQ1. *Plant J* 92: 787–795. doi:10.1111/tpj.13715

Seo YS, Lee SK, Song MY, Suh JP, Hahn TR, Ronald P and Jeon JS (2008) The HSP90-SGT1-RAR1 molecular chaperone complex: a core modulator in plant immunity. *J Plant Biol* 51: 1–10. doi:10.1007/BF03030734

Shabab M, Shindo T, Gu C, Kaschani F, Pansuriya T, Chintha R, Harzen A, Colby T, Kamoun S and van der Hoorn RAL (2008) Fungal effector protein AVR2 targets diversifying defense-related Cys proteases of tomato. *Plant Cell* 20: 1169–1183. doi:10.1105/tpc.107.056325

Shao ZQ, Xue JY, Wu P, Zhang YM, Wu Y, Hang YY, Wang B and Chen JQ (2016) Large-scale analyses of angiosperm nucleotide-binding site-leucine-rich repeat genes reveal three anciently diverged classes with distinct evolutionary patterns. *Plant Physiol* 170: 2095–2109. doi:10.1104/pp.15.01487

Shen QH, Saijo Y, Mauch S, Biskup C, Bieri S, Keller B, Seki H, Ulker B, Somssich IE and Schulze-Lefert P (2007) Nuclear activity of MLA immune receptors links isolate-specific and basal disease-resistance responses. *Science* 315: 1098–1103. doi:10.1126/science.1136372

Shirasu K (2009) The HSP90-SGT1 chaperone complex for NLR immune sensors. *Annu Rev Plant Biol* 60: 139–164. doi:10.1146/annurev.arplant.59.032607.092906

Sinapidou E, Williams K, Nott L, Bahkt S, Tör M, Crute I, Bittner-Eddy P and Beynon J (2004) Two TIR:NB:LRR genes are required to specify resistance to *Peronospora parasitica* isolate Cala2 in Arabidopsis. *Plant J* 38: 898–909. doi:10.1111/j.1365-313X.2004.02099.x

Singh PK, Ray S, Thakur S, Rathour R, Sharma V and Sharma TR (2018) Co-evolutionary interactions between host resistance and pathogen avirulence genes in rice-*Magnaporthe oryzae* pathosystem. *Fungal Genet Biol* 115: 9–19. doi:10.1016/j.fgb.2018.04.005

Skendžić S, Zovko M, Živković IP, Lešić V and Lemić D (2021) Impact of climate change on agricultural insect pests. *Insects* 12: 440. doi:10.3390/insects12050440

Slootweg E, Roosien J, Spiridon LN, Petrescu AJ, Tameling W, Joosten M, Pomp R, van Schaik C, Dees R, Borst JW, Smant G, Schots A, Bakker J and Goverse A (2010) Nucleocytoplasmic distribution is required for activation of resistance by the potato NB-LRR receptor Rx1 and is balanced by its functional domains. *Plant Cell* 22: 4195–4215. doi:10.1105/tpc.110.077537

Slootweg EJ, Spiridon LN, Roosien J, Butterbach P, Pomp R, Westerhof L, Wilbers R, Bakker E, Bakker J, Petrescu AJ, Smant G and Goverse A (2013) Structural determinants at the interface of the ARC2 and leucine-rich repeat domains control the activation of the plant immune receptors Rx1 and Gpa2. *Plant Physiol* 162: 1510–1528. doi:10.1104/pp.113.218842

Steinbrenner AD, Goritschnig S and Staskawicz BJ (2015) Recognition and activation domains contribute to allele-specific responses of an Arabidopsis NLR receptor to an oomycete effector protein. *PLoS Pathog* 11: e1004665. doi:10.1371/journal.ppat.1004665

Sun X, Lapin D, Feehan JM, Stolze SC, Kramer K, Dongus JA, Rzemieniewski J, Blanvillain-Baufume S, Harzen A, Bautor J, Derbyshire P, Menke FLH, Finkemeier I, Nakagami H, Jones JDG and Parker JE (2021) Pathogen effector recognition-dependent association of NRG1 with EDS1 and SAG101 in TNL receptor immunity. *Nat Commun* 12: 3335. doi:10.1038/s41467-021-23614-x

Takken FL and Goverse A (2012) How to build a pathogen detector: structural basis of NB-LRR function. *Curr Opin Plant Biol* 15: 375–384. doi:10.1016/j.pbi.2012.05.001

Tameling WI, Nooijen C, Ludwig N, Boter M, Slootweg E, Goverse A, Shirasu K and Joosten MH (2010) RanGAP2 mediates nucleocytoplasmic partitioning of the NB-LRR immune receptor Rx in the Solanaceae, thereby dictating Rx function. *Plant Cell* 22: 4176–4194. doi:10.1105/tpc.110.077461

Tameling WI, Vossen JH, Albrecht M, Lengauer T, Berden JA, Haring MA, Cornelissen BJ and Takken FL (2006) Mutations in the NB-ARC domain of I-2 that impair ATP hydrolysis cause autoactivation. *Plant Physiol* 140: 1233–1245. doi:10.1104/pp.105.073510

Thomas NC, Hendrich CG, Gill US, Allen C, Hutton SF and Schultink A (2020) The immune receptor Roq1 confers resistance to the bacterial pathogens *Xanthomonas*, *Pseudomonas syringae*, and *Ralstonia* in tomato. *Front Plant Sci* 11: 463. doi:10.3389/fpls.2020.00463

Toruño TY, Stergiopoulos I and Coaker G (2016) Plant-pathogen effectors: cllular probes interfering with plant defenses in spatial and temporal manners. *Annu Rev Phytopathol* 54: 419–441. doi:10.1146/annurev-phyto-080615-100204

van der Hoorn RAL and Kamoun S (2008) From Guard to Decoy: a new model for perception of plant pathogen effectors. *Plant Cell* 20: 2009–2017. doi:10.1105/tpc.108.060194

van Esse HP, Reuber TL and van der Does D (2020) Genetic modification to improve disease resistance in crops. *New Phytol* 225: 70–86. doi:10.1111/nph.15967

van Ooijen G, Mayr G, Kasiem MM, Albrecht M, Cornelissen BJ and Takken FL (2008) Structure-function analysis of the NB-ARC domain of plant disease resistance proteins. *J Exp Bot* 59: 1383–1397. doi:10.1093/jxb/ern045

van Wersch R, Li X and Zhang Y (2016) Mighty dwarfs: *Arabidopsis* autoimmune mutants and their usages in genetic dissection of plant immunity. *Front Plant Sci* 7: 1717. doi:10.3389/fpls.2016.01717

van Wersch S, Tian L, Hoy R and Li X (2020) Plant NLRs: the whistleblowers of plant immunity. *Plant Commun* 1: 100016. doi:10.1016/j.xplc.2019.100016

Voss M, Toelzer C, Bhandari DD, Parker JE and Niefind K (2019) Arabidopsis immunity regulator EDS1 in a PAD4/SAG101-unbound form is a monomer with an inherently inactive conformation. *J Struct Biol* 208: 107390. doi:10.1016/j.jsb.2019.09.007

Wagner S, Stuttmann J, Rietz S, Guerois R, Brunstein E, Bautor J, Niefind K and Parker JE (2013) Structural basis for signaling by exclusive EDS1 heteromeric complexes with SAG101 or PAD4 in plant innate immunity. *Cell Host Microbe* 14: 619–630. doi:10.1016/j.chom.2013.11.006

Wan L, Essuman K, Anderson RG, Sasaki Y, Monteiro F, Chung EH, Osborne Nishimura E, DiAntonio A, Milbrandt J, Dangl JL and Nishimura MT (2019) TIR domains of plant immune receptors are NAD(+)-cleaving enzymes that promote cell death. *Science* 365: 799–803. doi:10.1126/science.aax1771

Wang G, Roux B, Feng F, Guy E, Li L, Li N, Zhang X, Lautier M, Jardinaud M-F, Chabannes M, Arlat M, Chen S, He C, Noël LD and Zhou J-M (2015) The decoy substrate of a pathogen effector and a pseudokinase specify pathogen-induced modified-self recognition and immunity in plants. *Cell Host Microbe* 18: 285–295. doi:10.1016/j.chom.2015.08.004

Wang J, Hu M, Wang J, Qi J, Han Z, Wang G, Qi Y, Wang HW, Zhou JM and Chai J (2019a) Reconstitution and structure of a plant NLR resistosome conferring immunity. *Science* 364: eaav5870. doi:10.1126/science.aav5870

Wang J, Wang J, Hu M, Wu S, Qi J, Wang G, Han Z, Qi Y, Gao N, Wang HW, Zhou JM and Chai J (2019b) Ligand-triggered allosteric ADP release primes a plant NLR complex. *Science* 364: eaav5868. doi:10.1126/science.aav5868

Williams SJ, Sohn KH, Wan L, Bernoux M, Sarris PF, Segonzac C, Ve T, Ma Y, Saucet SB, Ericsson DJ, Casey LW, Lonhienne T, Winzor DJ, Zhang X, Coerdt A, Parker JE, Dodds PN, Kobe B and Jones JD (2014) Structural basis for assembly and function of a heterodimeric plant immune receptor. *Science* 344: 299–303. doi:10.1126/science.1247357

Williams SJ, Sornaraj P, deCourcy0Ireland E, Menz RI, Kobe B, Ellis JG, Dodds PN and Anderson PA (2011) An autoactive mutant of the M flax rust protein has a preference for binding ATP, whereas wild-type M protein binds ADP. *Mol Plant Microbe Interact* 24: 897–906. doi:10.1094/MPMI-03-11-0052

Wilton M, Subramaniam R, Elmore J, Felsensteiner C, Coaker G and Desveaux D (2010) The type III effector HopF2Pto targets Arabidopsis RIN4 protein to promote *Pseudomonas syringae* virulence. *Proc Natl Acad Sci USA* 107: 2349–2354. doi:10.1073/pnas.0904739107

Wirthmueller L, Zhang Y, Jones JD and Parker JE (2007) Nuclear accumulation of the Arabidopsis immune receptor RPS4 is necessary for triggering EDS1-dependent defense. *Curr Biol* 17: 2023–2029. doi:10.1016/j.cub.2007.10.042

Wu C-H, Belhaj K, Bozkurt TO, Birk MS and Kamoun S (2016) Helper NLR proteins NRC2a/b and NRC3 but not NRC1 are required for Pto-mediated cell death and resistance in *Nicotiana benthamiana*. *New Phytol* 209: 1344–1352. doi:10.1111/nph.13764

Wu Z, Li M, Dong OX, Xia S, Liang W, Bao Y, Wasteneys G and Li X (2019) Differential regulation of TNL-mediated immune signaling by redundant helper CNLs. *New Phytol* 222: 938–953. doi:10.1111/nph.15665

Wu Z, Tian L, Li X, Huang W, Zhang Y and Li X (2021) N-terminally truncated helper NLR NRG1C antagonizes immunity mediated by its full-length neighbors NRG1A and NRG1B. *Plant Cell* 34: 1621–1640. doi:10.1093/plcell/koab285

Xu F, Kapos P, Cheng YT, Li M, Zhang Y and Li X (2014) NLR-associating transcription factor bHLH84 and its paralogs function redundantly in plant immunity. *PLoS Pathog* 10: e1004312. doi:10.1371/journal.ppat.1004312

Yu D, Song W, Tan EYJ, Liu L, Cao Y, Jirschitzka J, Li E, Logemann E, Xu C, Huang S, Jia A, Chang X, Han Z, Wu B, Schulze-Lefert P and Chai J (2022) TIR domains of plant immune receptors are 2′,3′-cAMP/cGMP synthetases mediating cell death. *Cell* 185: 2210–2212. doi:10.1101/2021.11.09.467869

Zayan SA (2019) Impact of climate change on plant diseases and IPM strategies. In: (Ed. Topolovec-Pintarić S) *Plant Diseases – Current Threats and Management Trends*. IntechOpen, London, pp. 1–12. doi:10.5772/intechopen.87055

Zhang J, Li W, Xiang T, Liu Z, Laluk K, Ding X, Zou Y, Gao M, Zhang X, Chen S, Mengiste T, Zhang Y and Zhou J-M (2010) Receptor-like cytoplasmic kinases integrate signaling from multiple plant immune receptors and are targeted by a *Pseudomonas syringae* effector. *Cell Host Microbe* 7: 290–301. doi:10.1016/j.chom.2010.03.007

Zhang S and Klessig DF (1998) Resistance gene *N*-mediated *de novo* synthesis and activation of a tobacco mitogen-activated protein kinase by tobacco mosaic virus infection. *Proc Natl Acad Sci USA* 95: 7433–7438. doi:10.1073/pnas.95.13.7433

Zhang X, Bernoux M, Bentham AR, Newman TE, Ve T, Casey LW, Raaymakers TM, Hu J, Croll TI, Schreiber KJ, Staskawicz BJ, Anderson PA, Sohn KH, Williams SJ, Dodds PN and Kobe B (2017a) Multiple functional self-association interfaces in plant TIR domains. *Proc Natl Acad Sci USA* 114: E2046–E2052. doi:10.1073/pnas.1621248114

Zhang Y and Li X (2019) Salicylic acid: biosynthesis, perception, and contributions to plant immunity. *Curr Opin Plant Biol* 50: 29–36. doi:10.1016/j.pbi.2019.02.004

Zhang Z, Liu Y, Huang H, Gao M, Wu D, Kong Q and Zhang Y (2017b) The NLR protein SUMM2 senses the disruption of an immune signaling MAP kinase cascade via CRCK3. *EMBO Rep* 18: 292–302. doi:10.15252/embr.201642704

Zhao J and Song J (2021) NLR immune receptor RB is differentially targeted by two homologous but functionally distinct effector proteins. *Plant Commun* 2: 100236. doi:10.1016/j.xplc.2021.100236

Zhou JM and Chai J (2008) Plant pathogenic bacterial type III effectors subdue host responses. *Curr Opin Microbiol* 11: 179–185. doi:10.1016/j.mib.2008.02.004

Zhu Z, Xu F, Zhang Y, Cheng YT, Wiermer M, Li X and Zhang Y (2010) Arabidopsis resistance protein SNC1 activates immune responses through association with a transcriptional corepressor. *Proc Natl Acad Sci USA* 107: 13960–13965. doi:10.1073/pnas.1002828107

Zipfel C (2014) Plant pattern-recognition receptors. *Trends Immunol* 35: 345–351. doi:10.1016/j.it.2014.05.004

Zipfel C and Rathjen JP (2008) Plant immunity: AvrPto targets the frontline. *Curr Biol* 18: R218–R220. doi:10.1016/j.cub.2008.01.016

3 Modern Molecular Tools for Diagnostics of Plant Pathogens

Manjunath Hubballi and Anjali V.A.
University of Horticultural Sciences

Archana T.S.
Lovely Professional University

Chaitra C. Kulkarni
University of Horticultural Sciences

Nakkeeran S.
Tamil Nadu Agricultural University

3.1 INTRODUCTION

According to UN estimations, the existing population of 7.6 billion is anticipated to reach 9.8 billion by 2050, with a projected addition of 83 million people every year. The new population estimate clearly indicates the nine countries, viz., India, Nigeria, Pakistan, the Democratic Republic of the Congo, Ethiopia, the United Republic of Tanzania, Indonesia, Egypt and the United States of America, will contribute half of the projected population increase in 2050. This ever-increasing population not only puts pressure on natural resources but also on the food production system. On a global scale, it is estimated that there will be a 56% food gap and a land gap of 593 million ha between 2010 and 2050 (WRI, 2022). The enormous increase in population has created an environmental catastrophe by consuming more resources such as land, food, water, air, fossil fuels and minerals, as well as by producing waste products as a result of consumption. The global cropping area would reduce by 8%–20% owing to urbanization and the conversion of cropped area to non-food production (Nellemann et al., 2009). The FAO estimation indicates food production needs to be increased by 70% across the globe to meet these hungry mouths (MP, 2022), and this was one of the global challenges of the Millennium Project.

Plants, being the primary producers in the food chain, contribute to 80% of the food consumed by mankind and are also a key source of nutrition for livestock (FAO, 2022). It is very much necessary to have food and nutritional security in any country to have vibrant, productive and healthy societies (Savary et al., 2019). The state

DOI: 10.1201/9781003403142-3

of individuals having access to sufficient, safe, affordable and nutritious food at all times is considered as food security (Strange and Scott, 2005). Food security is an integral approach of One Health Project as well as an important pillar of the United Nations Sustainable Development Goals (SDGs) (Choffnes, 2012; UN, 2022).

Despite the unprecedented increase in food production over the last few decades through multifaceted research across the globe, nearly a billion people suffer from inadequate nutrition. According to the latest data from UNICEF, 50% of the total child deaths under 5 are due to undernutrition (UNICEF, 2022). Whether or not there is scope for increasing food production in many areas, it is critically important to safeguard the actual food produced. Thus, there is an increasing demand for improving food security by reducing crop losses and postharvest waste (Alexander et al., 2017). Although plant health is one part of the definition of One Health, plants have not been well integrated into the approaches. However, the health of plants is vital to sustaining human and animal life on earth. Recognizing the key role of plants in public health, the United Nations declared 2020 to be the International Year of Plant Health (IYPH).

It is estimated that plant diseases have made a remarkable impact both socially and economically on human life, as demonstrated by historical diseases like the late blight of potatoes in Ireland, the Bengal famine in India and coffee rust in Sri Lanka. Plant diseases, despite all measures and advanced management efforts, cause around 16% crop loss on a global basis (Oerke, 2006). Savary et al. (2012) reported that yield losses in crops due to diseases ranged from 20% to 40%. Furthermore, the postharvest losses owing to substandard quality are estimated to be 30%–40%. The overall economic losses incurred due to pathogens in the USA alone account for 40 billion dollars (Pimentel et al., 2005; Roberts et al., 2006). The yield loss caused by pathogens in major crops is given hereunder (Table 3.1).

It is important to minimize these huge losses being documented during crop growth, harvest and postharvest processing to enhance productivity and also to ensure the sustainability of agricultural production. In order to do so, advanced detection of diseases and prevention in crops are important. The correct identification and early detection of pathogens are the foremost steps in the management of any disease. The historical reliance on morphological and cultural characteristics for the identification of many pathogens is time-consuming and requires a lot of expertise in the field of taxonomy. Furthermore, many characteristics are subjective in nature, which leads to erroneous identification. Alternatively, molecular detection techniques offer a rapid opportunity to assess disease with greater objectivity (reliability, precision and accuracy). Furthermore, these techniques can give results with high accuracy and reproducibility. In the recent past, there has been great progress in the detection and diagnosis of pathogens using molecular techniques. These techniques have undergone two major breakthroughs in the last three decades. The first one was with respect to the use of antibodies, especially monoclonal antibodies, and ELISA assays for the detection of viruses and bacteria (Clark and Adams, 1977). The second breakthrough was the development of PCR-based techniques, which have totally revolutionized the field of detection (Miller and Martin, 1988). Thus, molecular techniques have led to hand in accurate diagnosis of the disease, thereby contributing to its management. The major modern molecular detection techniques and their implications for the diagnosis and detection of pathogens are discussed hereunder.

TABLE 3.1
Yield Losses Reported due to Diseases in Major Crops

Crop	Disease	Yield Loss (%)	References
	Rice blast	10–30	Asibi et al. (2019)
Rice	Bacterial leaf blight	50	Shekhar et al. (2020)
Wheat	Leaf rust	1–20	USDA (2022)
	Black stem rust	50–70	USDA (2022)
Maize	Turcicum leaf bight of maize	30–37	Shankara et al. (2017)
	Charcoal rot	63.5	Maize AICRP (2014)
Cowpea	Yellow mosaic virus, Cercospora leaf spot	10–100	Rana et al. (2016)
Pigeon pea	Sterility mosaic disease	95–100	Saxena et al. (2004)
Chickpea	Fusarium wilt, Ascochyta blight	50–100	Rana et al. (2016)
Field pea	Rust, wilt, downy mildew, powdery mildew	10–30	Rana et al. (2016)
Lentil	Rust, wilt, Sclerotia blight	20–70	Rana et al. (2016)
Potato	Late blight	70	Dowley et al. (2008)
	Potato virus X	10–40	Kolychikhina et al. (2021)
	Bacterial wilt	70	IIHR (2022)
	Tomato spotted wilt virus	70–100	IIHR (2022)
Onion	Soft rot	40–80	Gore and Rakhonde (2020)
Banana	Fusarium wilt	100	Nakkeeran et al. (2021)

3.2 POLYMERASE CHAIN REACTION-BASED ASSAYS

The polymerase chain reaction (PCR) was invented by Kary B. Mullis and was awarded the Nobel Prize for Chemistry in 1993 for his great contribution. This technology was patented in 1985. It is an instrument used for making millions of copies of a target piece of deoxyribonucleic acid (DNA), sometimes referred to as molecular photocopying. If the sequence of interest is known in DNA, it is possible to replicate the same very selectively in large copy numbers. Theoretically, the amount of matrix DNA is not at all a limiting factor in PCR, and therefore it makes it possible to amplify a very minute amount of DNA into a huge copy number.

PCR is an in vitro technology used for the amplification of DNA templates through various repeated cycles such as denaturation, annealing, extension and holding at various temperatures. Specific primers, deoxyribonucleotide triphosphates and Taq polymerases are used for the PCR reaction. In the process of amplification, three essential phases, viz., denaturation, hybridization and elongation, take place. Double-stranded DNA is broken down to single-stranded DNA by a process called denaturation. Annealing with primers is called as the hybridization phase, and the last phase is the elongation phase, where the strand grows. The products of each synthesis step serve as a template for the following steps; thus, exponential amplification is achieved (Pelt-Verkuil et al., 2008).

The discovery of PCR has virtually revolutionized the field of diagnosis of diseases caused by fungi, bacteria, nematodes, phytoplasma, viruses and viroids

(Ma and Michailides, 2007). The use of specific primers or universal primers that target specific pathogen species helps in proper identification and species-level confirmation. Furthermore, the identity of each isolate of a pathogen can be further confirmed by sequencing the amplified product and searching the homology in the NCBI GenBank database with the help of BLAST, the Basic Local Alignment Search Tool (Caetano Anolles, 2013). There have been a vast number of variants of this technology since its invention. The major types of PCR are discussed hereunder.

3.2.1 End-Point PCR

End-point PCR, where analysis happens after all cycles of PCR are completed. The products are visualized on an agarose gel to estimate size as well as quantity. It is used for many applications in molecular biology, such as genotyping, cloning and sequencing. Mostly end-point PCR used to detect the presence or absence of targets; however, it can also be used to estimate relative quantity. The detection is completed at the end of the temperature cycle. It is one of the most cost-effective molecular tools for the identification of various plant pathogens. This technique is widely used for the identification of various fungal plant pathogens. However, this technique is time-consuming, primer design is often difficult, and it is difficult to identify closely related species through this (Sikdar et al., 2014). Iturralde Martinez et al. (2019) used end-point PCR for the identification of *Ophiosphaerella*, which causes spring dead spots in Bermuda grass. The morphological identification of fungi is challenging because of the production of pseudothecium. Hence, species-specific diagnosis with specific primers helped the identification of fungal isolates. Tegli et al. (2010) developed a highly useful detection assay for *Pseudomonas savastanoi* pathovars causing olive knot disease by end-point PCR.

The end-point PCR assay found an effective tool for the diagnosis of pathogens from various symptomatic and asymptomatic plants. Arif et al. (2014) used an end-point PCR assay for the detection of *Phymatotrichopsis omnivora*, causing root rot in cotton. Specific primers were designed based on the nuclear ribosomal internal transcribed spacer (ITS) region. The assays effectively detected the pathogen in infected root samples of cotton as well as alfalfa. The current PCR-based assay was found very useful for rapid screening of infected plants, especially during various breeding programs.

Despite its large usage, the technique has its demerits. The major drawback of this technique is that it is time-consuming, as the procedure may not be completed in a day. The results obtained are based on a difference in the size of the amplicon, which is not accurate enough. The post-PCR process is cumbersome, and the agarose gel resolution is very poor. Thus, it is replaced by several advanced techniques.

3.2.2 Nested PCR

It is a modified end-point PCR method used for the identification of plant pathogens. Two sets of primers are used for the amplification of targeted DNA. Two successive rounds of PCR are mainly used to enhance specificity. During the first round of PCR, nonspecific primers are used, followed by species-specific primers in the next round, which increases the sensitivity of the reaction (Bhat and Browne, 2010). This is useful for rare templates or PCR with a high background. It is one of the most promising technologies for the detection of various phytoplasma diseases.

Pear decline phytoplasma, which belongs to the apple proliferation group, is one of the major constraints on pear decline. Phytoplasma detection by using nested PCR is one of the most preferable and reliable technologies for early detection. Other detection methods, such as serological methods, are less preferable because of a lesser amount of sensitivity (Garcia-Chapa et al., 2003). Srivastava et al. (2006) detected sugarcane grassy shoot phytoplasma presence in the leafhopper vector *Deltocephalus vulgaris* with the help of rRNA-specific primers of sugarcane grassy shoot phytoplasma. Furthermore, Gonzalez et al. (2011) detected phytoplasmas-causing X-disease in cherry by nested PCR assays.

Puig (2021) used a nested PCR assay for screening various germplasm against Cacao mild mosaic virus (CaMMV), a national and international quarantine-important virus. Low virus titer was the major problem for the detection of the virus. Nested PCR based on CaMMV-specific primers validated the correct detection from the samples with low virus titers and efficient germplasm screening.

Mudiyanselage et al. (2021) developed nested PCR for the detection of the downy mildew fungus *Peronospora sparsa*, especially from latently infected plants where systemic infection of the pathogen has started. This is a very effective and useful detection method for screening gooseberry mother plants as well as daughter plants from the nursery stages. Thus, this technique can be used for the detection of different types of pathogens with high reproducibility.

3.2.3 Multiplex PCR

Multiplex PCR is one of the novel detection techniques that helps in the simultaneous detection of multiple targets in a single reaction well by using different pairs of primers that specifically target each sample. The advancement of PCR detection systems with instantaneous multi-target detection has advanced research in many areas of plant pathology. The multiplex PCR assay works based on the principle of various pairs of primers, which helps in the amplification of several different plant pathogens simultaneously (Sint et al., 2012). Furthermore, the generated amplicon can be separated with the help of gel electrophoresis. The primers used in the corresponding reaction should be specific and have similar annealing temperatures (Zhao et al., 2014). Simultaneous detection of five viruses infecting pea, namely alfalfa mosaic virus, bean yellow mosaic virus, clover yellow vein virus, cucumber mosaic virus (CMV) and clover mottle virus in a single test tube and identification were carried out with the help of multiplex PCR product size (Bariana et al., 1994). Similarly, simultaneous detection of three wheat pathogens, such as *P. graminis* f.sp. *tritici*, *P. triticina* and *B. graminis* f.sp. *tritici*, during the early stage of infection was carried out with three specific sets of primers (Chen et al., 2015). Identification of *Fusarium solani* and *F. oxysporum*, which cause root rot and wilt disease, is challenging because of their complex morphological characteristics. Multiplex PCR helps in the identification of *Fusarium solani* and *F. oxysporum* in a single step more precisely and specifically (Villarino et al., 2021). Thus, this technique can be very effectively used in the detection of pathogens in complex diseases where they are caused by multiple pathogens. The major drawback with this is that the annealing temperature of all primers must be the same.

3.2.4 Quantitative PCR

Real-time polymerase chain reaction (RT-PCR), also known as quantitative PCR (qPCR), is one of the most important molecular biology-based detection techniques where the amplification of target DNA can be monitored during the PCR. It is mainly based on nonspecific fluorescent dyes or sequence-based specific DNA probes that intercalate with targeted DNA molecules and permit detection (Pal, 2022). qPCR allows the detection and quantification of targeted DNA or RNA sequences of phytopathogenic agents. The relative number of target DNA sequences can be valued by calculating the cycle threshold (Ct) value of the targeted samples with the help of sequence-specific primers (Balodi et al., 2017). The common fluorescent dyes used in the reaction mixture consist of SYBR Green, TaqMan (sequence-specific fluorescence-labeled reporter probes), Eva Green, or Molecular Beacons (Pal, 2022). The fluorescent signal produced during the amplification is directly proportional to the amount of amplicon generated (Alemu, 2014; Shao et al., 2022). The technology has been employed in the detection and diagnosis of many important pathogens in the past.

Didymella pinodella is one of the emerging pathogens causing root rot complex disease in pea. The wide host range of the pathogen and asymptomatic colonization in the host roots make detection methodologies difficult and challenging. The use of RT-PCR assays helps in the early and precise detection and quantification of *D. pinodella* with specific primers designed up on the TEF-1 alpha gene sequence (Šišić et al., 2022). Puertolas et al. (2021) employed qPCR for the detection of *Phytophthora* and *Pythium* in ornamental nursery plants. The major problem in ornamentals is proper detection and diagnosis at the early stages of growth. Hence, real-time amplification of oomycete DNA with specific primers such as ITS, trnM-trnP-trnM and atp9-nad9 helps in early detection as well as reducing the biosecurity risk associated.

Virus detection from apple rootstocks is mandatory to produce virus-free planting materials through vegetative propagation. The most precise diagnosis strategy to detect plant viruses from rootstocks is a RT-PCR assay. Heo and Chung (2021) tested the presence of *Apple chlorotic leaf spot virus* (ACLSV) and *Apple scar skin viroid* (ASSVd) through qPCR techniques. A singleplex RT-PCR assay with SYBR Green I and TaqMan probe targeting the conserved region of both ACLSV and ASSVd helps in the early detection of the virus.

3.2.5 Bio-PCR

BIO-PCR works based on a modified end-point PCR technique, where the pre-incubation step is required to increase the biomass of causal organisms. To improve detection and prevent nontarget microorganisms, the targeted pathogens are grown in specific media for detection purposes (Schaad et al., 1995). It is a very effective technique used for seed-borne fungal pathogens (Kumar et al., 2020). Detection and exclusion of pathogens from seed lots is one of the most difficult tasks in seedling production and disease management. The major challenge is that the number of infected seeds within the seed lot might be very low and asymptomatic, making decision-making difficult. To address these challenges, Hariharan and Prasannath (2021) developed the BIO-PCR technique, which selectively targets colonies used for the detection, selectively concentrates on targeted organisms and is successfully used for detecting seed-borne bacteria.

Detection of ring rot in potatoes, caused by *Clavibacter michiganensis* subsp. *sepedonicus*, is often difficult because of symptomless tubers. A BIO-PCR assay with specific primers for *C. michiganensis* subsp. *Sepedonicus* helped to detect the pathogen successfully and also to screen the symptomless potato tubers (Schaad et al., 1999).

Manulis et al. (1999) reported this method as an effective technique to screen disease-free planting materials in ornamentals. *Erwinia herbicola* pv. *gypsophilae* is one of the major constraints in *Gypsophila paniculata*, an ornamental used for cut flower production. Bio-PCR was employed in the detection of symptomless gypsophila cuttings and helps in effective disease-free mother stock production.

3.2.6 Magnetic-Capture Hybridization PCR

Magnetic-capture hybridization PCR (MCH-PCR) is another rapid and sensitive PCR technique where the extraction of purified DNA from the target organism is carried out with magnetic beads followed by PCR amplification (Jacobsen, 1995). Isolation and purification of DNA fragments were carried out with an ssDNA probe on magnetic beads. The purified DNA is further used for PCR amplification. It is a very specific method to inhibit plant extracts during the various stages of DNA isolation, which have negative impacts during PCR. The magnetic beads consist of a biotinylated, specific oligonucleotide sequence to target the DNA of interest (Walcott et al., 2004). This magnetic beads separation of DNA fragments helps in separation from complex inhibitors (Capote et al., 2012).

Agrobacterium vitis is an important bacteria causing crown gall disease in grapes. The bacterium mainly survives in grapevines and disseminates through propagating materials. Hence, disease indexing is very important to produce disease-free planting material. Johnson et al. (2013) developed a magnetic-capture hybridization assay for efficient and sensitive detection of *A. vitis* from grape tissue. This technique specifically overcame the effects of PCR inhibitors, which are present in plant tissues, and increased the specificity of the reaction. Furthermore, Ha et al. (2009) used this technology along with a RT-PCR assay for simultaneous detection of two important pathogens, *Acidovorax avenae* subsp. *citrulli* and *Didymella bryoniae*, causing bacterial fruit blotch and gummy stem blight in cucurbits.

Langrell and Barbara (2001) used MCH-PCR technology to remove the inhibitors present in the DNA extracted from apple tissues. Specific oligonucleotide-based magnetic beads were used for the selective isolation of DNA strands. This technology is very selective and reliable as compared with normal PCR techniques and increases the specificity in lignified and woody tissues, where the chance of additional inhibitors is higher.

Rapid and selective detection of botrytis neck rot in onions was carried out by using magnetic-capture hybridization and polymerase chain reaction (MCH-PCR) techniques. It was found to be very effective, with significant results within a reduced period. The magnetic beads used significantly allowed the detection of 1×10^3 conidia/mL. The technique was found to be more reliable and efficient as compared with conventional PCR and detected *B. aclada* from seed lots where natural infestation levels were 0.25% (Walcott et al., 2004).

This technology is also effective in reducing the effect of humic acid and other contaminants, which negatively affect the PCR reaction when targeting specific soil microorganisms. Magnetic beads with specific probes help in the species-specific isolation of DNA and remove other inhibitors present. Jacobsen (1995) used the MCH-PCR technique for detecting specific microorganisms in the rhizospheric soil of barley. Furthermore, Little and Rezaian (2006) used RT-PCR along with magnetic-capture hybridization for detecting the *grapevine leafroll-associated virus 1* (GLRaV-1) and effectively removed the RT-PCR inhibitors.

3.2.7 Serial Analysis of Gene Expression

Serial analysis of gene expression (SAGE) is a powerful high-throughput method that provides a gene expression profile of a cell type or tissue and also aids in the identification of specific genes by comparing gene profiles. In this technique, a set of unique sequence tags (9–10 bp) from individual mRNAs is isolated and concatenated into long DNA molecules for lump-sum sequencing (Yamamoto et al., 2001). The technique is mostly applied to study host-pathogen interactions and for transcriptional profiling infected plants. For instance, the interaction of the tomato yellow leaf curl ardinia virus with the host plant was studied employing long-term SAGE (Berardi et al., 2005). The technique enabled the quantative analysis of transcripts without prior knowledge of sequence information. A drawback faced by this method is the need for a large amount of mRNAs.

3.2.8 DNA Bar Coding

DNA bar coding is an emerging technique for plant pathogen diagnostics that relies on certain genes or genomic regions as taxon identifiers with significant sensitivity and reproducibility. Most short regions of the genome, approximately 600 bp, can be utilized as barcodes for identification. Although the technique has been mostly applied in diversity analysis, its application in plant pathogen detection depends on the use of species-specific small DNA sequences for identification. A DNA barcode should be universal with sufficient sequence variability, flanked by conserved regions. Most of the time, a combination of certain genes or genomic regions can be used as barcodes. For instance, detection of pathogens like Cercospora, Colletotrichum and Fusarium infecting Chinese masona was detected by targeting a combination of ITS, partial beta-tubulin and histone H3 (Hsieh et al., 2020). Although the ITS region has been established as a suitable DNA barcode for fungal identification, recent studies have implicated that the use of multiple housekeeping genes is expected to give more accurate results than the use of single gene-based barcodes. Similarly, the 16S rRNA region is the most frequently used barcode for bacterial identification. However, in a comparative study conducted by Links et al. (2012), it was found that the protein-coding gene cpn60, having a much wider barcode gap, was more robust for bacterial identification than the 16S rRNA region. The limitation faced by plant viruses is that they don't possess a universal gene that can be employed as a barcode for species identification. Thus, genes specific for a particular virus or a group of viruses can be used as targets for plant viral detection. For instance, for detecting the tomato brown rugose fruit virus, the small replicase subunit region was targeted to develop DNA barcodes (Alkowni et al., 2019). DNA bar coding has evolved in different ways since its development in recent years, but the main constraint is the detection of novel pathogens. Much more recent technologies like meta-bar coding and genome skimming have paved the way for more accurate diagnostics of plant pathogens.

3.3 ISOTHERMAL AMPLIFICATION-BASED METHODS

In the recent past, significant efforts have been made to develop detection methods for plant pathogens. However, the diagnostic method must meet the specific requirements in terms of selectivity, sensitivity and high throughput so as to have greater acceptability in the field of diagnosis (Shi et al., 2008). The most widely adopted amplification method is PCR. Despite its wide acceptance in diagnostic procedures, PCR carries inherent drawbacks. The major drawback is the time-consuming thermal cycling between different temperatures needed to obtain the target sequence amplification (Asiello and Baeumner, 2011). In order to address these drawbacks, alternative and promising thermal cycling procedures were developed. The isothermal amplification detects target DNA sequences in a streamlined and exponential manner and does not require thermal cycling. Since the DNA strands are not denatured, all isothermal methods depend on the polymerase with strand-displacement activity as an alternative approach to facilitate primer binding and initiation of amplification. Once the reaction is initiated, the polymerase must also separate the strand that is still annealed from the sequence of interest. Isothermal amplification chemistry has been applied to diagnostics with great success and is utilized in several commercial molecular diagnostic platforms. Various isothermal amplification methods used for the detection of plant pathogens are Loop-Mediated Isothermal Amplification (LAMP), Helicase-Dependent Amplification (HDA), Rolling Circle Amplification (RCA), Multiple Displacement Amplification (MDA), Recombinase Polymerase Amplification (RPA) and Nucleic Acid Sequence-Based Amplification (NASBA) (Zanoli and Spoto, 2013).

3.3.1 Rolling Circle Amplification

RCA is one of the promising isothermal amplification-based methods used for the detection of various plant pathogens. This assay exploits the RNA or DNA polymerase to generate ssDNA or RNA molecules. The techniques mainly require a circular template, deoxynucleotide triphosphates, DNA polymerase, homologous buffer and DNA/RNA primers (Gu et al., 2018). DNA amplification is carried out by using phi29 DNA polymerase, which has specific strand-displacement activity

(Van Emmerik et al., 2020). The displacement activity permits the newly synthesized DNA template to displace the previously synthesized DNA molecule to release single-stranded DNA (Bhat and Rao, 2020). A cascade of strand-displacement events generates 100–1,000 long ssDNA tandem repeats of the original targeted sequence (Kieser and Budowle, 2020). This assay using bacteriophage ϕ 29 DNA polymerase allows cheap, reliable detection and diagnosis of Gemini viruses from naturally infected host plants. The major advantage of this technique is that there is no need for additional expensive devices, it is easy to handle, it detects circular DNA components without prior knowledge of sequence information, and it has a low cost per reaction (Haible et al., 2006).

Davari et al. (2012) used the RCA technique for the rapid identification of pathogenic *Fusarium* fungi that cause head blight diseases in cereals. RCA was designed based on specific probes in the elongation factor 1-α (EF-1α) gene to detect the dominant *Fusarium* species, consisting of the *Fusarium graminearum* species complex (FGSC). It helped in the early detection of various Fusarium species, such as *Fusarium oxysporum*, *Fusarium incarnatum–equiseti* and *Fusarium tricinctum*, from infected cereals. Lin et al. (2018) used this assay for the detection of various pathogens such as *Neofabraea malicorticis*, *N. perennans*, *N. kienholzii* and *N. vagabunda* causing apple bull's-eye rot. Specific probes were designed by targeting *EF-1α* regions.

This is one of the quickest identification assays for the detection of various pathogens, as it is a very rapid and sensitive assay, and the specific probe designed helps in detecting the target pathogen. It is a very effective and useful tool for monitoring specific pathogens during quarantine and epidemiological studies (Javaheri Tehrani et al., 2014).

3.3.2 Loop-Mediated Isothermal Amplification

It is one of the latest and most promising techniques that has become a very popular diagnostic tool for detecting various plant pathogens (Le and Vu, 2017). The LAMP reaction consists of the initial step, cycling amplification and an elongation step (Panno et al., 2019). Two sets of internal primers, such as forward inner primer (FIP), backward inner primer (BIP), backward loop primer (B-Loop) and another set of two outer primers (F3 and B3), are used to identify six unique sequences on the targeted nucleic acid. FIP and BIP cover double distinct sequences matching to sense and anti-sense strands of the targeted DNA of an organism. Looping forward and backward with Bst polymerase helps to accelerate the LAMP reaction (Nagamine et al., 2002; Francois et al., 2011). The total reaction setup yields high exponential and isothermal amplification of about 10^9–10^{10} fold target DNA within 45–60 minutes at a reaction temperature of 60°–65° (Notomi et al., 2000; Chander et al., 2014).

The LAMP assay was developed for the detection of chili leaf curl disease caused by begomo viruses and transmitted by whiteflies. Although the PCR technique affords standards in the diagnostics of viruses from infected plants, it is time-consuming and requires sophisticated instruments. LAMP specifically targeting the AC1 and AC2 regions facilitated the detection of 1×10^{-1} ng DNA with 200-fold higher sensitivity than that of conventional PCR (Nagendran et al., 2022). Kim et al. (2022) developed a LAMP assay with specific primer sets for the detection of *Pea enation mosaic virus* (PEMV) infecting pea, chickpea, faba bean and lentil. The technique was found to be more reliable because there was less time required and no requirement for specific instruments.The technique confirmed the detection of a diluted PEMV up to 10^{-6} cDNA and was 10,000 times more sensitive than other detection techniques such as RT-nested PCR and RT-PCR.

Celik (2022) used reverse transcription LAMP (RT-LAMP) assay to target *prunus necrotic ringspot virus* (PNRSV) and *apple mosaic virus* (ApMV) based on the coat protein region of the virus. Warm Start 2X Master Mix was used to detect the color change from pink to yellowish, showing a positive result. This assay was found to be ten-fold higher than that of conventional PCR.

RT-LAMP assays were used for the detection of CMV and *Lily symptomless virus* (LSV). This approach captured viral antigens with anti-rabbit IgG antibodies for the RT-LAMP assay without the isolation of RNA. The detection limit was found to be 100 times greater than that of PCR. The use of SYBR Green dye made the results visible to the naked eye. The current essay is effective and helpful for detection where laboratory facilities and technical resources are limited (Zhang et al., 2022).

He et al. (2022) developed a LAMP assay for the specific detection of *Agroabcterium fabacearum*, causing crown gall disease. The staining of the amplified product with SYBR Green I provides visualization with naked eyes and easy detection of *A. fabacearum* from the roots and soil samples. Furthermore, Suzaki et al. (2022) suggested that this technique is also used for the effective detection of biovars 1 and 3 of *Pseudomonas syringae* pv. *actinidiae* with a specific set of primers.

The technique can also be used for detecting plant-parasitic nematodes. Omer et al. (2022) used a LAMP assay for detecting the root-knot nematode *Meloidogyne hapla*, which causes root-knot diseases in various plants. The primers are designed to be specific to nematodes and allow the detection of juveniles and soil samples. The assay was found to be fast, reliable and specific for detecting and quantifying *M. hapla* directly from the soil.

3.3.3 Nucleic Acid Sequence-Based Amplification

Nucleic acid amplification is another molecular technology used to produce numerous replicates of a specific segment of RNA/DNA. The amplified product of DNA/RNA is further used for detection purposes (Lamb et al., 2018). There are mainly two different types of amplification, including non-isothermal and isothermal amplification techniques, where the difference lies in the reaction temperature. Non-isothermal amplification produces several copies of the target RNA/DNA at different temperatures. Whereas isothermal amplification produces multiple copies at a constant temperature, Nucleic Acid Sequence-Based Amplification makes use of ssRNA and anneals the primers at 65°C; further amplifying occurs at a temperature of 41°C to yield numerous copies of ssRNA. To increase the efficacy and successful amplification of RNase H, Avian Myeloblastosis Reverse Transcriptase (AMV-RT) and RNA polymerase can be used (Malek et al., 1994).

It is commonly used for the amplification of ssRNA, which is converted to complementary DNA (cDNA) by reverse transcriptase (Chang et al., 2012). The current assay has the ability to generate 10^8 nucleic acid sequences within a short period of 30 minutes (Wernecke and Mullen, 2014), and detection can be carried out with the help of molecular beacons (Heo et al., 2019). The major advantage of this technology is that NASBA does not require a thermal cycler, and the amplification power is similar to that of RT-PCR assays. The reaction period is short, sensitive, reliable and not affected by any inhibitors (Honsvall and Robertson, 2017).

Detection of ASSVd based on NASBA along with real-time detection with molecular beacons was carried out by Heo et al. (2019). Viroid-specific primers were used for amplification of RNA, and molecular beacons help in the specific detection of viroid based on highly conserved regions. This assay allowed the detection of 1×10^4 to 1×10^{12} copies of viroid RNA/μL with high reproducibility and great precision. The combination of real-time quantitative assays and NASBA proved reliable, fast, sensitive and useful for routine detection and diagnosis, which greatly helped in the selection of viroid-free stock materials. Furthermore, the assay was found to be very useful for detecting low levels of viroid infection at early stages.

3.4 DNA OR RNA PROBE-BASED METHODS

The nucleic acids have unique sequences, and these sequences carry the identity of an organism. The probe-based assays mainly rely on the principle that single-stranded DNA or RNA specifically binds to a complementary sequence and forms a double-stranded hybrid. Thus, a piece of DNA or RNA sequence is used as a probe. A probe is a single-stranded sequence of DNA or RNA used to search for its complementary sequence in a sample genome. The probe is placed into contact with the sample under conditions that allow the probe sequence to hybridize with its complementary sequence. The probe is labeled with a radioactive or chemical tag that allows its binding to be visualized. In a similar way, labeled antibodies are used to probe a sample for the presence of a specific protein. Important probe-based methods are listed hereunder.

3.4.1 Southern Blotting Technique

It was invented by Edward Southern in 1975. It is a simple method for the detection of specific DNA molecules from the genome. In principle, this method involves the preparation of a DNA mixture using restriction enzymes and separation of DNA fragments by gel electrophoresis. The double-stranded DNA is denatured at this step, and the single-stranded DNA is transferred to the blotting membrane. The membranes are treated with a small DNA probe that is designed to have a complementary sequence in the genome. This allows the probe to hybridize, or bind, to a specific DNA fragment on the membrane. In addition, the probe has a label, which is typically a radioactive atom or a fluorescent dye. Thus, following hybridization, the probe permits the DNA fragment of interest to be detected from among the many different DNA fragments on the membrane. Grönberg et al. (2003) developed specific DNA probes targeting rDNA ITS of *Rhizoctonia* spp. And *Suillus ovines* for detection through the Southern dot blot technique. They reported that this technique was more discriminatoryas as target fungi were detected on the basis of combined probe hybridization-ITS capture and 5′ Cy-5 labeled ITS length polymorphism analysis.

3.4.2 Northern Blotting Technique

The northern blotting technique is yet another probe-based technique, with the target being RNA. Here, target RNA is purified from the tissue, loaded onto an agarose gel, electrophoresed and blotted onto a specialized filter paper. Furthermore, it is exposed to radioactively labeled probes, which then hybridize with their complementary sequences and are visualized using autoradiography. The technique provides low sensitivity, and the chances of RNA degradation during electrophoresis are higher. The probes for northern blotting are composed of nucleic acids with a complementary sequence to all or part of the RNA of interest. In situ hybridization or fluorescent in situ hybridization is much more accurate and specific compared to this method.

3.4.3 In Situ Hybridization

In situ hybridization is based on the hybridization of a labeled DNA or RNA probe with a complementary mRNA sequence. The method provides highly sensitive detection of mRNA contained in a single cell. The most commonly used probes in in situ hybridization are cDNAs, cRNAs and synthetically labeled nucleotides, which can be labeled with radioactive or non-radioactive nucleotides. Ellison et al. (2016) developed a non-radioactive in situ hybridization method for the detection of rust fungi in the host plants. A biotin-streptavidin reporter system was employed, targeting a 596-bp fragment of the 18S rRNA gene. Apart from detection, the technique also provides an avenue for observing the morphology of rust fungus in the host plant tissue. Moreover, in situ hybridization proves to be the classical technique for the validation of viral distribution in meristem tips. A non-radioactive, namely digoxygenin-labeled RNA probe, was employed in the detection of CMV in tobacco meristem tips for the validation of virus-free plantets (Mochizuki and Ohki, 2015). As the technique provides a direct visualization of viral RNA in plants, it is also a platform to study virus-plant interactions (Zou et al., 2022). A much more sensitive technique, using an RNAscope for in situ hybridization of target RNAs, was employed in the detection of viral as well as plant endogenous nucleic acids from cassava (Sheat et al., 2021).

3.4.4 Fluorescence In Situ Hybridization

Fluorescence in situ hybridization (FISH) is an important method in plant pathogen diagnostics as it aids in the visualization, identification, and enumeration of microbial cells in a given sample. An added advantage of this technique is that it also offers the detection of even unculturable organisms. For example, *Plasmopara obducens*, a biotrophic downy mildew-causing oomycete, was detected using the FISH technique by employing a species-specific 24-mer oligonucleotide probe targeting the ITS2 region (Salgado-Salazar et al., 2018). The major principle involved in this method is that nucleic acid sequences can be detected by using a fluorescently labeled probe that can hybridize to its specific complementary target sequence. Apart from aiding in the detection of plant pathogens, FISH can also be employed in colonization studies as well as to delineate the role of bacteria in pathogenesis. A specific FISH probe, KO 210 for *Xylella fastidiosa*, was designed to quantify the infection level and occlusion bodies in xylem vessels of petioles and branches of infected and healthy olive trees (Cardinale et al., 2018). Moreover, further accurate detection of pathogens, especially plant viruses, can be achieved with certain modifications to the in situ hybridization technique. For the sensitive detection of Cassava brown streak virus in plant tissues, the in situ hybridization technique was modified to include RNAscope technology. This technique comprises steps in the order of sample fixation and section preparation, pretreatment, hybridization, amplification, signal detection and mounting, and finally imaging. The imaging revealed the presence of viral RNA in the palisade, mesophyll and midrib tissues of infected plants, but no signals were received from non-infected plants. It also aids in the visualization of CBSV tropism during plant infection and disease development (Munganyinka et al., 2018).

3.5 POST-AMPLIFICATION TECHNIQUES

3.5.1 Microarrays

It is a technique for detecting the expression of many genes simultaneously. The DNA microarray is composed of small slides printed with tiny spots in defined positions. Each of these spots contains a DNA sequence, and these slides are referred to as DNA chips or gene chips (Robinson et al., 2000). The DNA molecules present on each slide act like probes in order to detect the expression of genes (Eshaque and Dixon, 2006). Since this technique does not require large-scale sequencing, it is very easy to use. Although this technique can monitor changes in gene expression as a whole, there are certain issues associated with it. The first issue is the requirement of a large amount of mRNA, and the other issue is cost and access (Singh and Kumar, 2013). Microarray technology has been used for the identification of the *Aspergillus candida* species (Singh and Kumar, 2013). The DNA microarray method is a sensitive technology with a broad spectrum of action. It can simultaneously detect thousands of diverse viruses and viroids (Martinelli et al., 2015).

3.5.2 Macroarray

The macroarray is otherwise called as the DNA array hybridization. In this method, there is no use of radioisotopes, and it only utilizes the sensitivity of DNA amplification (Singh and Kumar 2013). It is being used across the world as a standard molecular tool in diagnosis as well as epidemiological investigations. In this method, PCR amplification is combined with hybridization, and this combination gives increased sensitivity up to a thousandfold higher than PCR. Leinberger et al. (2005) used the macroarray technique in combination with PCR to detect the myobacterium at the species level. Furthermore, identification of *Alternaria alternata, Aspergillus fumigatus, Fusarium solani, Candida albicans* and *Cladosporium herbarum* has also been done using this method (Sato et al., 2010). The macroarray technique has been used for the identification of fungal and oomycete pathogens that cause disease in solanaceous crops (Zhang et al., 2008).

3.6 NEXT-GENERATION SEQUENCING

In next-generation sequencing (NGS), all or a subset of DNA present in a specimen, from an organism or host, is sequenced in parallel, generating millions of reads. Reads are basic elements produced by DNA sequencing and contain a sequence of sequential bases (Adenine, Glycine, Thymine, Cytosine) reflecting the DNA sequence of the input fragment. Assembling a complete genome sequence requires technologies that produce the longest reads, which typically range from 75 to 10,000 bp. NGS technologies allow for a generic (nonspecific method) approach to virus identification that does not require prior knowledge of the target pathogens but can provide a species/strain-specific result (Adams et al., 2009). Although they often only classify viruses at the genus level based on shared physical/biological characteristics. Sequencing (NGS) or high-flow sequencing (HTS) is a new approach for diagnostics. The development of NGS technologies has fueled innovative modalities for the detection and identification of plant pathogens (Chalupowicz et al., 2019). DNA isolation and fragmentation, library preparation, massively parallel sequencing, bioinformatics analysis, and variant/mutation annotation and interpretation are the main steps involved in DNA-based NGS (Qin, 2019). RNA sequencing (RNASeq) provides better coverage and higher resolution of the dynamic nature of the transcriptome. The Illumina Hi Seq platform is the most universally functional NGS platform for RNA Seq and has set the standard for NGS. The platform recently launched a desktop sequencer, Mi Seq (Kukurba and Montgomery, 2015).

Although it has been a decade since the introduction of NGS technologies, the applications of NGS sequencing platforms in plant pathology have been gaining popularity recently. It has been hailed as a much-improved method for the rapid, more efficient and cheaper detection of plant pathogens, including fungi, oomycetes, bacteria, nematodes and, more importantly, virus and viroids. Apart from detection, NGS platforms are also being applied in pathogen genome sequencing, in understanding its ecology and epidemiology and in the identification and characterization of prokaryotic and eukaryotic effector molecules. More importantly, it is now being successfully employed in the detection of novel viruses and viroids (Barba and Hadidi, 2015). For instance, NGS was employed in the discovery of negative sense RNA viruses in trees infected with Apple rubbery wood disease. The identified viral agents have been named as Apple rubbery wood virus (ARWV) 1 and 2, representing a new genus under the family *Phenuiviridae.* Numerous examples are available in the case of viruses identified using NGS platforms, namely Citrus yellow vein clearing virus, cassava brown streak virus, Cotton leafroll dwarf virus, Grapevine leaf roll associated virus-3, *Maize chlorotic mottle virus*, etc. (Varvara et al., 2018; Villamor et al., 2019). Nanopore sequencing, which is a single-molecule sequencing platform, enables the sequencing of long reads in negligible time with high-throughput data analysis. Chalupowicz et al. (2019) tested the sensitivity and ability of Oxford Nanopore sequencing platform in detecting several fungal, bacterial, viral and phytoplasmal plant pathogens. The sequencing technology was able to give rapid results, that is, within 1–2 hours of running of the nanopore sequencer, with the pathogenic organisms being identified to the genus or species level. Thus, it can serve as an excellent platform for routine laboratory diagnosis, as it is portable, low cost and has fast run times. NGS sequencing technologies have also gained popularity as an accurate and rapid method for indexing plant viral diseases. Recently, Prajapati et al. (2021) employed Illumina Hi Seq 2000 platform to fully sequence the Garlic virus A genome, comprising 8,793 nucleotides, excluding the poly (A) tail at the 3′ terminus with 5′ and 3′ untranslated regions (UTRs) of 193 and 123 nt, respectively. Further analysis of the genome revealed that the virus is a recombinant virus, with most of the genome segment derived from China GarV-E isolate as a minor parent and Russia ShV-X as a major parent. As garlic is propagated mainly through bulbs, the application of such sequencing technologies can aid in the indexing of planting materials, hence improving the quality of cultivation. Although NGS sequencing technologies put forth many advantages over conventional methods, the infrastructure required along with the expertise in data analysis can be taken as a major limitation posed by this platform.

3.7 VOLATILE SIGNATURES AS BIOMARKERS

Plants release various volatile organic compounds (VOCs) in response to different biotic and abiotic interactions. When the plants are subjected to either biotic or abiotic stresses, there is a considerable change in the amount and type of these plant VOCs released. These VOCs released by the plants are low molecular weight biomolecules that exist in the gaseous phase with high vapor pressure and a low boiling point. Thus, VOC profiling in plants as biomarkers is widely applied in disease detection, as many organic volatiles indicate the physiological health status of plants. Some examples of these volatile compounds include methyl salicylate, hexanol, 3 octanone, methyl jasmonates, phenylpropanoids, classes of terpenoids and indoles, etc. (Dudareva et al., 2006). These VOCs can be sampled either passively or actively. Active sampling involves the collection of immediate headspace surrounding the plant through high-capacity sorbent traps such as Carbopack D or TenaxTA (Martinelli et al., 2015). Passive sample pre-concentation involves the use of polymer traps like stir bar sorptive extraction (Baltussen et al., 2002) and solid-phase microextraction (Zhang et al., 1994; Pawliszyn, 1997). Mostly, two methods are used in the characterization and analysis of these compounds from plants: (1) gas chromatography-based analysis and (2) electronic nose systems. Several other spectrometry-based techniques, such as proton-transfer reaction mass spectrometry (PTR-MS) (Blake et al., 2009) and selected ion flow tube mass spectrometry (Smith and Spanel, 2005), are also employed in real-time profiling of VOC samples.

The application of electronic noses gains much popularity in postharvest diseases. For instance, an array of low-cost gas sensors could detect potato soft rot with up to 100% accuracy (Rutolo et al., 2018). Both GC-MS and PTR-MS analyses were carried out to distinguish healthy tubers from those infected with potato brown rot and ring rot. The products of bacterial metabolism, namely 2-propanol and 3-methylbutanoic acid, were used as specific markers for the detection of the pathogens

(Blasioli et al., 2014). Borowik et al. (2021) developed a low-cost electronic nose and successfully used it in the detection of *Fusarium oxysporum* and *Rhizoctonia solani.* Recently, Makarichian et al. (2022) utilized an electronic nose for the early detection of fungal infection in garlic, and the method proved to be non-destructive and rapid for the detection of pathogens like *Fusarium oxysporum* f.sp. *cepae, Alternaria embellisia* and *Botrytis alli* in garlic.

Application of VOCs in plant disease detection is a much recent innovation in diagnostics, for which instrumentation and equipment are still in development. Recognition of specific biomarker compounds and suitable methods for data analysis are essential components that need to be further advanced. Also, more sensitive sample collection methods that can detect ultra-low abundance compounds with lower detection rates also need to be developed.

3.8 CONCLUSION

Immigration of pathogens into the country and migration of the pathogens within the country create a major threat to reducing yield and productivity. On the other side, postharvest diseases cause at least 30% yield loss. In this juncture, it is imperative to detect the establishment of host-pathogen interaction well in advance, because it can aid in forewarning so as to make a decision on the management of yield loss imposed by inimical microbes. Hence, facilitating the availability of rapid detection kits using recent molecular tools at National Plant Protection Organization, quarantine stations at seaports and airports will strengthen the ability to intercept alien pathogens in the country. Furthermore, advances in the fields of metabolomics, proteomics and nucleic acid-based research have opened new vistas for developing rapid diagnostic kits that could benefit the seed industry and farmers by allowing them to distribute disease-free materials and monitor disease outbreaks and spread. Furthermore, in recent years, the occurrence of viruses in pulses, oilseeds, fruits and vegetables has reduced the yield and income of the farming community. At this juncture, advances in recent diagnostic tools will facilitate the supply of immunogenic, user-friendly kits to be used by scientists, private stakeholders and farmers to avoid yield loss. However, it would come into reality only when we sensitized policymakers and private stakeholders. Hence, policymakers should be advised to allocate an appropriate budget to promote the development of ready-to-use kits for early diagnosis, which would facilitate feeding millions of people across the world.

REFERENCES

Adams, I.P.; Glover, R.H.; Monger, W.A. et al. Next-generation sequencing and meta genomic analysis: a universal diagnostic tool in plant virology. *Mol. Plant Pathol.*, 2009, *10*(4), 537–545.

Alemu, K. Real-time PCR and its application in plant disease diagnostics. *Adv. Life Sci. Technol.*, 2014, *27*, 39–49.

Alexander, P.; Brown C.; Arneth, A. et al. Losses, inefficiencies and waste in the global food system, *Agric. Syst.*, 2017, *153*, 190–200.

Alkowni, R.; Alabdallah, O.; Fadda, Z. Molecular identification of tomato brown rugose fruit virus in tomato in Palestine. *J. Plant Pathol.*, 2019, *101*(3), 719–723.

Arif, M.; Dobhal, S.; Garrido, P.A. et al. Highly sensitive end-point PCR and SYBR green qPCR detection of *Phymatotrichopsis omnivora*, causal fungus of cotton root rot. *Plant Dis.*, 2014, *98*(9), 1205–1212.

Asibi, A.E.; Chai, Q.; Coulter, J.A. Rice blast: a disease with implications for global food security. *Agron.*, 2019, *9*(8), 451.

Asiello P.J.; Baeumner A.J. Miniaturized isothermal nucleic acid amplification, a review. *Lab Chip.*, 2011, *11*, 1420–1430.

Balodi, R.; Bisht, S.; Ghatak, A.; Rao, K.H. Plant disease diagnosis: technological advancements and challenges. *Indian Phytopathol.*, 2017, *70*(3), 275–281.

Baltussen, E., Cramers, C.A. and Sandra P. "Sorptive sample preparation–a review." *Analytical and bioanalytical chemistry*, 2002, 373, 3–22.

Barba, M; Hadidi A. An overview of plant pathology and application of next-generation sequencing technologies. *CAB Rev.*, 2015, *10*, 1–21.

Bariana, H.S.; Shannon, A.L.; Chu, P.W.G.; Waterhouse, P.M. Detection of five seedborne legume viruses in one sensitive multiplex polymerase chain reaction test. *Phytopathology*, 1994, *84*(10), 1201 1205.

Berardi, A; Lucioli, A.; Gatti, F.; Giuliano, G.; FIngerprint, S.P.B.P.M. Transcriptional profiling of tomato plants infected with the *Tomato yellow leaf curl sardinia virus* by long serial analysis of gene expression, *J Plant Pathol.*, 2005, *87*(4), 267–309.

Bhat, A.I.; Rao, G.P. "Rolling circle amplification (RCA)," In *Characterization of Plant Viruses – Methods and Protocols* (Humana, NY: Springer). 2020, 377–382.

Bhat, R.G.; Browne, G.T. Specific detection of *Phytophthora cactorum* in diseased strawberry plants using nested polymerase chain reaction. *Plant Pathol.*, 2010, *59*, 121–129.

Blake, R.S.; Monks, P.S.; Ellis, A.M. Proton-transfer reaction mass spectrometry. *Chem. Rev.*, 2009, *109*, 861–896.

Blasioli, S.; Biondi, E.; Samudrala, D. et al. Identification of volatile markers in potato brown rot and ring rot by combined GC-MS and PTR-MS techniques: study on *in vitro* and *in vivo* samples. *J. Agric. Food Chem.*, 2014, *62*(2), 337–347.

Borowik, P.; Adamowicz, L.; Tarakowski, R. et al. Development of a low-cost electronic nose for detection of pathogenic fungi and applying it to *Fusarium oxysporum and Rhizoctonia solani. Sensors*, 2021, *21*(17), 5868.

Caetano-Anolles, D. "Polymerase chain reaction," In *Brenner's Encyclopedia of Genetics*. Eds. S. Maloy and K. Hughes (San Diego: Academic Press). 2013, 392–395. doi:10.1016/B978-0-12-374984-0.01186-4

Capote, N.; Mara, A.; Aguado, A.; Snchez-Torres, P. "Molecular tools for detection of plant pathogenic fungi and fungicide resistance," In *Plant Pathology*. Ed. C.J.R. Cumagun (Glocester, RI: InTech). 2012, 151–202.

Cardinale, M.; Luvisi, A.; Meyer, J.B. et al. Specific fluorescence in situ hybridization (FISH) test to highlight colonization of xylem vessels by *Xylella fastidiosa* in naturally infected olive trees (Olea europaea L.). *Front. Plant Sci.*, 2018, *9*, 431.

Celïk, A. A novel technology for the one-step detection of prune dwarf virus: colorimetric reverse transcription loop-mediated isothermal amplification assay. *J. Crop Prot.*, 2022, *155*, 105910.

Chalupowicz, L.; Dombrovsky, A.; Gaba, V. et al. Diagnosis of plant diseases using the nanopore sequencing platform. *Plant Pathol.*, 2019, *68*(2), 229–238.

Chander, Y.; Koelbl, J.; Puckett, J. et al. A novel thermostable polymerase for RNA and DNA loop-mediated isothermal amplification (LAMP). *Front. Microbiol.*, 2014, *5*(395), 1–11.

Chang, C.C.; Chen, C.C.; Wei, S.C. et al. Diagnostic devices for isothermal nucleic acid amplification. Sensors, 2012, *12*(6), 8319–8337.

Chen, S.; Cao, Y.Y.; Li, T.Y; Wu, X.X. Simultaneous detection of three wheat pathogenic fungal species by multiplex PCR. *Phytoparasitica*, 2015, *43*(4), 449–460.

Choffnes, E.R.; Relman, D.A.; Olsen, L.A.; Hutton, R.; Mack, A. *Improving Food Safety Through a One Health Approach.* Forum on Microbial Threats, Institute of Medicine. (Washington, DC: National Academies Press). 2012.

Clark, M.F.; Adams, A.N. Characteristics of themicroplate method of enzyme-linked immunosorbent assay(ELISA) for the detection of plant viruses. *J. Gen. Virol.*, 1977, *34*, 475–483.

Davari, M.; van Diepeningen, A.D.; Babai-Ahari, A. et al. Rapid identification of *Fusarium graminearum* species complex using Rolling Circle Amplification (RCA). *J. Microbiol. Methods*, 2012, *89*(1), 63–70.

Dowley, L.J.; Grant, J.; Griffin, D. Yield losses caused by late blight (Phytophthora infestans (Mont.) de Bary) in potato crops in Ireland. *Irish J. Agric. Food Res.*, 2008, *47*, 69–78.

Dudareva, N.; Negre, F.; Nagegowda, D.A.; Orlova, I. Plant volatiles: recent advances and future perspectives. *CRC Crit. Rev. Plant Sci.*, 2006, *25*, 417–440.

Ellison, M.A.; McMahon, M.B.; Bonde, M.R. et al. In situ hybridization for the detection of rust fungi in paraffin embedded plant tissue sections. *Plant Methods*, 2016, *12*(1), 1–11.

Eshaque, B.; Dixon, B. Technology platforms for molecular diagnosis of cystic fibrosis. *Biotechnol. Adv.*, 2006, *24*(1), 86–93.

FAO. Food and Agriculture organization. 2022. https://www.fao.org/plant-health-2022. Accessed on March 2022.

Francois, P.; Tangomo, M.; Hibbs, J. et al. Robustness of a loop-mediated isothermal amplification reaction for diagnostic applications. *FEMS Immunol. Med. Microbiol.*, 2011, *62*, 41–48.

Garcia-Chapa, M.; Medina, V.; Viruel, M.A. et al. Seasonal detection of pear decline phytoplasma by nested-PCR in different pear cultivars. *Plant Pathol.*, 2003, *52*(4), 513–520.

Gonzalez, F.; Zamorano, A.; Pino, A.M. et al. Identification of phytoplasma belonging to X-disease group in cherry in Chile. *Bull. Insectol.*, 2011, *64*(Supplement), S235–S236.

Gore, P.M.; Rakhonde, R.I.P. Management of bulb rot of onion caused by *Erwinia carotovora* pv. *carotovora. J. Pharmacogn. Phytochem.*, 2020, *9*(4), 3294–3301.

Grönberg, H.; Paulin, L.; Sen, R. ITS probe development for specific detection of *Rhizoctonia* spp. and *Suillus bovinus* based on Southern blot and liquid hybridization-fragment length polymorphism. *Mycol Res.*, 2003, *107*, 428–438.

Gu, L.; Yan, W.; Liu, L. et al. Research progress on rolling circle amplification (RCA)-based biomedical sensing. *Pharmaceuticals*, 2018, *11*(2), 35.

Ha, Y.; Fessehaie, A.; Ling, K.S. et al. Simultaneous detection of *Acidovorax avenae* subsp. *citrulli* and *Didymella bryoniae* in cucurbit seedlots using magnetic capture hybridization and real-time polymerase chain reaction. *Phytopatholgy*, 2009, *99*(6), 666–678.

Haible, D.; Kober, S.; Jeske, H. Rolling circle amplification revolutionizes diagnosis and genomics of geminiviruses. *J. Virol. Methods*, 2006, *135*(1), 9–16.

Hariharan, G.; Prasannath, K. Recent advances in molecular diagnostics of fungal plant pathogens: a mini review. *Front. Cell. Infect. Microbiol.*, 2021, *10*, 829.

He, L.; Shi, J.; Zhao, Z. et al. First report of crown gall of kiwifruit (Actinidia deliciosa) caused by *Agrobacterium fabacearum* in China and the establishment of loop-mediated isothermal amplification technique. *Int. J. Mol. Sci.*, 2022, *23*(1), 207.

Heo, S.; Chung, Y.S. Rapid real-time detection method of ACLSV and ASSVd for apple quarantine field. *Plant Biotechnol. Rep.*, 2021, *15*(2), 187–195.

Heo, S.; Kim, H.R.; Lee, H.J. Development of a quantitative real-time nucleic acid sequence based amplification (NASBA) assay for early detection of apple scar skin viroid. *Plant Pathol. J.*, 2019, *35*(2), 164.

Honsvall, B.K.; Robertson, L.J. From research to lab to standard environmental analysis tool. Will NASBA make keep? *Water Res.*, 2017, *109*, 389–397.

Hsieh, C.W.; Chuang, Y.Y.; Lee, M.Z. et al. First Inventory of fungi in symptomless and symptomatic Chinese Mesonaindicates phytopathological threat. *Plant Dis.*, 2020, *104*(9), 2391–2397.

Iturralde Martinez, J.F.; Flores, F.J.; Koch, A.R.; Garzón, C.D.; Walker, N.R. Multiplex end-point PCR for the detection of three species of *Ophiosphaerella* causing spring dead spot of bermudagrass. *Plant Dis.*, 2019, *103*(8), 2010–2014.

Jacobsen, C.S. Microscale detection of specific bacterial DNA in soil with a magnetic capture-hybridization and PCR amplification assay. *Appl. Environ. Microbiol.*, 1995, *61*(9), 3347–3352.

Javaheri Tehrani, S.; Aliabadian, M.; Fata, A.; Najafzadeh, M.J. Rolling circle amplification (RCA): an approach for quick detection and identification of fungal species. *Mycol. Res.*, 2014, *1*(1), 55–62.

Johnson, K.L.; Zheng, D.; Kaewnum, S. et al. Development of a magnetic capture hybridization real-time PCR assay for detection of tumorigenic *Agrobacterium vitis* in grapevines. *Phytopathology*, 2013, *103*(6), 633–640.

Kieser, R.E.; Budowle, B. Rolling circle amplification: a (random) primer on the enrichment of an infinite linear DNA template. *WIREs Forensic Sci.*, 2020, *2*, e1359. doi:10.1002/wfs2.1359

Kim, J.H.; Lee, S.; Park, E.R.; Jang, W.C. Development of a highly sensitive and rapid detection method for Pea enation mosaic virus using loop-mediated isothermal amplification assay. *J. Virol. Methods*, 2022, *300*, 114427.

Kolychikhina, M.S.; Beloshapkina, O.O.; Phiri, C. Change in potato productivity under the impact of viral diseases. In *IOP Conference Series: Earth and Environmental Science* (Vol. 663, No. 1, p. 012035). IOP Publishing. 2021.

Kukurba, K.R.; Montgomery, S.B. RNA sequencing and analysis. *Cold Spring Harbor Protoc.*, 2015, *2015*(11), 951–969.

Kumar, R.; Gupta, A.; Srivastava, S. et al. "Diagnosis and detection of seed-borne fungal phytopathogens," In *Seed-Borne Diseases of Agricultural Crops: Detection, Diagnosis & Management*. Eds. R. Kumar and A. Gupta (Singapore: Springer). 2020, 107–142.

Lamb, L.E.; Bartolone, S.N.; Tree, M.O. et al. "Rapid detection of Zika virus in urine samples and infected mosquitos by reverse transcription-loop-mediated isothermal amplification". *Sci. Rep.*, 2018, *8*(1), 3803.

Langrell, S.R.H.; Barbara, D.J. Magnetic capture hybridisation for improved PCR detection of *Nectria galligena* from lignified apple extracts. *Plant Mol. Biol. Rep.*, 2001, *19*(1), 5–11.

Le, D.T.; Vu, N.T. Progress of loop-mediated isothermal amplification technique in molecular diagnosis of plant diseases. *Appl. Biol. Chem.*, 2017, *60*, 169–180.

Leinberger, D.M.; Schumacher, U.; Autenrieth, I.B.; Bachmann, T.T. Development of a DNA microarray for detection and identification of fungal pathogens involved in invasive mycoses. *J. Clin. Microbiol.*, 2005, *43*(10), 4943–4953.

Lin, H.; Jiang, X.; Yi, J. et al. Molecular identification of *Neofabraea* species associated with bull's-eye rot on apple using rolling-circle amplification of partial EF-1α sequence.*Can. J. Microbiol.*, 2018, *64*(1), 57–68.

Links, M.G.; Dumonceaux, T.J.; Hemmingsen, S.M.; Hill, J.E. The chaperonin-60 universal target is a barcode for bacteria that enables de novo assembly of metagenomic sequence data. *PLoS One*, 2012, *7*(11), 49755.

Little, A.; Rezaian, M.A. Improved detection of grapevine leafroll-associated virus 1 by magnetic capture hybridisation RT-PCR on a conserved region of viral RNA. *Arch. Virol.*, 2006, *151*(4), 753–761.

Ma, Z.; Michailides, T.J. Approaches for eliminating PCR inhibitors and designing PCR primers for the detection of phytopathogenic fungi. *Crop Prot.*, 2007, *26*, 145–161. doi:10.1016/j.cropro.2006.04.014

Maize AICRP. *Annual Progress Report Rabi*. Eds. O.P. Yadav, B. Kumar, G.K. Chikkappa et al. (New Delhi: Indian Institution of Maize Research). 2014.

Makarichian, A.; Chayjan, R.A.; Ahmadi, E.; Zafari, D. Early detection and classification of fungal infection in garlic (*A. sativum*) using electronic nose. *Comput. Electron. Agric.*, 2022, *192*, 106575.

Malek, L.; Sooknanan, R.; Compton, J. Nucleic acid sequence-based amplification (NASBA). *Methods Mol. Biol.*, 1994, *28*, 253–260.

Manulis, S., Kogan, N., Valinsky, L., Dror, O. and Kleitman, F. (1999). Detection of plant pathogens in flowers by the PCR technique. *Acta Hortic.* 482, 117–122. DOI: 10.17660/ActaHortic.1999.482.17.

Martinelli, F.; Scalenghe, R.; Davino, S. et al. Advanced methods of plant disease detection. A review. *Agron. Sustain. Dev.*, 2015, *35*(1), 1–25.

Millenium Project (MP). 2022. https://www.millennium-project.org/challenge-3/ Accessed on 10 March 2022.

Miller, S.A.; Martin, R.R. Molecular diagnosis of plant disease. *Annu. Rev. Phytopathol.*, 1988, *26*, 409–432.

Mochizuki, T.; Ohki, S.T. "Detection of plant virus in meristem by immunohistochemistry and in situ hybridization," In *Plant Virology Protocols* (New York: Humana Press). 2015, 275–287.

Mudiyanselage, A.H.; Jones, E.E.; Jaspers, M.V.; Walter, M.; Ridgway, H.J. A one step nested PCR method for detection of *Peronospora sparsa*, the downy mildew pathogen, in boysenberry (Rubus ursinus). *Eur. J. Plant Pathol.*, 2021, *160*, 1–9.

Munganyinka, E.; Margaria, P.; Sheat, S. et al. Localization of cassava brown streak virus in Nicotianarustica and cassava Manihotesculenta (Crantz) using RNAscope® in situ hybridization. *Virol. J.*, 2018, *15*(1), 1–11.

Nagamine, K.; Hase, T.; Notomi, T. Accelerated reaction by loop mediated isothermal amplification using loop primers. *Mol. Cell Probes*, 2002, *16*(3), 223–229.

Nagendran, K.; Kumari, S.; Kumar, R.; Pandey, K.K.; Singh, J. Loop-mediated isothermal amplification assay for quicker detection of tomato leaf curl Joydebpur virus infection in chilli. *J. Virol. Methods*, 2022, *302*, 114474.

Nakkeeran, S.; Rajamanickam, S.; Saravanan, R.; Vanthana, M.; Soorianatha sundaram, K. Bacterial endophytome-mediated resistance in banana for the management of *Fusarium* wilt. *Biotech*, 2021, *11*(6), 1–13.

Nellemann, C.; MacDevette, M.; Manders, T. et al. *The Environmental Food Crisis – The Environment's Role in Averting Future Food Crises. A UNEP Rapid Response Assessment* (Arendal, Norway: United Nations Environment Programme, GRID-Arendal). 2009.

Notomi, T.; Okayama, H.; Masubuchi, H. et al. Loop-mediated isothermal amplification of DNA. *Nucleic Acids Res.*, 2000, *28*(12), E63. doi:10.1093/nar/28.12.e63

Oerke, E.C. Crop losses to pests. *J. Agric. Sci.*, 2006, *144*, 31–43.

Omer, Z.S.; Wallenhammar, A.C.; Viketoft, M. Development of loop-mediated isothermal amplification assay for rapid detection and analysis of the root-knot nematode *Meloidogyne hapla* in soil. *Horticulturae*, 2022, *8*(2), 87.

Pal, A. "Real-time or quantitative PCR," In *Protocols in Advanced Genomics and Allied Techniques* (New York: Springer). 2022, 181–209.

Panno, S.; Ruiz-Ruiz, S.; Caruso, A.G. et al. Real-time reverse transcription polymerase chain reaction development for rapid detection of *Tomato brown rugose fruit virus* and comparison with other techniques. *Peer J.*, 2019, *7*, e7928.

Pawliszyn, J. *Solid Phase Microextraction-Theory and Practice* (Weinheim: Wiley-VCH). 1997.

Pelt-Verkuil, E.; Belkum, A.; John, P. "A brief comparison between *in vivo* DNA replication and in vitro PCR amplification," In *Principles and Technical Aspects of PCR Amplification* (Netherlands: Springer). 2008, 9–15.

Pimentel, D.; Zuniga, R.; Morrison, D. Update on the environmental and economic costs associated with alien-invasive species in the United States. *Ecol. Econ.*, 2005, *52*, 273–288.

Prajapati, M.R.; Manav, A.; Singh, J. et al. Identification of Garlic virus A infecting Allium sativum L. through next generation sequencing technology. *J. Hortic. Sci. Biotechnol.*, 2021, *97*, 1–10.

Puertolas, A.; Bonants, P.J.; Boa, E.; Woodward, S. Application of real-time PCR for the detection and quantification of oomycetes in ornamental nursery stock. *J. Fungi.*, 2021, *7*(2), 87.

Puig, A.S. Detection of *Cacao Mild Mosaic Virus* (CaMMV) using nested PCR and evidence of uneven distribution in leaf tissue. *Agronomy*, 2021, *11*(9), 1842.

Qin, D. Next-generation sequencing and its clinical application. *Cancer Biol. Med.*, 2019, *16*(1), 4.

Rana, D.S.; Dass, A.; Rajanna, G.A.; Kaur, R. Biotic and abiotic stress management in pulses. *Indian J. Agron.*, 2016, *61*, S238–S248.

Roberts, M.J.; Schimmelpfennig, D.E.; Ashley, E. et al. *The Value of Plant Disease Early-Warning Systems: A Case Study of USDA's Soybean Rust Coordinated Framework* (Washington, DC: United States Department of Agriculture, Economic Research Service). 2006.

Robinson, B.; Erle, D.; Jones, D. et al. Recent advances in molecular biological techniquesand their relevance to pulmonary research. *Thorax.*, 2000, *55*(4), 329–339.

Rutolo, M.F.; Clarkson, J.P., Covington, J.A. The use of an electronic nose to detect early signs of soft-rot infection in potatoes. *Biosyst. Eng.*, 2018, *167*, 137–143.

Salgado-Salazar, C.; Bauchan, G.R.; Wallace, E.C.; Crouch, J.A. Visualization of the impatiens downy mildew pathogen using fluorescence in situ hybridization (FISH). *Plant Methods*, 2018, *14*(1), 1–9.

Sato, T.; Takayanagi, A.; Nagao, K. et al. Simple PCRbased DNA microarray system to identify human pathogenic fungi in skin. *J. Clin. Microbiol.*, 2010, *48*(7), 2357–2364.

Savary, S.; Ficke, A.; Aubertot, J.; Hollier, C. Crop losses due to diseases and their implications for global food production losses and food security. *Food Secur.*, 2012, *4*, 519–537.

Savary, S.; Willocquet, L.; Pethybridge, S.J. et al. The global burden of pathogens and pests on major food crops. *Nat. Ecol. Evol.*, 2019, *3*(3), 430-439.

Saxena, K.B.; Lava Kumar, P.; Kulkarni, N.K. et al. Sterility mosaic disease the green plague of pigeonpea. *Plant Dis.*, 2004, *88*(5), 436–445.

Schaad, N.W.; Berthier-Schaad, Y.; Sechler, A.; Knorr, D. Detection of *Clavibacter michiganensis* subsp. *sepedonicus* in potato tubers by BIO-PCR and an automated real-time fluorescence detection system. *Plant Dis.*, 1999, *83*(12), 1095–1100.

Schaad, N.W.; Cheong, S.S.; Tamaki, S. et al. A combined biological and enzymatic amplification (BIO-PCR) technique to detect *Pseudomonas syringae* pv. *phaseolicola* in bean extracts. *Phytopathology*, 1995, *85*, 243–248. doi:10.1094/Phyto-85-243.

Shankara, K.; Pradhan, R.S.; Patole, S.P. Yield loss assessment due to turcicum leaf blight of maize caused by *Exserohilum turcicum. Int. J. Curr. Microbiol. App. Sci.*, 2017, *6*(10), 2888–2891.

Shao, Y.; Chen, A.; Chang, W. et al. Selection of appropriate reference genes in different tissues of *Vaccinium dunalianum* Wight by quantitative real-time PCR for gene expression studies. *Can. J. Plant Sci.*, 2022, *102*, 1–9.

Sheat, S.; Margaria, P.; Winter, S. Differential tropism in roots and shoots of resistant and susceptible cassava (*Manihot esculenta* Crantz) infected by cassava brown streak viruses. *Cells*, 2021, *10*(5), 1221.

Shi, L.; Perkins, R.G.; Fang, H.; Tong, W. Reproducible and reliable microarray results through quality control: good laboratory proficiency and appropriate data analysis practices are essential. *Curr. Opin. Biotechnol.*, 2008, *19*, 10–18.

Sikdar, P.; Okubara, P.; Mazzola, M.; Xiao, C.L. Development of PCR assays for diagnosis and detection of the pathogens *Phacidiopycnis washingtonensis* and *Sphaeropsis pyriputrescens* in apple fruit. *Plant Dis.*, 2014, *98*(2), 241–246.

Singh, A; Kumar, N. A review on DNA microarray technology. *Int. J. Curr. Res. Rev.*, 2013, *5*(22), 01–05.

Sint, D.; Raso, L.; Traugott, M. Advances in multiplex PCR: balancing primer efficiencies and improving detection success. *Methods Ecol. Evol.*, 2012, *3*(5), 898–905.

Šišić, A.; Oberhänsli, T.; Baćanović-Šišić, J.; Hohmann, P.; Finckh, M.R. A novel real time PCR method for the detection and quantification of *Didymella pinodella* in symptomatic and asymptomatic plant hosts. *J. Fungi*, 2022, *8*(1), 41.

Smith, D.; Spanel, P. Selected ion flow tube mass spectrometry (SIFT-MS) for on line trace gas analysis. *Mass Spectrom Rev.*, 2005, *24*, 661–700. doi:10.1002/mas.20033.

Srivastava, S.; Singh, V.; Gupta, P.S.; Sinha, O.K.; Baitha, A. Nested PCR assay for detection of sugarcane grassy shoot phytoplasma in the leafhopper vector *Deltocephalus vulgaris*: a first report. *Plant Path.*, 2006, *55*(1), 25–28.

Strange, R.N.; Scott, P.R. Plant disease: a threat to global food security. *Ann Rev Phytopathol.*, 2005, *43*, 83–116. 10. International Food Policy Research Institute. Food Security. https://www.ifpri. org/topic/food-security. Accessed 5 February 2021.

Sumit, S.; Diksha, S.; Anita, K. An overview of bacterial leaf blight disease of rice and different strategies for its management. *Int. J. Curr. Microbiol. App. Sci.*, 2020, *9*(04), 2250–2265.

Suzaki, K.; Sawada, H.; Kisaki, G. Loop-mediated isothermal amplification of bacterial effector genes to detect *Pseudomonas syringae* pv. *actinidiae* biovars 1 and 3. *J. Gen. Plant Pathol.*, 2022, *88*(1), 2–9.

Tegli, S.; Cerboneschi, M.; Libelli, I.M.; Santilli, E. Development of a versatile tool for the simultaneous differential detection of *Ps BMC Microbiol. eudomonas savastanoi* pathovars by End Point and Real-Time PCR. *BMC Microbiol.*, 2010, *10*(1), 1–14.

UNICEF.; Malnutrition in Children - UNICEF data 2022. https://data.unicef.org/topic/nutrition/malnutrition/#:~:text=While%20the%202021%20edition %20of,45.4%20million% 20suffered% 20from% 20wasting. Accessed on 10 March 2022.

United Nations. Sustainable development goals. https://www.un.org/sustainable development /sustainable-development-goals/. Accessed on 5 February 2022.

USDA. United State Department of Agriculture. https://www.ars.usda.gov//docs/cereal-rusts/wheat-leaf-rust. Accessed on 10 March 2022.

Van Emmerik, C.L.; Gachulincova, I.; Lobbia, V.R. et al. Ramified rolling circle amplification for synthesis of nucleosomal DNA sequences. *Anal. Biochem.*, 2020, *588*, 113469. doi:10.1016/j.ab.2019.113469.

Varvara, I.M.; Angelantonio, M.; Pasquale, S.; Ana, B. Recent advances on detection and characterization of fruit tree viruses using high-throughput sequencing technologies. *Viruses*, 2018, *10*, 436.

Villamor, D.E.V.; Ho, T.; Rwahnih, M.; Martin, R. R.; Tzanetakis, I.E. High throughput sequencing for plant virus detection and discovery. *Phytopathology*, 2019, *109*, 716–725.

Villarino, M.; De Cal, A.; Melgarejo, P.; Larena, I. Development of a multiplex PCR for the identification of *Fusarium solani* and *F. oxysporum* in a single step. *J. Plant Dis. Prot.*, 2021, *128*, 1–16.

Walcott, R.R.; Gitaitis, R.D.; Langston, D.B. Detection of *Botrytis aclada* in onion seed using magnetic capture hybridization and the polymerase chain reaction. *Seed Sci. Technol.*, 2004, *32*(2), 425–438.

Wernecke, M.; Mullen, C. "Molecular biology, molecular biology in microbiological analysis," In *Encyclopedia of Food Microbiology*. Eds. C.A. Batt and M.L. Tortorello (Oxford: Academic Press), 2014, 808–814.

WRI-World Resource Institute. 2022. https://www.wri.org/insights/how-sustainably-feed-10-billion-people-2050-21-charts. Accessed on 10 March 2022.

Yamamoto, M.; Wakatsuki, T.; Hada, A.; Ryo, A. Use of serial analysis of gene expression (SAGE) technology. *J. Immunol. Methods*, 2001 *250*(1–2), 45–66.

Yu, C.; Yang, C.; Song, S. et al. Development of a sensitive LuminexxMAP-based microsphere immunoassay for specific detection of Iris yellow spot virus. *Virol. J.*, 2018, *15*(1), 1–7.

Zanoli, L.M.; Spoto, G. Isothermal amplification methods for the detection of nucleic acids in microfluidic devices. 2013, *Biosensors*, *3*(1), 18–43.

Zhang, N..; McCarthy, M.L.; Smart, C.D. A macroarray system for the detection of fungal and oomycete pathogens of solanaceous crops. *Plant Dis.*, 2008, *92*(6), 953–960.

Zhang, Z.; Yang, M.J.; Pawliszyn, J. Solid-phase micro-extraction. *Anal. Chem.*, 1994, *66*, 844–853. doi:10.1021/ac00089a716

Zhang, Y.; Xie, Z.; Wang, Y. et al. Development of immunocapture reverse transcription loop-mediated isothermal amplification assays to detect both lily symptomless virus and cucumber mosaic virus. *Ann. Appl. Biol.*, 2022, *180*(1), 132–139.

Zhao, Y.Q.; Wu, Y.H;, Zhao, X.X.; An, M.N.; Chen, J.G. Study on the Taqman real-time PCR to the detection and quantification of *Rhizoctonia solani* AG-3 of tobacco target spot. *Adv. Mater. Res.*, 2014, *1010–1012*, 80–83.

Zou, J.Z., Liu, D.S., Tong, X., Zhang, X.P.; Wang, X.B. "RNA in situ hybridization of detecting cucumber mosaic virus in shoots of *Nicotiana benthamiana* plants," In *Plant Virology* (New York: Humana). 2022, 283–296.

4 Basic Research Advances on Bacterial T3SS-Effectors Mediated Blight Developments in Rice and Pomegranate

An Overview from India

Kalyan K. Mondal
ICAR-National Institute of Biotic Stress Management

4.1 INTRODUCTION

In India, the bacterial pathogens of crop plants cause substantial yield losses. Among the major bacterial diseases, bacterial blight (BB) has been a constant threat to the cultivation of rice and pomegranate (Mondal, 2011). The present disease management strategies through the use of bactericides are in place, but effective management through exploiting host resistance is not fully harnessed. Plant defence against pathogens has been categorized based on how resistance is manifested. Preferred cognates the presence of a pathogen by reading the pathogen-specific signatures (referred to as pathogen-associated molecular patterns or PAMP molecules) by its pattern recognition receptors (PRRs). Immediately after recognition, PRR transduces signals to its downstream members to activate defence via mitogen-activated protein kinase (MAPK) cascades. The activation of defence in plants is expressed through immune responses such as hypersensitive response (HR), callose deposition and accumulation of reactive oxygen species (ROS). Generally, PAMP molecules (such as bacterial flagellin or lipopolysaccharide) are recognized by plants through membrane-localized receptors and result in PAMP-triggered immunity (PTI) (Jones and Dangl, 2006; Schwessinger and Ronald, 2012). Some successful pathogens deploy effectors that interfere with PTI, resulting in effector-triggered susceptibility (ETS), thus contributing to the pathogen's virulence. The effectors are transported directly into the plant cytosol through a proteinaceous tubular channel, referred to as a type three secretion system (T3SS), and the secreted effectors are called T3SS-effectors. In bacteria, the components of the T3SS are encoded by *hrp* (HR and pathogenicity) genes (Alfano and Collmer, 1997). Thus, in the absence of functional T3SS, bacteria cannot infect and colonize the host plants (Alfano and Collmer, 2004; Mudgett, 2005).

 DOI: 10.1201/9781003403142-4

The bacteria employ T3SS effector arsenals to suppress PTI (Nurnberger et al., 2004; Ausubel, 2005; Jones and Dangl, 2006).

T3SS effectors manipulate various plant processes in such a way that they benefit the bacteria, such as the suppression of PTI (Büttner, 2016). In contrast, plants that recognize T3SS effectors through resistance (R) proteins initiate effector-triggered immunity (ETI) (Khan et al., 2016). ETI mostly manifests through HR, a programmed cell death that limits bacterial growth and multiplication in its hosts (Klement and Goodman, 1967). To overcome ETI, some pathogens evolve via natural selection by losing or modifying the recognized effector gene or by acquiring new effectors to suppress ETI, which results in plant susceptibility. In short, we can say that there is a tug of war between host and pathogen that either leads to resistance or susceptibility. The PTI and ETI are characterized by different defence-associated cellular or biochemical markers, including induction of MAP kinases, callose deposition, accumulation of ROS and a Ca^{2+}-burst (Meng and Zhang, 2013; Buscaill and Rivas, 2014; Cui et al., 2015). The widely accepted four-phased *Zig-Zag* model of Jones and Dangl (2006) represents how these two branches of the plant immune system interact. This chapter intends to provide an overview of the progress made by India in the field of basic strategic research, particularly in the areas of effector-driven susceptibility and resistance in plants and their sensible scope towards managing these two bacterial plant pathogens.

4.2 BACTERIAL BLIGHT OF RICE (*XANTHOMONAS ORYZAE* PV. *ORYZAE*)

BB of rice caused by *Xanthomonas oryzae* pv. *oryzae* (Xoo) is one of the important diseases of rice worldwide (Singh et al., 2011a,b; Mondal and Mani, 2012; Mondal et al., 2012). Xoo isolates from 13 states of north-west and north-eastern India were grouped into 6 races based on disease reaction to 12 near-isogenic lines (NILs). The distribution of the races varied across the states. It is interesting to note the co-existence of both less virulent and more virulent races in the Xoo population in a particular location. Race 4 was most virulent and predominant in the north-west states, including Punjab, Haryana, Uttar Pradesh, Uttarakhand and Himachal Pradesh (Mondal et al., 2020b), while races 2 and 3 were predominant in the north-eastern states, including Assam, Meghalaya and Manipur.

4.2.1 Detailed Understanding of T3SS Effectors of Xoo and Their Significance in Resistance Modulation

The sequences and composition of T3SS effectors vary widely across the strains of Xanthomonads, suggesting that T3SS effectors play important roles in bacterial pathogenicity, particularly in displaying host-specific virulence. The bacterial effectors are of two types: transcription activator-like (TAL) and non-TAL, or Xanthomonas outer protein (Xop) types. The effectors that are found across the species of phytopathogenic Xanthomonads are referred to as core (conserved or shared) effectors, while the non-shared group (specific effectors) are variable across the species. TAL effectors primarily promote plant susceptibility by inducing the expression of *S* genes (Susceptibility genes like *SWEET* genes) (Mondal et al., 2021). But, in some cases, plants have effectively evolved to recognize them and thereby utilize them as avirulence factors to trigger resistance, e.g., AvrXa7, AvrXa10 and AvrXa27 (Gu et al., 2005; Yang et al., 2005).

The composition and functional role of Xoo TAL effectors are well investigated in some strains. For example, there are 16 TAL genes in strain MAFF311018, 19 in strain PXO99A, and only 6 in strain KACC10331 of Xoo (Lee et al., 2005; Ochiai et al., 2005; Salzberg et al., 2008). Though the non-TAL or Xop effectors of Xoo were shown to play major virulence determinants, they are less extensively studied (Kim et al., 2008; White et al., 2009; Kumar and Mondal, 2013; Mondal, 2017). The functions of some effectors, such as XopN, XopD, XopJ, XopZ and XopR in Xoo, were predicted based on similarities with their known motifs or sequences as reported in other Xanthomonads (Kim et al., 2009; Bartetzko et al., 2009; Song and Yang, 2010; Akimoto-Tomiyama, 2012; Kumar and Mondal, 2013). XopN from *X. campestris* pv. *vesicatoria* has been demonstrated to suppress PTI by interacting with tomato atypical receptor-like kinase 1 (TARK1) and four tomato 14-3-3 isoforms (Kim et al., 2009). XopD from *X. campestris* pv. *vesicatoria* is predicted to function in the "de-SUMOlyzation" of plant proteins (Kim et al. 2008). XopJ from *X. campestris* pv. *Vesicatoria*, being structurally similar to the C55 group of cysteine protease, was proposed to be involved in the ubiquitin/proteasome system and thus capable of removing SUMO proteins, resulting in the destabilization or inactivation of SUMOlyated proteins (Bartetzko et al., 2009). XopZ from the Xoo Philippine strain PXO99A had been demonstrated to be involved in the suppression of PTI in rice (Song and Yang, 2010). Interestingly, two identical copies of $XopZ_{PXO99A}$ were shown to be essential for the full virulence of the strain PXO99A. Recently, Akimoto-Tomiyama et al. (2012) suggested that XopR from the Xoo Japanese strain MAFF311018 functions as a repressor of MAMP-induced plant innate immune responses.

The detailed characterization of Xoo race 4 revealed the presence of 21 Xop- and 15 TAL-type III effector families (Mondal et al., 2014). Among TALEs, both complete (tTALEs) as well as incomplete/pseudo/iTALEs are identified (Mondal et al., 2020b). The functional analysis based on loss-and-gain of effectors revealed that two, namely, XopF (Mondal et al., 2016) and XopR (Verma et al., 2018, 2019), contribute immensely during BB development. Xoo mutants (Xoo *ΔxopF* or Xoo *ΔxopR*) showed significantly reduced ability for *in planta* colonization, BB intensity but induced more callose deposition. Xoo mutants caused a significant fold increase in rice PTI marker transcripts. Both effectors localize to the plasma membrane. Recently, two rice interactors for XopF, namely photosystem I reaction subunit V (PSI-G) and cyclophilin II, were identified (Figure 4.1).

PSI-G interacts with many proteins of photosystem I, while cyclophilin II associates with proteins involved in protein folding, signal transduction and ubiquitination. This suggests that XopF interacts with the two interactors to fulfil its common goal of subverting plant immunity, either through interfering photosystem I or through destabilizing plant immune protection systems, like cyclophilin. Altogether, this insight into the Xoo-effectors vs rice exposes novel target gene(s) in rice for their sensible exploitation in the BB resistance programme.

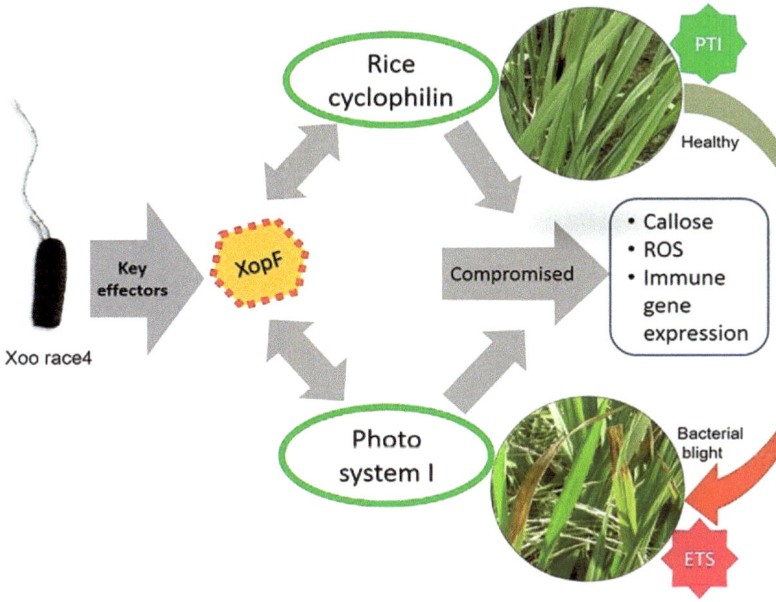

FIGURE 4.1 Schematic presentation of the underlying interaction between XopF T3SS-effector of Xoo and rice proteins leading to modulation of PTI to ETS via compromised plant defence.

A type III-deficient mutant of Xoo can induce a basal plant defence response like callose deposition and thereby immunize rice against subsequent Xoo infection. A T2S– T3S– double mutant exhibited a substantial reduction in the ability to evoke these responses (Jha et al., 2007).

Besides, disease management options using the biological control agent *Pseudomonas fluorescens* strain MBPF-01 and novel chemicals such as nanocopper have also been documented (Mondal et al., 2010). Resistance genes *Xa4* and *xa13*, which were found to be effective against the present races, could serve as potential candidates for disease resistance programmes (Mondal et al., 2014).

4.3 BACTERIAL BLIGHT OF POMEGRANATE (*XANTHOMONAS AXONOPODIS* PV. *PUNICAE*)

The recent year's BB of pomegranate has emerged in epidemic form, particularly in southern India, covering Karnataka, Andhra Pradesh, and Maharashtra (Mondal and Sharma, 2009; Kumar et al., 2016; Sharma et al., 2017). The disease, caused by *Xanthomonas axonopodis* pv. *punicae* (Xap), affects almost all parts of the plants, including fruits, reducing the exportable fruit quality. Phylogenetic analysis of 15 Xap strains representing Delhi, Maharashtra and Andhra Pradesh based on *gyrB* and 16S rDNA loci indicated that Xap shared 99.4% identity with *X. citri* subsp. *malvacearum* (Mondal et al., 2012). The Xap genome of 4.94 Mb contains 4,385 predicted coding regions and 50 tRNA and 3 rRNA genes. The G+C content in the

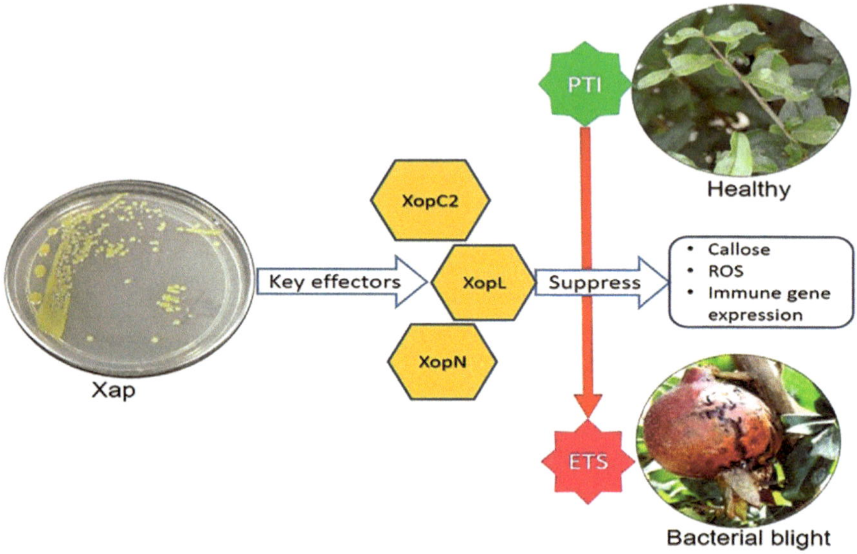

FIGURE 4.2 Schematic presentation of the underlying interaction between three T3SS-effectors of Xap and pomegranate leading to modulation of PTI to ETS via compromised plant defence.

Xap genome is 64.9% (Sharma et al., 2012). The presence of bacteria in the infected plant tissue (as low as 10^2 cells per g of leaf tissue) could now be detected through PCR-based amplification of 491 bp of *gyrB* gene regions (Mondal et al., 2012).

Disease management through application of bactericides, streptomycin and copper-based fungicides has limited success owing to their cost and inconsistent performance against the existing strains. New strategies using biological control agents and new-generation molecules such as nanocopper (Mondal and Mani, 2012) have shown some promise. However, novel strategies targeting genes involved in pathogenesis need to be explored. In this direction, recent studies on understanding the precise role of bacterial type III effectors in pathogenesis are pertinent. Xap contains six type III effector proteins (Kumar and Mondal, 2013). Functional investigation revealed that Xap depends on three major effectors, such as XopC2, XopL and XopN, for its own virulence and for modulating plant defence events such as callose deposition, ROS production, etc. (Kumar et al., 2016; Soni and Mondal, 2018; Mondal et al., 2020a) (Figure 4.2).

4.3.1 Suppression of Cell-Wall-Associated Immune Response by Effectors of Xap

Different physical as well as chemical defence mechanisms are adapted by the plants during pathogen attack. Callose, a β-1,3-glucan, is a high molecular weight polymer deposited at the point of invasion by the pathogen and acts as a matrix to restrict the spread of the pathogen. Callose deposition is thus considered a biochemical event related to basal defence like PTI and is usually triggered by PAMPs (Brown et al., 1998; Gomez-Gomez et al., 1999). The T3SS effectors of Xap, namely XopC2, XopL and XopN, were shown to suppress the callose accumulation in pomegranate leaves (Kumar and Mondal, 2013; Mondal et al., 2020a). The leaves inoculated with mutants lacking the individual effector gene showed higher callose deposition compared to leaves inoculated with wild strains. Suppression of callose deposition by T3SS effectors was observed by many workers. XopN from Xcv infecting tomatoes was reported to show reduced (PAMP)-induced gene expression and callose deposition in host tissue (Kim et al., 2009). XopD, another effector of Xcv that functions as a transcriptional repressor, was demonstrated to suppress SA-induced defence responses, such as callose deposition, that otherwise would limit Xcv growth in the host (Kim et al., 2008). XopJ from Xcv, when transiently expressed in *N. benthamiana* and in Arabidopsis, could suppress callose deposition, indicating that XopJ apparently can interfere with cell wall-based defence (Bartetzko et al., 2009). A quadruple mutant of *X. oryzae* pv. *oryzae* having mutated four Xop genes (XopN, XopQ, XopX and XopZ) was shown to trigger lipase/esterase A (LipA)-induced deposition of callose in rice (Sinha et al., 2013). The *xopQ* gene of Xoo when expressed in plasmids constitutively was found to suppress the innate immunity of rice, as observed by reduced callose deposition in leaves (Gupta et al., 2015). The above studies confirm that T3SS effectors play a major role in suppressing pathogen-induced callose deposition in host plants.

4.3.2 Role of Effector of Xap in Suppression of ROS Accumulation

During host-pathogen interactions, the accumulation of ROS is one of the earliest immune responses in plants. Superoxide (O_2^-), or its dismutation product hydrogen peroxide (H_2O_2), hydroxyl radical (OH^-) and the reactive nitrogen species, nitric oxide (NO), etc. are the most common molecules of ROS. They are highly reactive and toxic in nature, thus playing a role in several defence-related processes. ROS accumulation events in plants upon pathogen recognition are grouped into two phases: the first phase is categorized as low and transient accumulation, followed by a second phase with sustained higher accumulation leading to disease resistance (Lamb and Dixon, 1997). Virulent pathogens escape host recognition and induce only a low-amplitude transient phase. The main role of ROS is cross-linking glycoproteins for the strengthening of host cell walls (Bradley et al., 1992; Lamb and Dixon, 1997) or membrane damage and lipid peroxidation (Lamb and Dixon, 1997; Montillet et al., 2005). The histochemical analysis of ROS production in pomegranate leaves upon infection with Xap was studied (Kumar et al., 2016). A high level of H_2O_2 and O_2^- was observed in the form of brown or blue deposition on the infiltrated areas of pomegranate leaves inoculated with mutants lacking the individual effector gene in comparison to wild, suggesting the role of the effectors in reducing ROS accumulation and thus suppressing oxidative burst. The higher ROS accumulation in pomegranate leaves inoculated with a T3SS-defective mutant, Xap Δ*hrpG*, was also evident, confirming the key role of effectors on the regulation of ROS production. Comparatively higher ROS accumulation was observed when susceptible pepper plants were infiltrated with the Xcv *ΔxopB* strain than wild, indicating that XopB interferes with plant defence via targeting ROS accumulation (Priller et al., 2016). Further image-based quantification revealed an approximately 50% higher DAB intensity in infiltrated pepper leaves with mutants compared to wild. The T3SS effectors that have been described to impair the ROS burst but show normal MAPK signalling include AvrAC and AvrPphB (Zhang et al., 2010; Feng et al., 2012). These results corroborate the fact that T3SS effectors mediate inhibition of ROS production during pathogen invasion in host plants.

4.3.3 Identifying Cognate Interactors for XopC2, a Crucial Effector of Xap

Two pomegranate proteins, namely raffinose synthase-protein-probable galactinol sucrose galactosyltransferase (RS6) and serine protease inhibitor (SPI), were found to interact with the XopC2 effector. The interactors were named as PORSI (Pomegranate RS6-like Interactor) and POSPII (Pomegranate Serine Protease Inhibitor-like Interactor). Both proteins were validated for their interaction with XopC2 using the BiFC assay. The host protein-bacterial effector-protein interaction was further verified through protein docking analysis. Finally, a model to explain this two-way interaction that supports the blight bacterium was proposed to establish infection in its own host plant, pomegranate.

4.4 CONCLUSION

The ability of the pathogen to evade the plant immune system is probably a result of evolution in the bacterial genome, especially the genes governing pathogenicity. Plants immune system seems to have evolved from a simpler to a more complex system, keeping pace with the evolution of pathogens. Innate defence responses in plants rely only on surveillance and subsequent recognition of the PAMPs leading to PTI. However, the surveillance system is not as simple as it seems to be, as it is more dynamic and complex to remain successful, and that's why very few bacterial pathogens can invade plants. Pathogens have also been modified in such a way as to break the barriers of the plant immune system. The T3SS effector of bacterial pathogens is meant to disable the PTI and thus allow the bacteria to escape plant defences. Even though pathogens have successfully disabled PTI, plants have further evolved the ability to recognize its Avr proteins. To counteract this, the pathogens have also acquired many T3SS effectors, which are functionally redundant and thus overcome the recognition via PRRs. The endurance of the winners of this continued tug of war depends on the possession of exclusive and advanced weapons in their respective arsenals to cope with the situation. This necessitates understanding the plant immune responses operating to counter the pathogen's T3SS effectors. In this context, the recent insight into the T3SS effectors of bacterial pathogens and their functional analysis would lead to a better understanding of the molecular basis of pathogenesis. This understanding would certainly lead us to devise more appropriate strategies for managing these diseases.

REFERENCES

Akimoto-Tomiyama, C.; Furutani, A.; Tsuge, S.; Washington Erica, J.; Nishizawa, Y.; Minami, E. and Ochiai, H. XopR, a Type III effector secreted by *Xanthomonas oryzae* pv. *oryzae*, suppresses microbe-associated molecular pattern triggered immunity in *Arabidopsis thaliana. Mol. Plant Microbe Interact.* 2012, *25*: 505–514.

Alfano, J.R. and Collmer, A. The type III (Hrp secretion pathway of plant pathogenic bacteria: trafficking harpins, Avr proteins, and death. *J. Bacteriol.* 1997, *179*: 5655–5662.

Alfano, J.R. and Collmer, A. Type III secretion system effector proteins: double agents in bacterial disease and plant defense. *Annu. Rev. Phytopathol.* 2004, *42*: 385–414.

Ausubel, F.M. Are innate immune signaling pathways in plants and animals conserved? *Nat. Immunol.* 2005, *6*: 973–979.

Bartetzko, V.; Sonnewald, S.; Vogel, F.; Hartner, K.; Stadler, R.; Hammes, U.Z. and Bornke F. The *Xanthomonas campestris* pv. *vesicatoria* type III effector protein XopJ inhibits protein secretion: evidence for interference with cell wall-associated defense responses. *Mol. Plant Microbe Interact.* 2009, *22*: 655–664

Bradley, D.J; Kjellbom, I. and Lamb, C.J. Elicitor- and wound induced oxidative cross-linking of a plant cell wall proline-rich protein: a novel, rapid defense response. *Cell.* 1992, *70*: 21–30.

Brown, I.; Trethowan, J.; Kerry, M.; Mansfield, J. and Bolwell, G.P. Localization of components of the oxidative cross-linking of glycoproteins and of callose synthesis in papillae formed during the interaction between non-pathogenic strains of *Xanthomonas campestris* and french bean mesophyll cells. *Plant J.* 1998, *15*: 333–343.

Buscaill, P. and Rivas, S. Transcriptional control of plant defence responses. *Curr. Opin. Plant Biol.* 2014, *20*: 35–46.

Büttner, D. Behind the lines-actions of bacterial type III effector proteins in plant cells. *FEMS Microbiol. Rev.* 2016, *40*: 894–937.

Cui, H.; Tsuda, K. and Parker, J.E. Effector-triggered immunity: from pathogen perception to robust defense. *Annu. Rev. Plant Biol.* 2015, *66*: 487–511.

Feng, F.; Yang, F.; Rong, W.; Wu, X.; Zhang, J.; Chen, S.; He, C. and Zhou, J.M. A Xanthomonas uridine 5′-monophosphate transferase inhibits plant immune kinases. *Nature.* 2012, *485*: 114–118.

Gomez-Gomez, L.; Felix, G. and Boller, T. A single locus determines sensitivity to bacterial flagellin in *Arabidopsis thaliana. Plant J.* 1999, *18*: 277–284.

Gu, K.; Yang, B.; Tian, D.; Wu, L.; Wang, D.; Sreekala, C.; Yang, F.; Chu, Z.; Wang, G.L.; White, F.F. and Yin, Z. R gene expression induced by a type-III effector triggers disease resistance in rice. *Nature.* 2005, *435*: 1122–1125.

Gupta, M.K.; Nathawat, R; Sinha, D.; Haque, A.S.; Sankaranarayanan, R. and Sonti, R.V. Mutations in the predicted active site of *Xanthomonas oryzae* pv. *oryzae* XopQ differentially affect virulence, suppression of host innate immunity, and induction of the HR in a nonhost plant. *Mol. Plant Microbe Interact.* 2015, *28*(2): 195–206.

Jha, G.; Rajeshwari, R. and Sonti, R.V. Functional interplay between two *Xanthomonas oryzae* pv. *oryzae* secretion systems in modulating virulence on Rice. *Mol. Plant Microbe Interact.* 2007, *20*: 31–40.

Jones, J.D.G. and Dangl, J.L. Plant immune system. *Nature.* 2006, *444*: 323–329.

Khan, M.; Subramaniam, R. and Desveaux, D. Of guards, decoys, baits and traps: pathogen perception in plants by type III effector sensors. *Curr. Opin. Microbiol.* 2016, *29*: 49–55.

Kim, J.G.; Li, X.; Roden, J.A.; Taylor, K.W.; Aakre, C.D.; Su, B.; Lalonde, S.; Kirik, A.; Chen, Y.; Baranage, G.; McLane, H.; Martin, G.B. and Mudgett, M.B. *Xanthomonas* T3S effector XopN suppresses PAMP-triggered immunity and interacts with a Tomato atypical receptor-like kinase and TFT1. *Plant Cell.* 2009, *21*: 1305–1323.

Kim, J.G.; Taylor, K.W.; Hotson, A.; Keegan, M.; Schmelz, E.A. and Mudgett, M.B. XopD SUMO protease affects host transcription, promotes pathogen growth, and delays symptom development in *Xanthomonas*-infected tomato leaves. *Plant Cell.* 2008, *20*: 1915–1929.

Klement, Z. and Goodman, R. The hypersensitive reaction to infection by bacterial plant pathogens. *Annu. Rev. Phytopathol.* 1967, *5*: 17–44.

Kumar, R. and Mondal, K.K. XopN T3SS effector is required for growth and pathogenicity by *Xanthomonas axonopodis* pv. *punicae*, the causal agent of bacterial blight of pomegranate. *Physiol. Mol. Plant Pathol.* 2013, *84*: 36–43.

Kumar, R.; Soni, M. and Mondal, K.K. XopN-T3SS effector of *Xanthomonas axonopodis* pv. *punicae* localizes to the plasma membrane and modulates ROS accumulation events during blight pathogenesis in pomegranate. *Microbiol. Res.* 2016, *193*: 111–120.

Lamb, C. and Dixon, R.A. The oxidative burst in plant disease resistance. *Annu. Rev. Plant Physiol. Plant Mol. Biol.* 1997, *48*: 251–275.

Lee, B.M.; Park, Y.J.; Park, D.S.; Kang, H.W.; Kim, J.G.; Song, E.S.; Park, I.C.; Yoon, U.H.; Hahn, J.H.; Koo, B.S.; Lee, G.B.; Kim, H.; Park, H.S.; Yoon, K.O.; Kim, J.H.; Jung, C.; Koh, N.H.; Seo, J.S. and Go, S.J. The genome sequence of *Xanthomonas oryzae* pathovar *oryzae* KACC10331, the bacterial blight pathogen of rice. *Nucleic Acids Res.* 2005, *33*: 577–586.

Meng, X. and Zhang, S. MAPK cascades in plant disease resistance signaling. *Annu. Rev. Phytopathol.* 2013, *51*: 245–266.

Mondal, K.K. *Plant Bacteriology*, Kalyani Publishers, New Delhi, 2011, 190 pp. ISBN:978-93-272-1631-8.

Mondal, K.K. Dialogue between Xop-T3SS effectors of *Xanthomonas oryzae* pv. *oryzae* and rice during bacterial blight pathogenesis. *Indian Phytopath.* 2017, *70*(3): 269–274. DOI:10.24838/ip.2017.v70.i3.72486.

Mondal, K.K.; Bhar, L.M. and Mani, C. Combined efficacy of *Pseudomonas fluorescens* strain MBPF-01 and nanocopper against bacterial leaf blight in rice. *Indian Phytopath.* 2010, *63*: 266–268.

Mondal, K.K. and Mani, C. Investigation of the antibacterial properties of nanocopper against *Xanthomonas axonopodis* pv. *punicae*, the incitant of pomegranate bacterial blight. *Ann. Microbiol.* 2012, *62*: 889–893.

Mondal, K.K.; Meena, B.R.; Junaid A.; Verma, G.; Mani, C.; Majumder, D.; Manju, K.; Kumar, S. and Banik, S. Pathotyping and genetic screening of type III effectors in Indian strains of *Xanthomonas oryzae* pv. *oryzae* causing bacterial leaf blight of rice. *Physiol. Mol. Plant Pathol.* 2014, *86*: 98–106.

Mondal, K.K.; Rashmi, E.R. and Ghoshal, T. Rice SWEET proteins: the key targets by the bacterial blight pathogen, *Xanthomonas oryzae* pv. *oryzae* to ensure nutrition in the rice apoplast during infection. *Indian Phytopathol.* 2021, *74*: 323–331. DOI:10.1007/s42360-021-00369-5.

Mondal, K.K. and Sharma, J. Bacterial blight: an emerging threat to pomegranate export. *Indian Farming.* 2009, *59*(8): 22–23.

Mondal, K.K.; Soni, M.; Verma, G.; Kulshreshtha, A.; Mrutyunjaya, S. and Rishikesh, K. *Xanthomonas axonopodis* pv. *punicae* depends on multiple non-TAL (Xop) T3SS effectors for its coveted growth inside the pomegranate plant through repressing the immune responses during bacterial blight development. *Microbiol. Res.* 2020a, *240*: 126560. DOI:10.1016/j.micres.2020.126560.

Mondal, K.K.; Verma, G.; Kulshreshtha, A.; Rajrana, Y.; Mani, C.; Soni, M.; Reddy, K.K.; Ghoshal, T.; Lakshmi, A. and Kalaivanan, N.S. Complete genome sequence of Indian race 4 of Xanthomonas oryzae pv. oryzae, the causal agent of bacterial blight of rice. *Mol. Plant Microbe Interact.* 2020b, *33*(4): 573–575. DOI:10.1094/MPMI-12-19-0335-A.

Mondal, K.K.; Verma, G. and Mani, C. Phylogenetic relatedness of *Xanthomonas axonopodis* pv. *punicae*, the causal agent of bacterial blight of pomegranate based on two loci, 16S rRNA and gyrB. *Ann. Microbiol.* 2012, *63*: 801–804. DOI:10.1007/s13213-012-0498-4.
Mondal, K.K.; Verma, G.; Manju; Junaid and Mani, C. Rice pathogen *Xanthomonas oryzae pv. oryzae* employs inducible *hrp*-dependent XopF type III effector protein for its growth, pathogenicity and for suppression of PTI response to induce blight disease. *Eur. J. Plant Pathol.* 2016, *144*: 311–323. DOI:10.1007/s10658-015-0768-7.
Montillet, J.L.; Chamnongpol, S.; Rusterucci, C.; Dat, J.; van de Cotte, B.; Agnel, J.P.; Battesti, C.; Inze, D.; Van Breusegem, F. and Triantaphylides, C. Fatty acid hydroperoxides and H_2O_2 in the execution of hypersensitive cell death in tobacco leaves. *Plant Physiol.* 2005, *138*: 1516–1526.
Mudgett, M.B. New insights to the function of phytopathogenic bacterial type III effectors in plants. *Annu. Rev. Plant Biol.* 2005, *56*: 509–531.
Nurnberger, T.; Brunner, F.; Kemmerling, B. and Piater, L. Innate immunity in plants and animals: striking similarities and obvious differences. *Immunol. Rev.* 2004, *198*: 249–266.
Ochiai, H.; Inoue, Y.; Takeya, M.; Sasaki, A. and Kaku, H. Genome sequence of *Xanthomonas oryzae* pv. *oryzae* suggests contribution of large numbers of effector genes and insertion sequences to its race diversity. *Jpn. J. Agric. Res. Q.* 2005, *39*: 275–287.
Priller, J.P.R.; Reid, S.; Konein, P.; Dietrich, P. and Sonnewald, S. The *Xanthomonas campestris* pv. *vesicatoria* Type-3 Effector XopB Inhibits plant defence responses by interfering with ROS production. *PLoS One*. 2016, *11*(7): e0159107.
Salzberg, S.L.; Sommer, D.D.; Schatz, M.C.; Phillippy, A.M.; Rabinowicz, P.D.; Tsuge, S.; Furutani, A.; Ochiai, H.; Delcher, A.L.; Kelley, D.; Madupu, R.; Puiu, D.; Radune, D.; Shumway, M.; Trapnell, C.; Aparna, G.; Jha, G.; Pandey, A.; Patil, P.B.; Ishihara, H.; Meyer, D.F.; Szurek, B.; Verdier, V.; Koebnik, R.; Dow, J.M.; Ryan, R.P.; Hirata, H.; Tsuyumu, S.; Won Lee, S.; Seo, Y.S.; Sriariyanum, M.; Ronald, P.C.; Sonti, R.V.; Van Sluys, M.A.; Leach, J.E.; White, F.F. and Bogdanove, A.J. Genome sequence and rapid evolution of the rice pathogen *Xanthomonas oryzae* pv. *oryzae* PXO99A. *BMC Genom.* 2008, 9: 204.
Schwessinger, B. and Ronald, P.C. Plant innate immunity: perception of conserved microbial signatures. *Annu. Rev. Plant Biol.* 2012, *63*: 451–482.
Sharma, J.; Sharma, K.K.; Kumar, A.Q.; Mondal, K.K.; Thalor, S.; Maity, A.; Gharate, R.; Chinchure, S. and Jadhav, V.T. Pomegranate bacterial blight: symptomatology and rapid inoculation technique for *Xanthomonas axonopodis* pv. *punicae. J. Plant Pathol.* 2017, *99*(1): 109–119. DOI:10.4454/jpp.v99i1.3825.
Sharma, V.; Midha, S.; Ranjan, M.; Pinnaka, A.K. and Patil, P.B. Genome sequenceof Xanthomonas axonopodis pv. punicae strain LMG 859. *J. Bacteriol.* 2012, *194*: 2395.
Singh, D.; Mondal, K.K. and Kishun, R. Current status and future research strategies in plant bacteriology in India. In *Plant Pathology in India: Vision 2030*, pp. 115–120, Indian Phytopathological Society, New Delhi, India. 2011a; 294 pp.
Singh, U.D.; Gogoi, R. and Mondal, K.K. Molecular approach for augmenting disease resistance in cereals: rice and maize. In *Plant Pathology in India: Vision 2030*, pp. 21–30, Indian Phytopathological Society, New Delhi, India. 2011b; 294 pp.
Sinha, D.; Gupta, M.K.; Patel, H.K.; Ranjan, A. and Sonti, R.V. Cell wall degrading enzyme induced rice innate immune responses are suppressed by the type 3 secretion system effectors XopN, XopQ, XopX and XopZ of *Xanthomonas oryzae* pv. oryzae. *Plos One.* 2013, *8*: e75867. DOI:10.1371/journal.pone.0075867.
Song, C. and Yang, B. Mutagenesis of 18 Type III effectors reveals virulence function of XopZpxo99 in *Xanthomonas oryzae* pv.*oryzae*. *Mol. Plant Microbe Interact.* 2010, *23*: 893–902.
Soni, M. and Mondal, K.K. *Xanthomonas axonopodis* pv. *punicae* employs XopL effector to suppress pomegranate immunity. *J. Integrat. Plant Biol.* 2018, *60*: 341–357. DOI:10.1111/jipb.12615.

Verma, G.; Mondal, K.K.; Kulshreshtha, A. and Sharma, M. XopR T3SS-effector of Xanthomonas oryzae pv. oryzae suppresses cell death-mediated plant defense response during bacterial blight development in rice. *3 Biotech.* 2019, 9: 272–281. DOI:10.1007/s13205-019-1802-9.

Verma, G.; Sharma, M. and Mondal, K.K. XopR TTSS-effector regulates *in planta* growth, virulence of Indian strain of *Xanthomonas oryzae* pv. *oryzae* via suppressing reactive oxygen species production and cell wall-associated rice immune responses during blight induction. *Funct. Plant Biol.* 2018, *45*: 561–574. DOI:10.1071/FP17147.

White, F.F.; Potnis, N.; Jones, J.B. and Koebnik, R. The type III effectors of *Xanthomonas*. *Mol. Plant Pathol.* 2009, *10*: 749–766.

Yang, B.; Sugio, A. and White, F.F. Avoidance of host recognition by alterations in the repetitive and C-terminal regions of AvrXa7, a type III effector of *Xanthomonas oryzae* pv. *oryzae*. *Mol. Plant Microbe Interact.* 2005, *18*: 142–149.

Zhang, J.; Li, W.; Xiang, T.; Liu, Z.; Laluk, K.; Ding, X.; Zou, Y.; Gao, M.; Zhang, X.; Chen, S.; Mengiste, T.; Zhang, Y. and Zhou, J. M. Receptor-like cytoplasmic kinases integrate signaling from multiple plant immune receptors and are targeted by a *Pseudomonas syringae* effector. *Cell Host Microbe.* 2010, *7*: 290–301.

5 Small RNA *Understanding and Prospects in Crop Protection*

Ashish Marathe
ICAR-National Institute of Biotic Stress Management

Deepak Vishwanath Pawar
ICAR-Directorate of Weed Research

Senthilkumar K. Muthusamy
ICAR-Central Tuber Crops Research Institute

Palaiyur Nanjappan Sivalingam and Jagdish Kumar
ICAR-National Institute of Biotic Stress Management

The biotic stresses, which include viruses, bacteria, fungi, insect pests and nematodes, are the major cause of the reduction of global food production and have adversely affected food security in the past and present. Approximately one-third of the edible food losses annually are estimated to be 750 billion US dollars every year (Mesterházy et al., 2020). The growing human population globally is going to utilize the 100 Mha of land additionally in the next three decades (Pareek et al., 2020), which will lead to a decrease in natural resources. Therefore, several environmentally friendly and sustainable plant protection strategies are needed. To avoid the negative impacts of pesticides on human health, alternate strategies like utilizing small RNAs (sRNAs) for crop protection are one of the most promising options. sRNA is the driver of the RNAi machinery that ensures downregulation of the target genes (Kaur et al., 2021). The various approaches utilizing sRNA for the management of crop pests and their understanding are compiled in this chapter.

5.1 sRNA-MEDIATED RESISTANCE IN INSECTS

RNAi and other sRNA-based genetic engineering are considered as one of the important strategies for the development of insect-resistant cultivars in crop plants. Moreover, the inclusion of these eco-friendly technologies would be helpful in reducing the consumption of pesticides. Several groups have made numerous efforts to engineer insect resistance in crop plants (Grover et al., 2019; Abdellatef et al., 2021;

DOI: 10.1201/9781003403142-5

Jain et al., 2021). Feeding of transgenic rice lines overexpressing artificial miRNA (amiRNA) affected the growth of the rice pest striped stem borer (*Chilo suppressalis*) and delayed the pupation of the larve (Jiang et al., 2017). Similarly, striped stem borer larve feeding on the transgenic rice lines expressing the amiRNA-csu-novel-miR53 resulted in reduced larval growth, increased pupation duration, and decreased pupal weight and eclosion rate (Liu et al., 2022). *Lymantria dispar* larvae feeding on transgenic poplar plants that are expressing dsRNAs of *insect Cytochrome P450 monooxygenase* (LdCYP6B53) gene inhibit the development of the larve and delay its overall growth (Sun et al., 2022).

Adult flies of *B. tabaci* feeding on transgenic tobacco lines that are expressing dsRNAs of *B. tabaci* acetylcholinesterase (*BtAChE*) and ecdysone receptor (*BtEcR*) genes showed increased mortality than the flies feeding on wild-type tobacco plants (Malik et al., 2016). *Spodoptera littoralis* larvae that were feeding on the transgenic tobacco lines expressing dsRNA targeting an immunosuppressive gene resulted in increased impairment in the hemocyte-mediated encapsulation response and became more susceptible to the *Bacillus thuringiensis*-based insecticide (Di Lelio et al., 2022). Downregulation of *ferritin 1 Heavy Chain (NlFer1)* and *ferritin 2 Light Chain (NlFer2)* genes in *Nilaparvata lugens* caused underdeveloped ovaries, resulting in extremely low fecundity and a zero-hatching rate in RNAi-treated individuals (Shen et al., 2021).

Also, silencing of the detoxification genes *DcGSTe2* and *DcGSTd1* of Asian citrus psyllid *Diaphorina citri* Kuwayama increases the susceptibility of the insects to the pesticides thiamethoxam and fenpropathrin (Yu and Killiny, 2018). Silencing of *nuclear export factor 1 (NXF1)* in *Locust amigratoria* increased the mortality rate of the dsRNA-treated locusts (Xie et al., 2019). In *Tetranychus urticae*, silencing of any one of the vital genes, viz., T-COPB2 (coatomer protein complex, subunit beta 2), T-M1MP (protease m1 zinc metalloprotease), T-RPS4 (40S ribosomal protein S4) and T-VATPase (vacuolar-type H^+-ATPase), increased the mortality rate of the insect. Mao et al. (2022) showed that feeding siRNA and co-silencing of the two genes, viz., vacuolar-type H^+-ATPase and β subunit of the coatomer protein complex, T-COPB2, resulted in reduced weight gain and survival rate of the larvae of the cotton bollworm, *Helicoverpa armigera.* These studies show both genetic engineering of sRNA and the introduction of siRNAs externally would be a potential approach for controlling insects in plants (Table 5.1).

5.2 sRNAS IN PLANT–BACTERIA INTERACTION

Plant immune systems are strictly controlled by an endogenous sRNA mechanism that silences or suppresses specific genes associated with plant hormone biosynthesis or antimicrobial activities. Biogenesis and/or buildup of all these sRNAs is modulated in response to pathogens, which fine-tunes phytohormone concentrations and the regulation of gene expression associated with plant defense (Figures 5.1a and 5.2a). Plants are infected by bacteria, fungus, and oomycetes without immediate genomic and RNA association with the host plant's RNA suppression system.

TABLE 5.1
List of Some sRNA Studies Mediated Insect Resistance in Crops

Insect	Small RNA	Crop (Transgenic Line)	Target Gene	Remarks	References
Striped stem borer (*Chilo suppressalis*)	Artificial miRNA (amiRNA) csu-novel-miR15	Rice	Endogenous genes	Delayed pupation	Jiang et al. (2017)
Striped stem borer (*Chilo suppressalis*)	Artificial miRNA (amiRNA) csu-novel-miR53	Rice	DN90065_c0_g12 RNA-directed DNA polymerase	Inhibited larval growth, delayed pupation time, and reduced pupal weight and eclosion rate	Liu et al. (2022)
Chilo suppressalis and Cnaphalocrocis medinalis	miRNA Csu-novel-260	Rice	Disembodied (dib) gene	Inhibits the pupation of larva	Wen et al. (2021)
Chilo suppressalis	RNAi silencing ds*CssHsp*	Rice	Small heat shock protein gene (*CssHsp*)	Increased mortality	Mao et al. (2022)
Lymantria dispar	RNAi silencing CYP6B53	Poplar	CYP6B53	Inhibited larval feeding but also delayed overall growth	Sun et al. (2022)
Bemisia tabaci	RNAi silencing *TPS* genes	Tobacco	BtTPS1 BtTPS2	Retarded growth, low hatchability increased mortality and decreased the fecundity in whitefly adults	Gong et al. (2021)
Bemisia tabaci	RNAi silencing v-ATPase gene	Tobacco	v-ATPase A	Increased mortality	Thakur et al. (2014)
Bemisia tabaci	Agroinfiltration (transient expression of dsRNA in plants)	Tobacco	dsRNAs of *whitefly genes* viz., acetylcholinesterase (*BtAChE*) and ecdysone receptor (*BtEcR*)	Increased mortality	Malik et al. (2016)

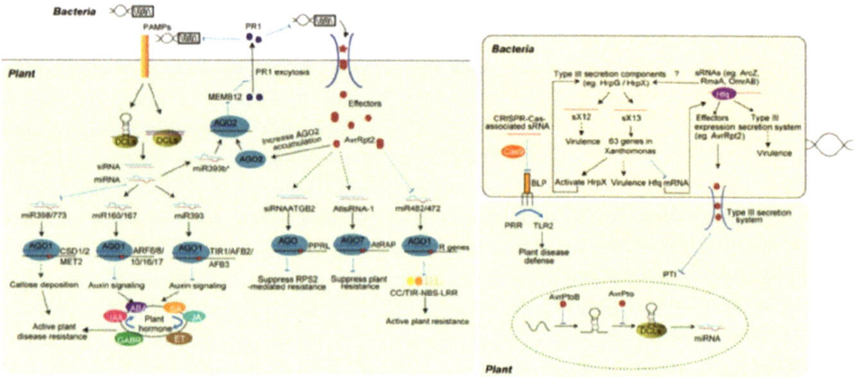

FIGURE 5.1 Role of sRNAs in plant–bacteria interaction. (a) Plant sRNAs defend bacteria from attack by fine-tuning plant hormones and disease resistance activity. Upon infection, plants detect PAMPs and modulate the accumulation of miRNA and siRNA. miRNAs, such as miR393, miR160, and miR167, regulate disease resistance by fine-tuning plant hormone networks, while other miRNAs regulate the activation of R protein (miR482/miR472) or the slicing of genes inhibiting plant immunity (miR398/miR773). miR393b∗, the pairing strand of miR393, increases plant immunity by promoting the exocytosis of antimicrobial proteins. siRNAs, including siRNAATGB2 and AtlsiRNA-1, are induced by bacteria effectors and enhance ETI by silencing genes that negatively regulate plant disease resistance. (b) Bacteria non-coding sRNAs (ncRNAs) regulate bacteria gene expression to improve virulence. Through imperfect base-pairing of short regions (10- to 25-nt), bacteria ncRNAs bind to target mRNAs and guide the suppression of genes or proteins that are involved in virulence. ncRNAs can regulate bacterial virulence by inhibiting proteins that trigger host defense (BLP) or affecting the expression of effectors (AvrRpt2). Bacteria effectors translocate into host plant cells and inhibit the regulation of plant sRNA (bottom). The AvrPtoB effector specifically represses the accumulation of miR393 at the transcriptional level, while AvrPto reduces the processing of miR393 (Adapted from Huang et al., 2016).

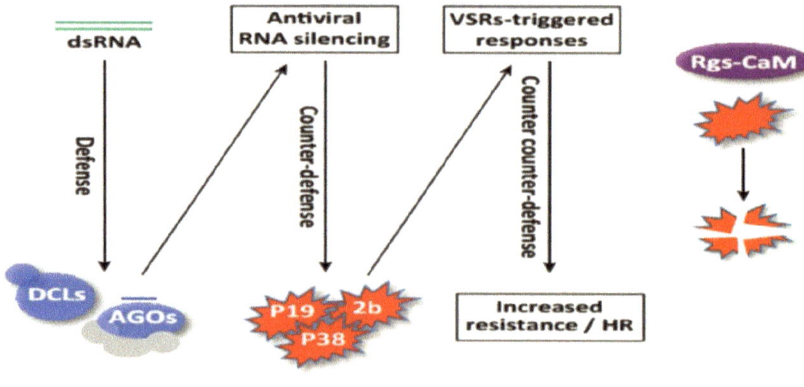

FIGURE 5.2 RNA silencing interplay during plant–virus interactions.

The plant's immunity must be fine-tuned in order for the host plant to be resistant to such infections. Antibacterial immunity is mediated by a variety of plant miRNAs and siRNAs (Figure 5.1a). The miRNA miR393 was discovered to play a role in antibacterial defense for the first time. Exposure to the unique N-terminal portion of flagellin, flg22, or inoculation with the bacterial pathogen *Pseudomonas syringae* pv. tomato (Pst) DC3000 increases the buildup of miR393 (Navarro et al., 2006). By adversely modulating the expression of F-Box auxin receptors such as Transport Inhibitor Response 1 (TIR1), Auxin signaling F-Box proteins 2 (AFB2), and 3 (AFB3), miR393 improves host tolerance to Pst DC3000 (Navarro et al., 2006). Auxin response factor (ARF) group transcription factors are also targeted by miR160 and miR167, which are increased by infection with Pst DC3000 hrcC, a strain with a mutant type III secretion system, to enhance antibacterial resistance (Fahlgren et al., 2007). Further investigations revealed that MiR160a as well as 15 additional miRNAs are activated by flg22 administration. Whereas MiR398b, miR773, and 9 additional miRNAs are suppressed after being exposed to flg22 (Li et al., 2010). Increased levels of miR398b and miR773 suppress PTI through inhibition of flg22 or pathogen-induced callose accumulation, implying that miRNAs play a vital role in immune function. Two sRNAs, miR393 and miR393b, engage with AGO1 and AGO2, respectively, to modulate different hormonal networks and boost resistance in plants in tandem (Navarro et al., 2006; Zhang et al., 2011a). A further fascinating result concerning miRNA in bacterial defense has been that, depending on the timing and intensity of defensive reactions, the same miRNA can address both positive and negative regulation of resistance. In primary infection, miR863-3p promotes plant defense by suppressing *ARLPK1* and *ARLPK2*, but subsequent infection negatively regulates defense by SE gene suppression (Niu et al., 2016).

Plants induce effector-triggered immunity (ETI) in order to fight pathogenic effector proteins. The ETI response is strictly controlled by siRNAs and miRNAs because ETI is much more persistent and generally causes a hypersensitivity reaction (HR). Pst DC3000 effector AvrRpt2-induced siRNA nat-siRNAATGB2 improves ETI by downregulating the pentatricopeptide repeats (PPR) protein-like gene (PPRL) and preventing PPRL's negative impact on the resistance mechanism controlled by RPS2, the gene that recognizes effector AvrRpt2. Also, suppressing the activity of *AtRAP*, a repressor of plant immune function, AtlsiRNA-1, that is also produced by *AvrRpt2*, enhances resistance to disease (Katiyar-Agarwal et al., 2007). In addition to these sRNAs, genome-wide sRNA deep sequencing indicates that the accumulation of more than 20 miRNAs and various nat-siRNAs is significantly altered upon ETI. Furthermore, genome-wide sRNA and deep sequencing demonstrated the accumulation of about 20 miRNAs and numerous nat-siRNAs that are radically changed upon ETI (Zhang et al., 2011b; Zhang et al., 2012). Essential genes associated with hormone synthesis and signal transduction pathways involved in plant defense are among the miRNAs' targets. TE-siR815, a TE-siRNA derived from the intron of WRKY45-1, suppresses ST1 and so lowers WRKY45-mediated immunity against *Xanthomonas oryzae* pv. Oryzae, resulting in the WRKY45-1 and WRKY45-2 functions being reversed (Zhang et al., 2016). Host plant's sRNAs have a role in ETI not only by influencing the activation of resistance genes but also by modulating the engagement of R proteins effectively. RPP4 and SNC1, two R genes in the RPP5 region, appear to be involved in immune function in response to bacterial and fungal infections, respectively (Baldrich et al., 2014, 2015).

The discovery and profiling of miRNAs' targets would aid in the identification of novel components in innate immunity networks. If such bacteria-regulated miRNAs operate as positive moderators of resistant bacteria, the constitutive accumulation of miRNA or amiRNA could be used to create genetically modified crops with increased resistance. Whenever genetically modified crops overexpress miRNAs that serve as suppressors, they become far more susceptible to pathogens. Upregulation of their target loci, which can be accomplished by overexpressing amiRNA-tolerant variant of their target or by utilizing amiRNA target mimics, could be an effective technique for enhancing plant stress resilience in such circumstances (Franco-Zorrilla et al., 2007).

5.3 sRNAS IN PLANT–FUNGI AND PLANT–OOMYCETES INTERACTION

PTI as well as ETI are indeed modulated by host sRNAs in response to fungus and oomycete infections. In an attempt to discover *Phytophthora* effectors with downregulation of suppressor action, it was revealed that the two effectors, *Phytophthora Suppressor of RNA Silencing 1* (*PSR1*) and *PSR2*, could inhibit transgene-governed *GFP* gene downregulation in genetically modified Nicotiana benthamiana by impairing the biosynthetic of sRNAs (Qiao et al., 2013). PSR1 binding to the RNA helicase PSR1-Interacting Protein 1 (PINP1) inhibits miRNA and siRNA synthesis, perhaps through the disassembling of slicing complexes (PSR1) (Qiao et al., 2015). In another study, it was found that phasiRNA production regulated by miR161 (and not by miR173) is critical for *Arabidopsis* defense during *Phytophthora* infection (Wang et al., 2016). PSR2, which has also been found to have suppressor silencing function due to its association with host plant DRB4 protein (Hou et al., 2019). *Verticillium dahliae* is an intrusive systemic fungal pathogen (Fradin and Thomma, 2006). After infection with *V. dahlia* and *Pseudomonas syringae*, the transcription of four miRNAs was shown to be significantly elevated in *Arabidopsis*; these four miRNAs target ARF10, NAC1, PHV, and ARF6 genes (Jian et al., 2018; Deleris et al., 2006).

Botrytis cinerea causes gray mold disease in tomatoes. According to sRNA sequencing reports, numerous phasiRNAs were found to be overexpressed in tomatoes infected with *B. cinerea* (Wu et al., 2017). Likewise, when oilseed rape was infected with *Sclerotinia sclerotiorum*, the buildup of 51 miRNAs was found to be increased. Upon treatment with *Magnaporthe oryzae* elicitor, the synthesis of rice miRNA528/miR1879 was found to be reduced, causing overexpression of its oxidative stress-responsive genes (Baldrich et al., 2015). In addition, elicitors of the fungal pathogen *Fusarium oxysporum* have been shown to control a group of sRNAs in A. thaliana, particularly miR156, miR165/166, miR170, and miR172 (Baldrich et al., 2014). Furthermore, fungal elicitors were discovered to activate miR823 and siRNA415, which are also engaged in RNA-directed DNA methylation (RdDM) (Baldrich et al., 2014). MiR159, miR164, miR167, miR171, miR444, miR408, miR1129, and miR1138, which govern three different defense response processes, are considerably elevated at the preliminary phase of *Puccinia graminis* f.sp. tritici

(Bgt) invasion, although not at the later phase. As a result, such miRNAs could play an essential role in HR during the initiation of pathogen infection (Gupta et al., 2012).

However, the roles of these miRNAs' in induced genes, on the other hand, are obscure (Jian et al., 2018). In these instances, it's still unclear which effectors or PAMPs are effective in inducing sRNA synthesis in plants. In the powdery mildew fungus *Blumeria graminis* f.sp. hordei (Bgh), the involvement of sRNA in plants innate immunity has been further established (Liu et al., 2014). The miRNA group miR9863 targets the Mildew resistance locus a (*Mla*), which encodes a set of Bgh-responsive CC-NBS-LRR proteins. In the *N. benthamiana* expression system, miR9863 has been found to break down *Mla1* transcripts and also suppress the buildup of MLA1 polypeptides. In diverse plants, the suppression of NBS-LRR genes through these unique 22-nt miRNAs, as well as their activation after miRNA down-regulation due to bacterial, fungal, or viral infections, has been extensively explored (He et al., 2008; Xin et al., 2010; Zhai et al., 2011; Li et al., 2012; Shivaprasad et al., 2012; Zhu et al., 2013; Boccara et al., 2014; Liu et al., 2014; Fei et al., 2015).

Host plant sRNAs are clearly important in controlling the transcription of genes encoded in the host defense response. Nevertheless, every sRNA serves a different purpose with plant immune reactions because sRNA synthesis and activity are pathogen-dependent. As a result, the functionality of much more sRNAs needs to be further researched in order to ascertain the comprehensive involvement of sRNAs in host plant immunity. Also, signal transduction in sRNA biogenesis, as well as the actual role of sRNA target genes, must be given due attention.

5.4 sRNAS IN PLANT–FUNGI–VIRUS INTERACTION

sRNA is imperative in mediating several developmental roles in the plant system, including the regulation of gene expression at both the epigenetic and post-transcriptional levels. The various classes of sRNA involved in regulating eukaryotic gene expression include the following: small-interfering RNA (siRNA), microRNAs (miRNAs), natural antisense transcript-derived siRNAs (nat-siRNAs), heterochromatic siRNAs (hcsiRNAs) or repeat-associated siRNAs (rasiRNAs), and long siRNAs (lsiRNAs). In addition to developmental regulation, sRNAs have also proven their activity in mediating tolerance to various biotic stresses. Although required in other biotic stresses too, sRNAs are notably very important in the case of viral infections since viruses are obligate parasites and most of them are systemic in nature. Several research groups have demonstrated the role of sRNAs as a major factor in disease outcome during host–virus interaction (Ramesh et al., 2014; Pradhan et al., 2015; Chen et al., 2017). In response to virus infections, dynamics in the levels of sRNA accumulation determine pathogenicity or impart a resistant nature to the host (Prasad et al., 2019). The antiviral sRNAs organize a cascade of molecular events by regulating the host defense machinery and are a complex phenomenon (Guidotti and Iannacone, 2015). Host–virus interaction mainly involves changes in the levels of sRNAs produced by plants as a response to invading viruses. Upon infection, the plant sRNAs that are 21–22 nt in length are known to increase in their levels, while those that are 24 nt long decrease in number. These perturbations in the levels of sRNAs have been explored to understand the biochemical and molecular processes in virus infections in several crops. miR162 expression was observed to have increased in response to Cotton Leaf Role Dwarf Virus (CLRDV) (Silva et al., 2011). In the case of Rice Stripped Virus (RSV) infection, miR144 expression was observed to have increased, while miR528 and miR396 were downregulated (Wang et al., 2106; Wu et al., 2017). Besides miRNAs and siRNAs, virus-induced siRNA (vasiRNA) plays an important role in the RNA silencing mechanism in plants. vasiRNAs, a separate class of sRNAs encoded by the host exons and 24 nt in length, are produced by the combined activity of DCL4 and RDR1. Several host factors, such as phospholipid flippases and vasiRNAs, are regarded as important players for the antiviral response in plants (Cao et al., 2014; Guo et al., 2017). In recent times, sRNAs have been widely explored for triggering host-induced gene silencing, owing to their movement. RNA molecules, including sRNAs, have been found to be mobile between cells and tissues. The mobile nature of sRNAs was demonstrated in *Arabidopsis* by grafting wild-type shoots to *dcl* mutant roots, showing the downward movement of sRNAs (Molnar et al., 2010). sRNAs are also known to be involved in trans-kingdom gene silencing.

5.5 sRNA-INDUCED GENE SILENCING

Ever since its discovery, RNAi technology has been deployed for imparting virus resistance in several crops utilizing partial or complete coding sequences yielding functional viral products. On the basis of their mechanism of silencing, sRNA-induced silencing can be categorized into various types: sense gene-induced post-transcriptional gene silencing (S-PTGS), hairpin RNA-induced PTGS (hp-PTGS), and amiRNA-induced PTGS (AMIR).

S-PTGS can be considered as the first generation of RNAi silencing and is successful in generating viral resistance, where viral proteins are overexpressed in the transgenic host plants (Lindbo et al., 1993). However, the level of resistance depends on the accumulation of viral siRNA transcripts in the absence or prior to viral infection (Guo et al., 1998; Lindbo et al., 1993).

Hp-PTGS is a second-generation technology developed after recognizing dsRNA-triggered RNA silencing. In this technology, researchers constructed silencing vectors with fragments of both sense and antisense viral cDNA sequences under control of plant promoters and terminators. The potential of Hp-PTGS was first demonstrated in tobacco against Potato virus Y (PVY). Later, this technology was explored in several crops against various viruses to generate resistant immune transgenics.

AMIR, considered the third-generation technology in the area of RNAi silencing, emerged as a potential technology overcoming the limitations of the first- and second-generation technologies and was demonstrated in *Arabidopsis* and tobacco (Niu et al., 2006; Qu et al., 2007).

5.6 USE OF sRNAs FOR IMPARTING VIRUS RESISTANCE

Biotechnological applications have explored the possibility of using amiRNA to impart virus resistance through a transgenic approach. *Nicotiana benthamiana* transgenic lines expressing *Oryza sativa* miR528 as the backbone were generated, expressing amiRNAs targeting the RNA-dependent RNA polymerase (RdRp) gene of Cymbidium mosaic virus (CymMV) and Odontoglossum ringspot virus (ORSV) (Petchthai et al., 2018). Transgenic tomato plants expressing amiRNA targeting the ATP binding domain of AC1 are resistant to Tomato leaf curl disease (Sharma and Prasad, 2020). miRNA controls post-transcriptional events negatively and is involved in the degradation or repression of the viral genome. Recent reports have demonstrated that overexpression of miR7695 enhances the innate immune response against *M. oryzae* through high iron and increased phytoalexin accumulation in rice (Perisperis et al., 2017).

REFERENCES

Abdellatef, E., Kamal, N.M., and Tsujimoto, H. 2021. Tuning beforehand: a foresight on RNA interference (RNAi) and in vitro-derived dsRNAs to enhance crop resilience to biotic and abiotic stresses. *International Journal of Molecular Sciences*, *22*(14), 7687.

Baldrich, P., Campo, S., Wu, M.T., Liu, T.T., Hsing, Y.I., and San Segundo, B. 2015. MicroRNA-mediated regulation of gene expression in the response of rice plants to fungal elicitors. *RNA Biology*, *12*, 847–863.

Baldrich, P., Kakar, K., Sire, C., Moreno, A.B., Berger, A., Garcia-Chapa, M., et al. 2014. Small RNA profiling reveals regulation of *Arabidopsis* miR168 and heterochromatic siRNA415 in response to fungal elicitors. *BMC Genomics*, *15*, 1083.

Boccara, M., Sarazin, A., Thiebeauld, O., Jay, F., Voinnet, O., Navarro, L., et al. 2014. The *Arabidopsis* miR472-RDR6 silencing pathway modulates PAMP- and effector-triggered immunity through the post-transcriptional control of disease resistance genes. *PLoS Pathogens*, *10*, e1003883.

Cao, M., Du, P., Wang, X., Yu, Y.Q., Qiu, Y.H., Li, W., et al. 2014. Virus infection triggers wide spread silencing of host genes by a distinct class of endogenous siRNAs in Arabidopsis. *Proceedings of the National Academy of Sciences of the United States of America*, *111*, 14613–14618.

Chen, H., Adam Arsovski, A., Yu, K., and Wang, A. 2017. Deep sequencing leads to the identification of eukaryotic translation initiation factor 5a as a key element in Rsv1-mediated lethal systemic hypersensitive response to Soybean mosaic virus infection in Soybean. *Molecular Plant Pathology*, *18*, 391–404.

Deleris, A., Gallego-Bartolome, J., Bao, J., Kasschau, K.D., Carrington, J.C., and Voinnet, O. 2006. Hierarchical action and inhibition of plant Dicer-like proteins in antiviral defense. *Science*, *313*, 68–71.

Di Lelio, I., Barra, E., Coppola, M., Corrado, G., Rao, R., and Caccia, S. 2022. Transgenic plants expressing immunosuppressive dsRNA improve entomopathogen efficacy against Spodoptera littoralis larvae. *Journal of Pest Science*, 95, 1–16.

Fahlgren, N., Howell, M.D., Kasschau, K.D., Chapman, E.J., Sullivan, C.M., Cumbie, J.S., et al. 2007. High-throughput sequencing of *Arabidopsis* microRNAs: evidence for frequent birth and death of MIRNA genes. *PLoS One*, 2, e219.

Fei, Q., Li, P., Teng, C., and Meyers, B.C. 2015. Secondary siRNAs from medicago NB-LRRs modulated via miRNA-target interactions and their abundances. The *Plant Journal*, *83*, 451–465.

Fradin, E.F. and Thomma, B.P. 2006. Physiology and molecular aspects of *Verticillium* wilt diseases caused by V. dahliae and V. albo-atrum. *Molecular Plant Pathology*, *7*, 71–86.

Franco-Zorrilla, J.M., Valli, A., Todesco, M., Mateos, I., Puga, M.I., Rubio-Somoza, I., et al. 2007. Target mimicry provides a new mechanism for regulation of microRNA activity. *Nature Genetics*, *39*, 1033–1037.

Gong, C., Yang, Z., Hu, Y., Wu, Q., Wang, S., Guo, Z., et al. 2021. Silencing of the BtTPS genes by transgenic plant-mediated RNAi to control *Bemisia tabaci* MED. *Pest Management Science*, *78*(3), 1128–1137.

Grover, S., Jindal, V., Banta, G., Taning, C.N.T., Smagghe, G., and Christiaens, O. 2019. Potential of RNA interference in the study and management of the whitefly, Bemisia tabaci. *Archives of Insect Biochemistry and Physiology*, *100*(2), e21522.

Guidotti, L.N. and Iannacone, M. 2015. Editorial overview: viral pathogenesis. *Current Opinion in Virology*, *11*, v–vii.

Guo, H.S., Cervera, M.T., and Garcia, J.A. 1998. Plum pox potyvirus resistance associated to transgene silencing that can be stabilized after different number of plant generations. *Gene* 206, 263–272.

Guo, Z., Lu, J., Wang, X., Zhan, B., Li, W., and Ding, S.W. 2017. Lipid flippases promote antiviral silencing and the biogenesis of viral and host siRNAs in Arabidopsis. *Proceedings of the National Academy of Sciences of the United States of America*, *114*, 1377–1382.

Gupta, O.P., Permar, V., Koundal, V., Singh, U.D., and Praveen, S. 2012. MicroRNA regulated defense responses in *Triticum aestivum* L. during *Puccinia graminis* f.sp. tritici infection. *Molecular Biology Reports*, *39*, 817–824.

He, X., Fang, Y., Feng, L., and Guo, H. 2008. Characterization of conserved and novel microRNAs and their targets, including a TuMV-induced TIR-NBS-LRR class R gene-derived novel miRNA in *Brassica*. *FEBS Letters*, *582*, 2445–2452.

Hou, Y., Zhai, Y., Feng, L., Karimi, H.Z., Rutter, B.D., Zeng, L., et al. 2019. A *Phytophthora* effector suppresses trans-kingdom RNAi to promote disease susceptibility. *Cell Host and Microbe*, *25*, 153–165.

Huang, J., Yang, M., Lu, L., and Zhang, X. 2016. Diverse functions of small RNAs in different plant-pathogen communications. *Frontiers Microbiology*, *7*, 1552.

Jain, R.G., Robinson, K.E., Asgari, S., and Mitter, N. 2021. Current scenario of RNAi-based hemipteran control. *Pest Management Science*, *77*(5), 2188–2196.

Jian, H., Ma, J., Wei, L., Liu, P., Zhang, A., Yang, B., et al. 2018. Integrated mRNA, sRNA, and degradome sequencing reveal oilseed rape complex responses to *Sclerotinia sclerotiorum* (Lib.) infection. *Scientific Reports*, *8*, 10987.

Jiang, S., Wu, H., Liu, H., Zheng, J., Lin, Y., and Chen, H., 2017. The overexpression of insect endogenous small RNAs in transgenic rice inhibits growth and delays pupation of striped stem borer (*Chilo suppressalis*). *Pest Management Science*, *73*(7), 1453–1461.

Katiyar-Agarwal, S., Gao, S., Vivian-Smith, A., and Jin, H. 2007. A novel class of bacteria-induced small RNAs in *Arabidopsis*. *Genes Development*, *21*, 3123–3134.

Kaur, R., Choudhury, A., Chauhan, S., Ghosh, A., Tiwari, R. and Rajam, M.V. 2021. RNA interference and crop protection against biotic stresses. *Physiology and Molecular Biology of Plants*, *27*, 2357–2377.

Li, F., Pignatta, D., Bendix, C., Brunkard, J.O., Cohn, M.M., Tung, J., et al. 2012. MicroRNA regulation of plant innate immune receptors. *The Proceedings of the National Academy of Sciences United States of America*, *109*, 1790–1795.

Li, Y., Zhang, Q., Zhang, J., Wu, L., Qi, Y., and Zhou, J.M. 2010. Identification of microRNAs involved in pathogen-associated molecular pattern-triggered plant innate immunity. *Plant Physiology*, *152*, 2222–2231.

Lindbo, J.A., Silva-Rosales, L., Proebsting, W.M., and Dougherty, W.G. 1993. Induction of a highly specific antiviral state in transgenic plants: implications for regulation of gene expression and virus resistance. *Plant Cell*, 5, 1749–1759.

Liu, H., Shen, E., Wu, H., Ma, W., Chen, H., and Lin, Y., 2022. Trans-kingdom expression of an insect endogenous microRNA in rice enhances resistance to striped stem borer *Chilo suppressalis*. *Pest Management Science*, *78*(2), 770–777.

Liu, J., Cheng, X., Liu, D., Xu, W., Wise, R., and Shen, Q.H. 2014. The miR9863 family regulates distinct Mla alleles in barley to attenuate NLR receptor-triggered disease resistance and cell-death signaling. *PLoS Genetics*, *10*, e1004755.

Malik, H.J., Raza, A., Amin, I., Scheffler, J.A., Scheffler, B.E., Brown, J.K., et al. 2016. RNAi-mediated mortality of the whitefly through transgenic expression of double-stranded RNA homologous to acetylcholinesterase and ecdysone receptor in tobacco plants. *Scientific Reports*, *6*(1), 1–11.

Mao, C., Zhu, X., Wang, P., Sun, Y., Huang, R., Zhao, M., et al. 2022. Transgenic double-stranded RNA rice, a potential strategy for controlling striped stem borer (*Chilo suppressalis*). *Pest Management Science*, *78*(2), 785–792.

Mesterházy, A., Judit Oláh, J., and Popp, J. 2020. Losses in the grain supply chain: causes and solutions. *Sustainability*, *12*, 2342.

Molnar, A., Melnyk, C.W., Bassett, A., Hardcastle, T.J., Dunn, R., and Baulcombe, D.C. 2010. Small silencing RNAs in plants are Mobile and direct epigenetic modification in recipient cells. *Science*, *328*(5980), 872–875.

Navarro, L., Dunoyer, P., Jay, F., Arnold, B., Dharmasiri, N., Estelle, M., et al. 2006. A plant miRNA contributes to antibacterial resistance by repressing auxin signaling. *Science*, *312*, 436–439.

Niu, D., Lii, Y.E., Chellappan, P., Lei, L., Peralta, K., Jiang, C., et al. 2016. miRNA863-3p sequentially targets negative immune regulator ARLPKs and positive regulator SERRATE upon bacterial infection. *Nature Communications*, *7*, 11324.

Niu, Q.W., Lin, S.S., Reyes, J.L., Chen, K.C., Wu, H.W., Yeh, S.D., et al. 2006. Expression of artificial microRNAs in transgenic *Arabidopsis thaliana* confers virus resistance. *Nature Biotechnology*, *24*, 1420–1428.

Pareek, A., Dhankher, O.P., and Foyer, C.H. 2020. Mitigating the impact of climate change on plant productivity and ecosystem sustainability. *Journal of Experimental Botany*, *71*, 451–456.

Perisperis, C., Serracardona, A., Sanchezsanuy, F., Campo, S., Ariño, J., and San Segundo, B. 2017. Two NRAMP6 isoforms function as Iron and manganese transporters and contribute to disease resistance in Rice. *Molecular Plant-Microbe Interactions*, *30*, 385–398.

Petchthai, U., Yee, C.S.L., and Wong, SM. 2018. Resistance to CymMV and ORSV in artificial microRNA transgenic Nicotiana benthamiana plants. *Science Reports*, *8*, 9958.

Pradhan, B., Naqvi, A.R., Saraf, S., Mukherjee, S.K., and Dey, N. 2015. Prediction and characterization of Tomato leaf curl New Delhi virus (ToLCNDV) responsive novel microRNAs in Solanum lycopersicum. *Virus Research*, *195*, 183–195.

Prasad, A., Sharma, N., Muthamilarasan, M., Rana, S., and Prasad, M. 2019. Recent advances in small RNA mediated plant-virus interactions. *Critical Reviews in Biotechnology*, *39*(4), 587–601.

Qiao, Y., Liu, L., Xiong, Q., Flores, C., Wong, J., Shi, J., et al. 2013. Oomycete pathogens encode RNA silencing suppressors. *Nature Genetics*, *45*, 330–333.

Qiao, Y., Shi, J., Zhai, Y., Hou, Y., and Ma, W. 2015. Phytophthora effector targets a novel component of small RNA pathway in plants to promote infection. Proceedings of the National Academy of Science *United States of America*, *112*, 5850–5855.

Qu, J., Ye, J., and Fang, R. 2007. Artificial microRNA-mediated virus resistance in plants. *Journal of Virology*, *81*, 6690–6699.

Ramesh, S.V., Ratnaparkhe, M.B., Kumawat, G., Gupta, G.K., and Husain, S.M. 2014. Plant miRNAome and antiviral resistance: a retrospective view and prospective challenges. *Virus Genes*, *48*, 1–14.

Sharma, N. and Prasad, M. 2020. Silencing AC1 of Tomato leaf curl virus using artificial microRNA confers resistance to leaf curl disease in transgenic tomato. *Plant Cell Reports*, *39*, 1565–1579.

Shen, Y., Chen, Y.Z., and Zhang, C.X. 2021. RNAi-mediated silencing of ferritin genes in the brown planthopper *Nilaparvata lugens* affects survival, growth and female fecundity. *Pest Management Science*, *77*(1), 365–377.

Shivaprasad, P.V., Chen, H.M., Patel, K., Bond, D.M., Santos, B.A., and Baulcombe, D.C. 2012. A microRNA superfamily regulates nucleotide binding site-leucine-rich repeats and other mRNAs. *Plant Cell*, *24*, 859–874.

Silva, T.F., Romanel, E.A., Andrade, R.R., Farinelli, L., Østerås, M., Deluen, C., et al. 2011. Profile of small interfering RNAs from cotton plants infected with the polerovirus Cotton leafroll dwarf virus. *BMC Molecular Biology*, *12*, 40.

Sun, L., Gao, Y., Zhang, Q., Lv, Y., and Cao, C. 2022. Resistance to *Lymantria dispar* larvae in transgenic poplar plants expressing CYP6B53 double-stranded RNA. Annals of Applied Biology, 181(1), 40–47.

Thakur, N., Upadhyay, S.K., Verma, P.C., Chandrashekar, K., Tuli, R., and Singh, P.K. 2014. Enhanced whitefly resistance in transgenic tobacco plants expressing double stranded RNA of v-ATPase A gene. *PloS One,* 9(3), e87235.

Wang, H.C., Jiao, X.M., Kong, X.Y., Hamera, S., Wu, Y., Chen, X.Y., et al. 2016. A signaling cascade from miR444 to RDR1 in rice antiviral RNA silencing pathway. *Plant Physiology*, *170*, 2365–2377.

Wen, N., Chen, J., Chen, G., Du, L., Chen, H., Li, Y., et al. 2021. The overexpression of insect endogenous microRNA in transgenic rice inhibits the pupation of Chilo suppressalis and Cnaphalocrocis medinalis. *Pest Management Science*, *77*(9), 3990–3999.

Wu, F., Chen, Y., Tian, X., Zhu, X., and Jin, W. 2017. Genome-wide identification and characterization of phased small interfering RNA genes in response to Botrytis cinerea infection in Solanum lycopersicum. *Scientific Reports*, *7*, 3019.

Xie, J., Li, S., Zhang, W., and Xia, Y. 2019. RNAi-knockdown of the Locusta migratoria nuclear export factor protein results in insect mortality and alterations in gut microbiome. *Pest Management Science*, *75*(5), 1383–1390.

Xin, M., Wang, Y., Yao, Y., Xie, C., Peng, H., Ni, Z., et al. 2010. Diverse set of microRNAs are responsive to powdery mildew infection and heat stress in wheat (Triticum aestivum L.). *BMC Plant Biology*, *10*, 123.

Yu, X. and Killiny, N. 2018. RNA interference of two glutathione S-transferase genes, Diaphorina citri DcGSTe2 and DcGSTd1, increases the susceptibility of Asian citrus psyllid (Hemiptera: Liviidae) to the pesticides fenpropathrin and thiamethoxam. *Pest Management Science*, *74*(3), 638–647.

Zhai, J., Jeong, D.H., De Paoli, E., Park, S., Rosen, B.D., Li, Y., et al. 2011. MicroRNAs as master regulators of the plant NB-LRR defense gene family via the production of phased, trans-acting siRNAs. *Genes and Development*, *25*, 2540–2553.

Zhang, H., Tao, Z., Hong, H., Chen, Z., Wu, C., Li, X., et al. 2016. Transposon-derived small RNA is responsible for modified function of WRKY45 locus. *Nature Plants*, *2*, 16016.

Zhang, M., Zhao, H., Xie, S., Chen, J., Xu, Y., Wang, K., et al. 2011. Extensive, clustered parental imprinting of protein-coding and noncoding RNAs in developing maize endosperm. Proceedings of the National Academy of Science *United States of America*, *108*, 20042–20047.

Zhang, X., Xia, J., Lii, Y., Barrera-Figueroa, B., Zhou, X., Gao, S., et al. 2012. Genome-wide analysis of plant nat-siRNAs reveals insights into their distribution, biogenesis and function. *Genome Biology*, *13*, R20.

Zhang, X., Zhao, H., Gao, S., Wang, W., Katiyar-Agarwal, S., Huang, H., et al. 2011. Arabidopsis argonaute 2 regulates innate immunity via miRNA393*-mediated silencing of a golgi-localized SNARE gene, MEMB12. *Molecular Cell*, *42*, 356–366.

Zhu, Q., Fan, L., Liu, Y., Xu, H., Llewellyn, D., and Wilson, I. 2013. miR482 regulation of NBS-LRR defense genes during fungal pathogen infection in cotton. *PLoS One*, *8*, e84390.

6 Genome Editing Strategies

A Way Forward for Developing Resistance Against Plant Viruses

Divya K.
ICAR-Central Tuber Crops Research Institute

Ying Zhai
Washington State University

Makeshkumar T.
ICAR-Central Tuber Crops Research Institute

Enrique Moriones
Instituto de Hortofruticultura Subtropical
y Mediterránea "La Mayora"

Hanu R. Pappu
Washington State University

6.1 INTRODUCTION

Biotic stress induced by a wide range of pests and diseases continues to be a major threat to crop production. Among the various plant pathogens, viruses cause 10%–15% yield loss every year, and the cost of managing viral diseases is estimated to be over US $1 billion on a global scale. The impact of climate change on disease dynamics continues to evolve, which could potentially lead to more frequent and severe outbreaks of viral diseases. This could be manifested through the expansion of viral vectors to new geographic areas, and the spread of viruses along with the vectors, the emergence of new and more virulent viral strains, and host adaptation leading to the expansion of the host range.

Due to the lack of effective chemical control options, the management of viral diseases requires an integrated approach that involves a multitude of tactics that need to be adapted from the pre-planting stage, through the production season, and during

DOI: 10.1201/9781003403142-6

storage of the produce. One of the most effective tools for viral disease management is growing virus-resistant cultivars. Conventional breeding techniques were successfully used for developing resistant varieties of numerous crop species. While this approach has made a significant contribution toward reducing crop losses, some of the inherent disadvantages of hybridization techniques include linkage drag, loss of variability, lack of sources of genetic resistance, and time from initial crosses to releasing a variety. Mutation breeding using artificial mutagens is another approach to creating disease-resistant varieties. Examples include resistance to powdery mildew conferred by a loss-of-function mutation of the mildew resistance locus o (mlo) in barley, *Arabidopsis*, tomato, and pea. However, this approach is also time-consuming and costly. Moreover, the effect of such random mutagenesis is often unpredictable, and the chances of getting undesirable traits or loss of desirable traits are high.

The advent of recombinant DNA technology facilitated the creation of transgenic plants expressing genes encoding desirable traits through stable and heritable integration of transgenes. Subsequently, other molecular approaches, such as RNA interference, artificial micro-RNAs, transacting small-interfering RNAs, hairpin RNAs, and direct repeats, were shown to impart virus resistance (Figure 6.1). Several virus-resistant transgenic crops were developed, including papaya, tomato, watermelon, wheat, and rice. However, some of the disadvantages of transgenic crops include, besides meeting the biosafety and regulatory requirements of countries, the lack of consistency of transgenes over generations, non-specific integration events, and the low frequency of single-copy integration events. Most importantly, the biosafety concerns about such genetically modified plants hindered their broad range/universal field-level adaptation. Several countries approved the commercial cultivation and consumption of transgenic food crops, while others, including India, restricted them, citing ecological and biosafety concerns.

Over the last two to three decades, gene-editing techniques such as ZFNs, TALENs, and CRISPR/Cas have emerged as a potential alternative strategy for modifying crop genomes. Among these, the recently identified CRISPR/Cas system gained much popularity owing to its simplicity, specificity, ease of multiplexing, efficiency, and

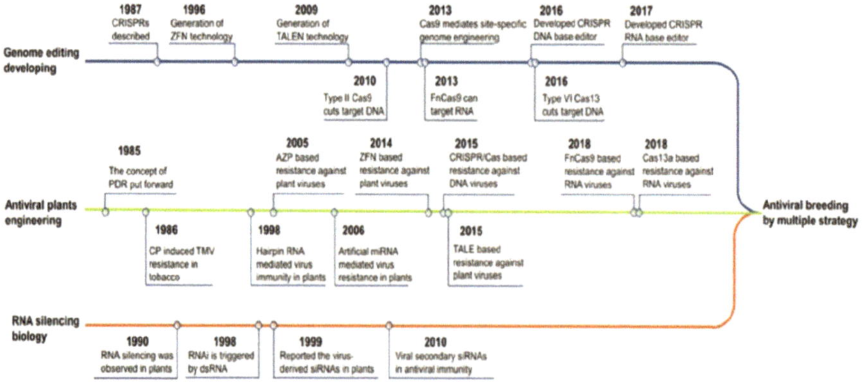

FIGURE 6.1 A historical context showing the development of antiviral strategies for virus resistance (Zhao et al. 2020).

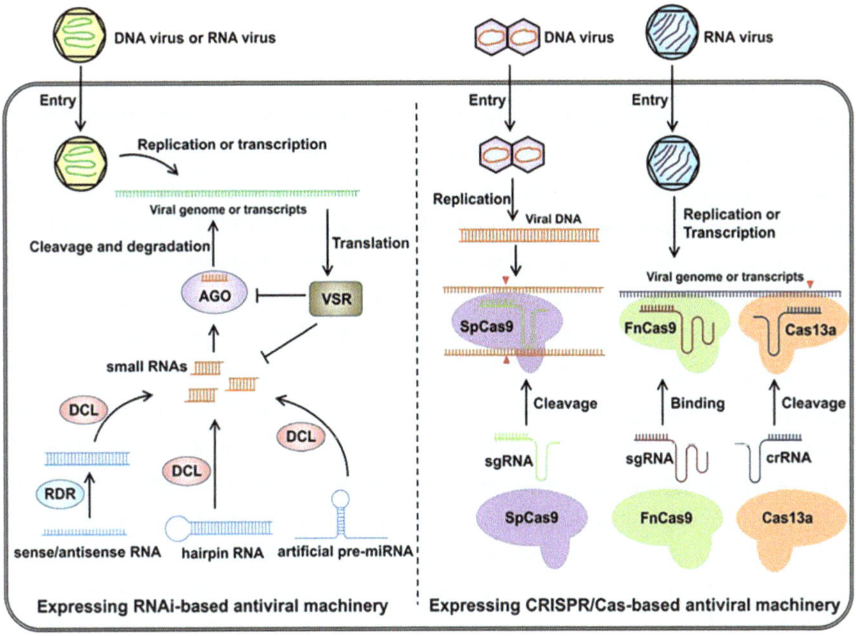

FIGURE 6.2 Diagrammatic representation of the mechanism of action of RNAi-based and CRISPR/Cas-based antiviral machinery (Zhao et al. 2020).

reproducibility. Numerous publications document the application of this technique for conferring resistance to plant viruses. It has been shown that CRISPR/Cas and its modified forms provide a robust platform for genetic manipulation with versatile editing options that could confer crop species with durable, effective, and broad-spectrum resistance against pathogenic viruses. Moreover, non-transgenic editing options significantly enhanced the acceptability of CRISPR-edited crops. The mechanism of action of RNAi-based and CRISPR-based strategies for inducing resistance against viruses is illustrated in Figure 6.2.

6.2 GENOME EDITING IN CROPS: BRIEF OUTLINE OF MECHANISM AND APPLICATIONS

Genome editing is the modification of genomic sequences in a site-specific manner, mostly with the help of site-specific nucleases (SSNs). SSNs create double-stranded breaks in the target genome, which are then repaired by the cell's intrinsic DNA repair pathways, non-homologous end joining, or homology-directed repair. The former is an error-prone repair mechanism that helps to induce point mutations in the target genomic site, whereas the latter is a high-fidelity pathway facilitating replacement/insertion of the desired gene segment at a specific target site. ZFNs and TALENs were developed as second-generation gene-editing tools following the meganucleases. Both of them consist of customizable DNA-binding protein domains coupled with the FokI endonuclease domain (Pabo et al. 2001; Boch et al. 2009). Although certain species have been edited using ZFNs and TALENs, their application has been limited due to the complexity of designing target-specific protein domains and lower efficiency rates (Gaj et al. 2013). CRISPR/Cas is an RNA-guided engineered nuclease in which a single-stranded RNA segment guides the transient expression of CRISPR constructs to induce desired edits. The *Agrobacterium*-mediated delivery results in the inheritance of the CRISPR construct along with the induced mutations. Through genetic segregation, Cas null segregants with the desired mutation can be selected as non-transgenic lines, and transgenic ones carrying the Ti plasmid can be eliminated. This non-transgenic characteristic of edited plants is one of the most important attributes of the technique over existing transgenic strategies. CRISPR-edited plants are likely to be exempted from GMO-related regulations, for example, the guidance issued by the United States Department of Agriculture.

CRISPR/Cas systems are classified into 2 classes, 6 types, and 33 subtypes, each of which has a different set of Cas proteins for performing different functions (Makarova et al. 2020). Of these, the class2 system consisting of a single Cas protein, present in *Streptococcus pyogenes* (*Sp*) is the most commonly used. Cas identifies a 3′ NGG and cleaves the target sequence three base pairs upstream to the PAM with its two nuclease domains, i.e., HNH and RuvC (Jinek et al. 2012). Modified CRISPR/Cas systems with mutations in H840A in the HNH domain and D10A in the Ruv C domain lose their cleavage activity but still retain guide RNA- and DNA-binding properties. Fusion of this 'dead' Cas with FokI nuclease domain can facilitate efficient nicking of the target site, which is specifically useful for gene insertion purposes as well as for inducing long deletions. Apart from editing, the 'dead' Cas fused with different transcription factors, activators, or repressors can be employed for transcriptional regulation (Moradpour and Abdulah 2020). Other CRISPR systems with variant Cas that possess robust target recognition ability and have high-efficiency and high-flexible editing features irrespective of PAM specificities have been identified in recent years. CRISPR/Cpf1 or Cas12a is a class 2 type V system with a comparatively smaller size and an inbuilt single-guide RNA (sgRNA), which identifies a T-rich PAM at the 5′ end of the target and produces staggered end cuts. This will facilitate gene knock-ins easily compared to blunt end cuts by Cas9. Its ribonuclease activity facilitates the self-processing of crRNAs, which enables multiple loci editing with a single cassette comprising sgRNAs specific to each target (Kim et al. 2017). Cas12f is another variant with similar properties. Another one is Cas13a, a class2, type VI CRISPR system that is an RNA-guided RNA editing system. The Cas13a endonuclease targets single-stranded RNAs, and it lacks a DNA endonuclease domain, and possesses two higher eukaryotic and prokaryotic nucleic acid-binding endo RNase domains (HEPN), independent of specific PAM site requirements (Abudayyeh et al. 2016, 2017).

A general workflow involved in the genome editing of plants for virus resistance (Figure 6.3) includes developing appropriate CRISPR cassettes and introduction into the host through agro-infiltration or any other suitable method. This is followed by the identification of the edited plants containing the desired trait modification through various molecular approaches. The precise editing takes place either by the near homologous end joining (NHEJ) or homologous recombination (HR) method. These CRISPR systems together provide a broad range of editing strategies for

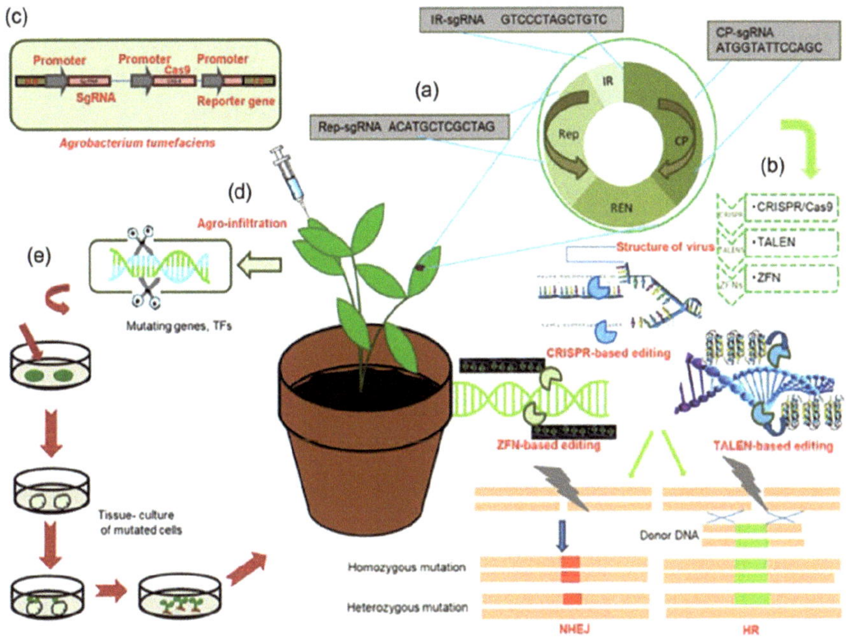

FIGURE 6.3 General workflow of gene-editing technologies to engineer virus resistance in crops (Mushtaq et al. 2019).

enhancing resistance to both DNA and RNA viruses. CRISPR has the advantage of specificity over RNAi because it can effectively target viruses located in plant nuclei or viruses with strong gene silencing suppressors since RNAi is less effective than viruses located in plant nuclei or encoding strong gene silencing suppressors. CRISPR can be used to enhance virus resistance by combining with RNAi. A large number of species, ranging from major staple crops such as rice and wheat to horticultural crops including tomato, grapes, cucumber, and cassava have been successfully edited using CRISPR. Recent advancements such as base editing and prime editing further enhance its applicability and acceptability for crop improvement, particularly in terms of specificity and non-transgenic characteristics.

6.3 CRISPR/CAS GENOME EDITING FOR VIRUS RESISTANCE IN CROP PLANTS

Soon after its establishment as a potential tool for plant genome editing in 2013, different groups of researchers verified the feasibility of using CRISPR/Cas for developing virus resistance in plants. Initial attempts were solely focused on viral genome silencing strategies, i.e., targeted mutagenesis of vital viral genes to inhibit their replication and movement, which in turn conferred resistance to the respective virus. Most of these studies used geminiviruses, which cause significant damage to crop plants. Genes essential for their replication and movement, i.e., the replication-associated gene and the coat protein gene, were the common targets for editing.

The intergenomic region of bipartite geminiviruses was identified as a suitable candidate for editing to obtain broad-spectrum resistance, as this sequence is conserved among several geminivirus species. The effectiveness of CRISPR/Cas in imparting resistance to RNA virus has also been evaluated. Although most of the studies are confined to model plant systems such as *Nicotiana benthamiana* and *Arabidopsis thaliana* as proof of concept, the effectiveness of the CRISPR/Cas editing in conferring virus resistance was shown in certain crop species as well. However, when we target viral genomes, the possibility of viruses escaping the editing is high, owing to their wide range of adaptability and evolutionary potential. Thus, targeting host susceptibility factors was proposed as a more efficient strategy, and a few studies were reported in this regard, which will be discussed in the following sections.

6.3.1 Transgenic Approaches Deploying CRISPR/CAS to Confer Virus Resistance

The feasibility of application of the CRISPR/Cas tool for imparting virus resistance was first demonstrated by Ji et al. (2015) against a geminivirus, Beet severe curly top virus (BSCTV), in *N. Benthamiana* and *A. thaliana*. It was achieved by targeted mutagenesis of viral genomic sequences that are critical for replication and survival in the host. BSCTV replication and accumulation were inhibited by targeting Rep, CP, and IR using specific gRNAs. The efficiency and degree of virus inhibition were correlated with the expression levels of Cas. The level of resistance in successive generations of Cas-sgRNA-overexpressing plants was also evaluated in this study.

The same strategy was employed by Baltes et al. (2015) to confer resistance to another geminivirus, Bean yellow dwarf virus (BeYDV), in *N. benthamiana*. Six regions of the BeYDV genome were targeted: RBS, hairpin, nonanucleotide sequence, and three Rep motifs essential for rolling circle replication (motifs I, II, and III). This impaired the replication as well as the systemic movement of the virus in the host and thereby conferred resistance. The nature of NHEJ-induced INDELs at each target site was also analyzed. The additive effect of multiple sgRNAs targeting the same viral genome for conferring robust virus resistance was assessed. Transgenic *N. benthamiana* constitutively expressing Cas9 and sgRNAs showed reduced symptoms relative to control plants.

Tomato yellow leaf curl virus (TYLCV) is a ssDNA begomovirus that causes severe crop losses, and existing control measures are inadequate. The linkage of the resistance locus with poor fruit quality genes renders conventional resistance breeding inefficient for TYLCV resistance, and RNA silencing strategies also failed to impart significant resistance. Ali et al. (2015a,b) demonstrated the efficacy of CRISPR/Cas for targeting the viral genome to develop resistance against this ssDNA begomovirus. Target-specific sgRNAs were delivered to Cas-overexpressing *N. benthamiana* lines. Different coding and non-coding sequences of TYLCV, viz., IR, CP, and Rep (RCR II motif), were targeted and resulted in a considerable reduction in virus titer. IR was found to be a better target for preventing virus accumulation in comparison with others (CP and Rep). Targeted mutagenesis of IR caused a significant reduction in virus replication and accumulation, whereas Rep and CP-targeted transgenic lines showed attenuated symptoms. Feasibility of multiplexing based on polycistronic – tRNA – gRNA system was analyzed with CP- and IR-targeting gRNAs. Simultaneous delivery of these two gRNAs induced targeted mutations in both sites and resulted in a greater reduction of virus titer and recovery from disease symptoms in the host compared to those caused by individual gRNAs. The authors also succeeded in conferring broad-spectrum resistance by using IR-sgRNA containing the invariant TAATATTAC sequence common to all geminiviruses. Resistance against TYLCV, mermia mosaic virus (MeMV), and BCTV in *N. benthamiana* plants was achieved in this manner.

Subsequently, in 2018, Tashkandi et al. showed resistance to TYLCV by targeting CP and Rep, and the resistance imparted was found to be stably inherited across generations. This study also showed that although targeting Rep could induce mutations, the ORF remained functional, and therefore replication and systemic spread were not impaired. Contrary to this, IR nonanucleotide targeting gRNA effectively prevented viral infection and limited the generation of escape mutants. But tomato plants with mutated IR regions failed to regenerate. Thus, targeting CP appeared to be a more viable way to confer TYLCV resistance (Tashkandi et al. 2018). Multiplex editing and paired nickases are likely to provide robust resistance.

In 2016, Ali et al. further investigated the potential of CRISPR/Cas-edited viral genomes to evade the CRISPR/Cas system. Resistance to cotton leaf curl Kokhranvirus (CLCuKov), MeMV, and TYLCV geminiviruses was imparted using CRISPR/Cas targeting coding and non-coding sequences in *N. benthamiana.* Even though mutations induced in coding regions such as RCRIII and CP are capable of inhibiting viral replication, targeting of such regions was found to have comparatively higher probabilities of evading the CRISPR/Cas system by evolving mutant variants that can continue to replicate and spread systemically. But targeting non-coding sequences such as IR was found to permanently interfere with viral replication and systemic spread, thereby providing durable resistance (Ali et al. 2016).

Cotton leaf curl disease (CLCuD) causes significant yield loss in the Indian subcontinent. Most of the methods employed for CLCuD resistance so far have failed to confer complete resistance, particularly when there are chances of multiple infections. This could be because the helper virus was targeted while the satellites associated with it (alpha satellite and beta satellite) played an important role in the pathogenesis. Iqbal et al. (2016) proposed a multiplex editing strategy for achieving broad-spectrum resistance to CLCuD-associated begomoviruses (CABs) and their associated satellites. Cotton leaf curl Alabad virus (CLCuAlV), cotton leaf curl Multan virus (CLCuMuV), cotton leaf curl Bangalore virus (CLCuBaV), CLCuKoV-Bu (Burewala strain), cotton leaf curl Kokhran virus (CLCuKoV), and cotton leaf curl Rajasthan virus (CLCuRaV) constitute the CLCuD complex. sgRNAs specific to CABs (16), alpha satellites (8), and CLCuMuB (12) were designed, and efficiency was evaluated *in silico.* When sgRNAs specific to CABs (16), alpha satellites (8), and CLCuMuB (12) were designed and their efficiency was evaluated *in silico*, high off-target effects for most of the sgRNAs were predicted, pointing to the need for employing more stringent strategies such as nickases to obtain efficient resistance to the entire CLCuD complex. A multiplex editing of IR and Rep of the CABs, IR of CLCuMuB, and Rep of diverse alpha satellites was proposed to be a potential strategy to trigger efficient broad-spectrum resistance in cotton.

Khan et al. (2019) assessed the efficiency of dCas9 as a DNA-binding protein to suppress CLCuV replication by targeting the nonanucleotide sequence of CLCuV. Transgenic *N. benthamiana* plants harboring CRISPR/dCas9 constructs were inoculated with CLCuV, and viral load and symptoms were evaluated. Although partial inhibition of CLCuV replication reduced symptoms, and virus accumulation in transgenic plants, but results suggested that dCas9 is less efficient at inhibiting virus replication when compared to Cas9.

Yin et al. (2019) demonstrated the effectiveness of dual gRNA-mediated editing against CLCuMuV). Replication-associated protein (Cl) and IR regions were selected as targets, and gRNA with the highest efficiency and activity was designed for the respective targets. Large deletions were created in transgenic *N. benthamiana* plants by dual gRNAs, and this conferred considerable resistance to CLCuMuV. Edited *N. benthamiana* lines showed resistance marked by a lack of any symptoms. A reduction in viral load as well as double-stranded DNA forms indicated the efficient inhibition of CLCuMuV replication by dual gRNAs.

The wheat dwarf virus infects both wheat and barley, causing severe yield loss, and sources of natural host plant resistance are lacking. Kis et al. (2019) used CRISPR/Cas to develop WDV resistance in *Hordeum vulgare*. The overlapping region of the MP and CP coding sequences, the sequence encoding the N-terminus of the Rep/RepA protein, the LIR region, and the sequence encoding the C-terminus of Rep were selected as editing targets, and sgRNAs were designed and showed improved resistance and reduced symptoms in edited barley lines. Sequence analysis of edited lines revealed that the sgRNA targeting sequence encoding the N-terminus of Rep/RepA was the most efficient in generating mutations in the target site for conferring resistance, whereas other sgRNAs failed to induce any useful mutations.

Banana streak virus is a complex of pararetroviruses whose genome gets integrated within the B genome derived from *Musa balbisiana* and is referred to as endogenous banana streak virus (eBSV). The eBSV becomes a functional episome, causing infection under stress conditions and during in vitro culture for propagation and hybridization. Thus, eBSV in the B genome of *M. balbisiana* is a barrier to banana breeding programs, despite its elite trait attributes. CRISPR/Cas could inactivate the eBSV in the B genome of plantain by specifically mutating the genomic sequence of the eBSV. Three gRNAs targeting ORF1, ORF2, and ORF3 of BSV strain Obinol'Ewai (BSOLV), distributed in Africa, were used for multiplex editing. It was found that simultaneous editing of all three ORFs is the most efficient way to combat eBSV, rather than targeting them independently. Of these, targeting ORF3 is an important target since it encodes most of the proteins required for infection and symptom development, such as movement protein, coat protein, aspartic protease, reverse transcriptase, and ribonuclease H. CRISPR/Cas-induced mutations efficiently inactivated eBSV and thereby rendered the B genome safe for breeding and hybrid dissemination (Tripathi et al. 2019).

6.3.1.1 CRISPR/CAS-Mediated Resistance to Cassava Mosaic Virus: Case Study of Editing Resistant Mutant Evolution

African cassava mosaic virus (ACMV) is an economically important virus affecting cassava production in several parts of the world. Mehta et al. (2018) developed cassava lines by targeting two coding sequences of DNA A of the ACMV bipartite genome: the AC2 gene coding for the multifunctional TrAP protein, which is involved in gene activation, virus pathogenicity, and suppression of gene silencing, and the AC3 gene coding for replication enhancer protein (REn), which is involved in replication enhancement. Contrary to the results obtained with other viruses, they observed no significant difference in viral load, symptom severity, and disease incidence in transgenic plants compared to controls, which can be attributed to the evolution of Cas cleavage-resistant variants harboring a conserved mutation in the AC2 gene.

Sequence analyses of the viral genome isolated from transgenic cassava lines harboring gRNA: Cas showed substitution and indel mutations at target sites of AC2 and AC3. In genomes with edits, both open reading frames stopped prematurely. Amino acid sequence analysis of AC2 and AC3 gene products revealed the insertion of a single T nucleotide in H54Q of the ACMV AC2 protein, thereby creating a premature stop codon and resulting in premature termination of the AC2 ORF. This T insertion event and substitution occurred in the AC2 gRNA seed region and was observed in 33%–48% of mutant lines, suggesting its selection in the population as it gave rise to a new open reading frame for AC2, which makes ACMV-AC2 H54Q mutant lines resistant to CRISPR/Cas-mediated cleavage (Mehta et al. 2019). Even though this cleavage-resistant virus variant is incapable of causing infection independently as it requires AC2 and AC3 encoded by wild-type viruses, the possibility of the evolution of potentially virulent mutants cannot be ruled out. The use of highly efficient Cas or simultaneous targeting of multiple viral sequences is supposed to be effective for mitigating the probability of evolution of mutant viral strains.

6.3.1.2 Editing Viral Genome Using Variant CRISPR/CAS Systems for Virus Resistance

Genome editing-based disease resistance strategy is not only confined to target DNA viruses but has also extended to target RNA viruses, using RNA targeting Cas variants. Zhang et al. (2018) used Cas9 from the *Francisella novicida* species, *Francisella novicida*, which specifically targets RNA to confer resistance against positive sense ssRNA viruses, cucumber mosaic virus (CMV), and tobacco mosaic virus (TMV) in *N. benthamiana* and *Arabidopsis*. Stable inheritance of resistance was also analyzed over generations (up to T6 in *Arabidopsis*). A total of 23 candidate sites in CMV were targeted, of which three sites in ORF 1A, 3C, and 30UTR-A, were identified as the most efficient for imparting resistance. Similarly, three sites in the coding region of TMV were selected as targets. PAM sequence-independent target identification by FnCas9 was evaluated in this study. RNA-binding capacity, arginine-rich motif region (for RNA-binding), and cytoplasmic localization of FnCas9 were proposed to be important. RNA-binding activity was found to be more inevitable for virus interference than endonuclease activity. FnCas9 binds genomic RNA and inhibits replication and translation, which in turn prevents infection. Since it does not cleave the target virus genome, FnCas9 limits the possibility of the evolution of CRISPR-imparted resistance-evading mutant viruses, thereby providing more durable resistance, unlike Cas9, which cleaves the viral genome and creates mutations that may lead to the evolution of cleavage-resistant mutants.

A Class II, Type VI CRISPR system (C2c2) with a single effector, Cas13a, was used to develop resistance to a TuMV by Aman et al. (2018). Both CP and HC-Pro regions were targeted with suitable guide RNAs, and HC-Pro targeting crRNA was found to be most efficient. As mentioned earlier, Cas13a lacks the DNase catalytic site but possesses RNase activity. Additionally, Cas13a has pre-crRNA processing ability, which enables the generation of multiple crRNAs for multiplexed targeting of the same or multiple RNA/DNA viruses. The optimal protospacer length (crRNA) for Cas13a is 28 nucleotides, along with a 28-nucleotide direct repeat. Even though it is independent of PAM specificity, a protospacer flanking site (PFS) flanking 3′ of protospacer was found to affect the activity of Cas13a.

Mahas et al. (2019) attempted to compare the efficiency of different Cas13 variants to interfere with RNA viruses. The efficiency and RNAi potential of three Cas13 variants, namely, LwaCas13a, PspCas13b, and Cas13d orthologue CasRx from *Leptotrichiawadei*, *Bergeyellazoohelcum*, and *Prevotella* sp. P5–125 and *Ruminococcus flavefaciens* XPD3002, respectively, were assessed by transient expression assays using the tobacco rattle virus (TRV) system and *in planta* using TMV RNA-based overexpression system. Although all three variants showed efficient RNAi, CasRx was found to be the ideal candidate.

In 2019, Zhang et al. reported the use of engineered Cas13a from *Leptotrichiashahii* (LshCas13a) to confer resistance against TMV in *N. benthamiana*. Five TMV genomic sites were targeted, and a significant reduction in viral titer was observed. In addition to proving the efficacy of Cas13a in conferring resistance to RNA viruses, this study also emphasized the involvement of HEPN domains in the cleavage process and the significance of the cleavage activity of Cas13a in inducing targeted mutations.

In addition to dicot systems, the efficacy of CRISPR/Cas13a in conferring virus resistance was demonstrated in monocot systems. In rice, CRISPR/Cas13a was used against southern rice black-streaked dwarf virus (SRBSDV) and rice stripe mosaic virus (RSMV) by Zhang et al. (2019). Southern rice black-streaked dwarf virus (SRBSDV) has a double-stranded RNA genome, whereas RSMV possesses a single-stranded RNA genome. Both viruses affect rice cultivation in East Asian countries and South China, respectively. Three genomic targets from both SRBSDV and RSMV were cleaved by Cas13a and resulted in a considerable reduction in virus load as well as symptom severity. Stable inheritance of resistance conferred by CRISPR/Cas13a was also analyzed.

The potyviral membrane protein P3 is involved in virus replication and systemic infection; CI forms the cytoplasmic inclusion bodies involved in virus movement; RNA-dependent RNA polymerase NIb; and coat protein-encoding CP regions were selected as targets for editing. Potato virus Y (PVY) infection was suppressed in transgenic potato lines expressing Cas13a/sgRNA constructs. Plants exhibited reduced PVY accumulation and disease symptoms.

The level of virus resistance was correlated with the expression levels of the Cas13a/sgRNA construct in the plants. Transgenic lines with a low level of Cas13a/sgRNA expression had mild mosaic symptoms, whereas those with a high level of Cas13a/sgRNA expression do not produce any symptoms. It was also revealed that appropriately designed sgRNAs can specifically interfere with multiple PVY strains (Zhan et al. 2019).

Sweet potato virus disease (SPVD), caused by the combined infection of one potyvirus (genus *Potyvirus* in the Potyviridae family), the sweet potato feathery mottle virus (SPFMV), and the criniviru (genus *Crinivirus* in the Closteroviridae family), sweet potato chlorotic stunt virus (SPCSV), poses a major threat to sweet potato cultivation and production by significantly reducing the tuber yield. Management of SPVD is difficult due to limitations in conventional control measures and the complex nature of the causal agents. CRISPR-Cas13a was engineered to target one of its important pathogenesis-related factors, i.e., SPCSV-RNase3. Targeting of SPCSV-RNase3 with two sgRNAs using the LwaCas13a system inhibited its RNA silencing suppressor activity and recovered the RNA silencing activity in N. *benthamiana* leaf cells. Transgenic *N. benthamiana* plants carrying an RNase3-targeting CRISPR/LwaCas13a system exhibited enhanced resistance against TuMV and CMV (CMV-RNase3) co-infection. It was also shown that transgenic sweet potato plants carrying an RNase3-targeted RfxCas13d system with three sgRNAs exhibited substantially improved SPVD resistance. This study revealed the potential of the CRISPR/Cas13a system for the development of SPVD-resistant sweet potato varieties (Yu et al. 2022).

6.3.2 Transgene-Free Approaches for Virus Resistance by CRISPR/CAS-Induced Mutations

6.3.2.1 Targeting Host Susceptibility Factors Using CRISPR/CAS to Confer Virus Resistance

Editing or targeted mutagenesis of host susceptibility genes is effective in imparting recessive resistance against viruses. Chandrasekaran et al. (2016) developed CRISPR/Cas-mediated broad-spectrum resistance to an ipomovirus, cucumber vein yellowing virus (CVYV), and two potyviral species, zucchini yellow mosaic virus (ZYMV) and papaya ringspot virus (PRSV)-Win cucumber. Unlike previous studies that targeted the viral genome, this was achieved by targeted mutagenesis of the host genomic sequence, the eukaryotic translation initiation factor eIF4e, with which the viral protein Vpg interacts to manifest infection and thereby acts as a negative regulator of plant immunity (Wang and Krishnaswamy 2012). Two gRNAs targeting exon 1 and exon 3 of the eIF4E gene were employed to induce mutations in the C and N termini of eIF4E. The former gRNA disrupts translation and prevents eIF4E protein formation, whereas the latter allows the translation of two-thirds of the protein product, although intact eIF4E is not produced. Targeted mutations, including 20 nucleotide deletions, were induced by these gRNAs without any detectable off-target effects and resulted in recessive resistance in cucumber. Homozygous *eif4e* mutants exhibited complete resistance to CVYV. Although *eif4e* mutants showed resistance to ZYMV and PRSV, later they developed mild symptoms with low viral titers. However, mild symptoms did not affect the plant's health and development. Production of non-transgenic virus-resistant cucumber (*Cucumis sativus* L.) plants (from T3 generation) was demonstrated.

Non-transgenic *Arabidopsis* plants showing stable and heritable resistance to another potyvirus, TuMV, were developed using the same strategy as above by Pyott et al. (2016). The loss-of-function mutation of eIF(iso)4E was achieved by targeting the 5′ end, i.e., the translation initiation site of eIF(iso)4E locus, without any off-target effects. Transgene-free Cas null segregant T3 plants with a homozygous mutation in eIF(iso)4E showed resistance without compromising the plant's vigor. As this approach results in the complete absence of eIF(iso)4E, it is expected to confer more durable resistance and unlikely leave a chance of breakdown of recessive resistance. However, the durability of this LOF-induced recessive resistance and the possibility of resistance breaking need to be evaluated.

Gomez et al. (2019) employed CRISPR/Cas-mediated genome editing to develop resistance to an RNA virus belonging to the Potyviridae family, cassava brown streak virus (CBSV), which causes cassava brown streak disease, a major constraint to cassava production in Africa. CBSV, like other potyviruses, interacts with the eukaryotic translation initiation factor 4E (eIF4E) isoforms to establish infection. Cassava encodes four eIF4E isoforms in cassava eIF(iso)4E-1, eIF(iso)4E-2, novel cap-binding protein-1 (nCBP-1), and nCBP-2. CBSV VPg specifically interacts with novel cap-binding protein-1 (nCBP-1) and nCBP-2. So, these two nCBPs act as recessive resistance sources. Five different regions of both genes were selected as targets, and gRNA constructs were designed. *Agrobacterium*-mediated transformation of friable embryogenic callus was followed by the regeneration of plants bearing 1–16 nt insertions and deletions up to 128 bp at target sites. Ncbp-1/ncbp-2 mutants displayed a significant reduction in areal tissue and storage root symptom severity and disease onset. Although leaf viral RNA load did not show much difference from wild-type, viral titer was considerably reduced in storage roots. However, factors such as affinity for other eIF4E isoforms and unpredictable splice variations created by mutagenesis may hinder the achievement of complete resistance to CBSD.

A similar strategy was employed in tomatoes and efficient resistance to the pepper mottle virus (PepMoV) was achieved. Guide RNAs targeting the first exon of eIF4e created deletions ranging from 11 to 43 bp and 38 bp insertion in the region spanning the gRNA target and, in certain cases, beyond the target region, most of which resulted in premature stop codons and truncated eIF4E protein (Yoon et al. 2020). Stable inheritance of induced mutations was observed in T-DNA-free non-transgenic tomato plants, and they exhibited resistance to PepMoV, which was evidenced by a significant reduction in viral accumulation in edited lines upon PepMoV inoculation. However, this strategy failed to confer resistance to another potyvirus infecting tomatoes, tobacco etch virus, which can be attributed to the redundancy of eIF4E homologs in the plant.

6.3.2.2 Engineering Specific Mutations in the Host Defense Factors to Impart Virus Resistance

Zhang et al. (2020) observed that the modification of the secondary metabolite profile in soybeans is coupled with the virus resistance phenotype. Multiplex editing of three genes, i.e., GmF3H1, GmF3H2, and GmFNSII-1, which are involved in the synthesis of isoflavone, resulted in an increase in isoflavone content in hairy roots, leaves, and seeds. Isoflavone content was doubled in homozygous triple mutants, while mutagenesis efficiency was only 44%. Interestingly, the increase in leaf isoflavone content significantly reduced the accumulation of soybean mosaic virus coat protein. Hence, it was suggested that multiplex editing of isoflavone synthesis genes could result in the dual benefits of enhancing secondary metabolite production and imparting virus resistance.

CRISPR/Cas was successfully used to confer resistance to PVY by editing the gene encoding the nuclear protein coilin (Makhotenko et al. 2019). CRISPR/Cas editing constructs in the form of gold or chitosan microparticles on which Cas endonuclease and a short gRNA were immobilized were delivered to the apical meristem cell by bioballistics and vacuum infiltration methods, respectively. Editing of coilin gene alleles significantly enhanced the resistance to PVY infection (Table 6.1).

6.4 CRISPR/CAS GENOME EDITING FOR INSECT CONTROL

The CRISPR/Cas genome editing system is increasingly being applied to gene function studies and genetic modifications of economically important genes in insects (Taning et al. 2017). As a classic insect model for genetics and molecular biology research, *Drosophila melanogaster* has been used to develop multiple CRISPR/Cas knockout lines for both euchromatic and heterochromatic genes (Yu et al. 2013). Conditional CRISPR/Cas-induced mutagenesis can be achieved by using *D. melanogaster* strains expressing Cas specifically in the germline (Gratz et al. 2013). The cross between a UAS-Cas/gRNA transgenic line and a tissue/development-specific Gal4 line induces CRISPR/Cas-mediated mutagenesis in the corresponding tissues or developmental stages (Xue et al. 2014). The efforts to create CRISPR/Cas mutants have been expanded to other insect models such as *Bombyx mori*, *Aedes aegypti*, and *Tribolium castaneum*, with various gene-editing efficiencies (Taning et al. 2007). To achieve the best practices for CRISPR/Cas editing of insect genomes, particular considerations should be kept in mind, such as designing high-specificity guide RNAs, effective and efficient delivery of Cas and gRNA constructs, and optimizing Cas mRNA and gRNA dosages (Taning et al. 2007).

Some insect species are not compatible with standard CRISPR/Cas gene-editing protocols. For example, members of the silverleaf whitefly (*Bemisia tabaci*) cryptic species Middle East-Asia Minor I (MEAM1), an insect pest and vector of numerous plant viruses causing devastating crop loss worldwide, have small embryos that are technically challenging to inject and exhibit high mortality rates even after successful injections. Heu et al. (2020) developed an improved protocol for switching the injection targets from embryos to vitellogenic adult females, which is much easier to carry out without the need for specialized equipment. In this protocol, they used an ovary-targeting peptide ligand (BtKV) to fuse with Cas. After injection, BtKV was able to translocate the ribonucleoprotein complex to MEAM1 germline, which led to efficient and heritable editing of their offspring genomes. This adult-targeting and germline-translocation strategy can be used to edit other insect species that are difficult for embryo injection.

TABLE 6.1
List of Plants Edited Using CRISPR/Cas System for Virus Resistance

Plant Species	Target Region	Resistance to Virus/Disease	CRISPR System Used	References
Arabidopsis /N. benthamiana	IR, CP, Rep	Beet severe curly top virus	Cas9	Ji et al. (2015)
Cucumis sativus	Host factor eIF4E	CVYV, ZYMV, and PRSV	Cas9	Chandrasekaran et al. (2016)
A. thaliana	Host factor eIF(iso)4E	Turnip mosaic virus (TuMV)	Cas9	Pyott et al. (2016)
Oryza sativa var. indica cv. IR64	Host factor eIF4G	Rice tungro spherical virus	Cas9	Macovei et al. (2018)
Rice		Southern rice black-streaked dwarf virus; Rice stripe mosaic virus	Cas13a	Zhang et al. (2019)
Barley	MP/CP, Rep/RepA, LIR	Wheat dwarf virus	Cas9	Kis et al. (2019)
Nicotiana benthamiana	RBS, IR hairpin, 3x Rep	Bean yellow dwarf virus	Cas9	Baltes et al. (2015)
	IR, CP, RCRII of Rep	Tomato yellow leaf curl virus (TYLCV), BCTV, and MeMV	Cas9	Ali et al. (2015a,b)
	IR, CP, RCRII of Rep	Gemini viruses (CLCuKoV, MeMV, TYLCV)	Cas9	Ali et al. (2016)
	IR, C1	Cotton leaf curl Multan virus (CLCuMuV)	Cas9	Yin et al. (2019)
	CP, Hc-Pro	Turnip mosaic virus (TuMV)	Cas13a > RNA	Aman et al. (2018)
	Nonanucleotidesequence of CLCuV	Cotton leaf curl virus (CLCuV)	dCas9	Khan et al. (2019)
		Tobacco mosaic virus (TMV)	Cas13a	Zhang et al. (2019)
Potato	P3, CI, Nib, CP	Potato Virus Y (PVY)	Cas13a > RNA	Zhan et al. (2019)
	Nuclear protein coilin	PVY	Cas9	Makhotenko et al. (2019)
Cassava	AC2 and AC3	African cassava mosaic virus	Cas9	Mehta et al. (2019)
	eIF4E isoforms nCBP-1 and nCBP-2	Cassava brown streak virus	Cas9	Gomez et al. (2019)
Tomato	CP, Rep	Tomato yellow leaf curl virus	Cas9	Tashkandi et al. (2018)
	Host factor eIF4E1	Pepper mottle virus	Cas9	Yoon et al. (2020)
Musa balbisiana	ORF 1, 2 & 3 of BSV genome	Endogenous banana streak virus	Cas9	Tripathi et al. (2019)
Soybean	Isoflavone biosynthesis gene GmF3H1, GmF3H2 and GmFNSII-1	Soybean mosaic virus	Cas9	Zhang et al. (2020)

In addition to editing insect genes, the CRISPR/Cas platform has been successfully adopted for developing virus-resistant insect strains. For example, Dong et al. (2016) optimized the CRISPR/Cas system to cleave the immediate early gene *ie-1* in *B. mori* nucleopolyhedrovirus (BmNPV). *ie-1* encodes a key protein required for BmNPV replication. Its disruption effectively reduced BmNPV proliferation in *B. mori* cells. This virus-inducible CRISPR/Cas system can be rapidly activated upon BmNPV infection and, at the same time, minimize unwanted off-target effects.

Improvement of plant resistance to insect pests can be achieved by genome editing efforts on either the plant or the insect side (Bisht et al. 2019). To date, multiple plant-pest insects have been engineered to reduce their injury to plants (Bisht et al. 2019). The targeted genes include those that encode chitin synthases, cytochrome P450 enzymes, hydroxylases, and receptors. Similar to the transformation of other insect species, embryo injection/microinjection is still the most popular method for the delivery of CRISPR/Cas apparatus to plant-pest insect cells. The editing events result in gene knockout, substitution, or insertions/deletions (Indels). Host susceptibility genes (S) and genes encoding intracellular resistance receptor (R) proteins are common plant targets for engineering insect resistance.

6.5 CRISPR/CAS-BASED DIAGNOSIS OF PLANT VIRUSES

CRISPR/Cas genome editing has been adopted for virus detection (Sharma et al. 2021). The first CRISPR-based detection of viruses in non-plant hosts was performed using Cas9 to recognize double-stranded DNA (dsDNA) (Zhang et al. 2017). Cas12a (Chen et al. 2018), Cas13a (Gootenberg et al. 2017), and Cas14 (Harrington et al. 2018) were then adopted to establish improved CRISPR-based detection systems. Nucleic acid sequence-based amplification (NASBA)-CRISPR cleavage (NASBACC) involves specific recognition of nucleic acid targets (Pardee et al. 2016). Cas9-mediated cleavage requires the PAM sequence. Both the endonuclease Cas12 and the RNA-guided RNase Cas13 possess collateral or trans-cleavage activity, which enables readouts through the addition of reporter nucleic acids. In both Cas12-based DNA endonuclease-targeted CRISPR trans reporter (DETECTR) (Chen et al. 2018) and Cas13-based specific high-sensitivity enzymatic reporter unlocking (SHERLOCK) (Kellner et al. 2019) systems, fluorescence signals are generated via the separation of the quencher from the fluorophore, resulting in target recognition and reporter cleavage.

NASBACC needs specific recognition of nucleic acid targets (Pardee et al. 2016). After NASBA in the isothermal amplification condition, Cas9 only cleaves the PAM-dependent target, which was then detected by a toehold sensor for the readout (Figure 6.4, left). Both Cas12 and Cas13 possess collateral or trans-cleavage activity. Cas12 is an endonuclease guided by crRNA to find the target DNA sequence, and it then cuts both strands and collateral DNA. Cas13 is an RNA-guided RNase that cleaves single-stranded and collateral RNA. The collateral activity of Cas12 and Cas13 enables readouts through the addition of reporter nucleic acids. In both Cas12-based DETECTR (Chen et al. 2018) and Cas13-based SHERLOCK (Kellner et al. 2019) systems, fluorescence signals are generated via the separation of the quencher from the fluorophore resulting from target recognition and reporter cleavage (Figure 6.4, right).

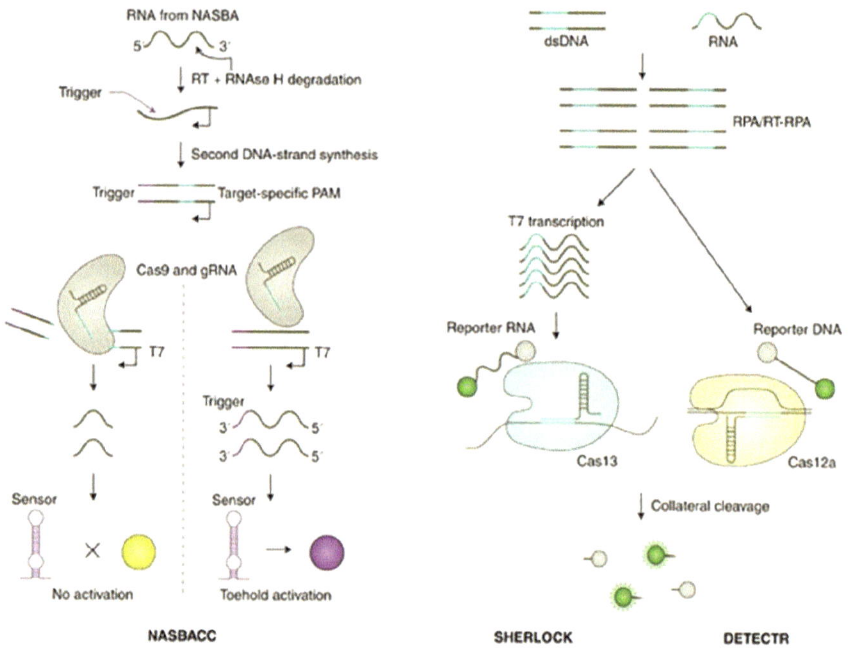

FIGURE 6.4 Comparative cleavage patterns of gene-editing using Cas9, Cas12a and Cas13 modules.

The CRISPR/Cas12a module tends to be the most popular diagnostic platform for plant viruses (Alon et al. 2021), which is often coupled with traditional rapid virus detection tools such as recombinase polymerase amplification (RPA) (Aman et al. 2020; Jiao et al. 2021; Ramachandran et al. 2021) and loop-mediated isothermal amplification (LAMP) (Mahas et al. 2021; Zhu et al. 2022). In an example of CRISPR/Cas12a-based diagnosis coupled with regular RT-PCR, Alon et al. (2021) designed specific CRISPR RNAs (crRNAs) to differentially detect tomato brown rugose fruit virus (ToBRFV) and its closely related tomato mosaic virus (ToMV). Both ToBRFV and ToMV are emerging tobamoviruses (genus *Tobamovirus*, family Virgaviridae) affecting tomato production.

Coupling with RPA or LAMP further improves the sensitivity and specificity of CRISPR-based plant virus detection protocols. Aman et al. (2020) developed a simple, rapid, efficient, but also highly sensitive and specific RT-RPA-coupled CRISPR/Cas12a diagnostic tool for the detection of multiple economically damaging plant RNA viruses, including potexvirus (genus *Potexvirus*, family Alphaflexiviridae), potyvirus (genus *Potyvirus*, family Potyviridae), and tobamovirus (genus *Tobamovirus*, family Virgaviridae). This one-step, 30-minute method can be performed in the field at a single temperature. Its fluorescence visualizer is commercially available with a reasonably inexpensive cost per sample. Jiao et al. (2021) optimized a CRISPR/Cas12a-RPA diagnostic protocol to detect individual apple-infecting RNA virus/viroid directly from crude leaf extracts. The protocol is specific enough to simultaneously distinguish apple necrotic mosaic virus (ApNMV), apple stem pitting virus (ASPV), apple stem grooving virus (ASGV), apple chlorotic leaf spot virus (ACLSV), and apple scar skin viroid (ASSVd) via visible oligonucleotide-conjugated gold nanoparticles. Their diagnostic results were validated by field tests. Ramachandran et al. (2021) adopted CRISPR/Cas12a-RPA for the detection of beet necrotic yellow vein virus (BNYVV), which causes rhizomania disease in sugar beet roots.

With regard to CRISPR-LAMP, Mahas et al. (2021) coupled CRISPR/Cas12a with LAMP for rapid and sensitive detection of DNA viruses. The diagnostic targets were two single-stranded DNA geminiviruses (genus *Geminivirus*, family Geminiviridae), tomato yellow leaf curl virus (TYLCV), and tomato leaf curl New Delhi virus (ToLCNDV). Similarly, they utilized a simple, low-cost fluorescence visualizer to provide visual readouts. Zhu et al. (2022) recently developed a novel protospacer adjacent motif (PAM)-free LAMP assay assisted by CRISPR/Cas12a cleavage (Cas-PfLAMP). Cas-PfLAMP was used to simultaneously detect three rice disease agents: the blight-causing bacterium *Xanthomonas oryzae* pv. *Oryzae* (XOO) as well as two viral pathogens, rice stripe virus (RSV) and rice black-streaked dwarf virus (RBSDV). The two-layer sequence recognition by AMP primer sets and FnCas12a/sgRNA in the Cas-PfLAMP design provides high specificity and sensitivity of as low as three copies. The Cas-PfLAMP platform also features solid-phase nucleic acid extraction, a lateral flow strip (LFS) test, and visual on-spot detection.

6.6 VIRUSES AS FRIENDS: CRISPR-CAS GENOME EDITING OF PLANTS USING VIRUS VECTORS

CRISPR/Cas has resulted in an open window for unprecedented genome engineering in plants. However, to obtain plants with heritable modifications, transgenic lines that stably express Cas and gRNA should be generated, from which progeny with targeted modifications are then recovered in subsequent generations. This requires intensive and time-consuming tissue culture approaches. Moreover, the need for tissue culture has additional drawbacks as it can only be applied to a limited number of species and genotypes and may cause unintended changes to the genome and epigenome. The use of viral vectors offers an alternative to expressing CRISPR/Cas at the whole plant level without the need for stable transformation. Thus, "virus-induced genome editing" (VIGE) can be used to deliver one or more sgRNAs using RNA or DNA virus vectors, usually in transgenic plants that stably express the Cas nuclease.

The availability of tools for transformation-free genome editing of plants has been reported recently using plant virus-derived vectors that are able to move systemically through a plant expressing Cas nuclease and the corresponding guide RNA (Ariga et al. 2020; Ma et al. 2020). The latter alternative has been improved by using two compatible viral vectors (tobacco etch virus, TEV, genus *Potyvirus*, and potato virus X, PVX, genus *Potexvirus*) that replicate in the same cells and coordinately move systemically through the whole plant to express the Cas nuclease and the guide RNA (Uranga et al. 2021b). Thus, the possibility of delivering several CRISPR-Cas components simultaneously can be extended, which could be useful for more sophisticated approaches in plant genome editing and gene expression regulation studies. Selection of adequate virus vectors with large host ranges enlarges the possibilities for genome editing to plant host species recalcitrant to transformation. Also, significant editing levels can be recovered in the next generation using RNA viral vectors, especially by designing sgRNA that could gain better access to the germline (Ali et al. 2015a,b; Ellison et al. 2020). Interestingly, one of the advantages of virus-mediated delivery of gRNA molecules is the possibility to simultaneously target (multiplex editing) several loci by expression of several guide RNAs (Ali et al. 2015a,b; Cody et al. 2017; Ellison et al. 2020; Uranga et al. 2021a,b). Thus, a virus-based system enlarges the toolbox for DNA-free VIGE, providing a means for easy recovery of seeds with the desired modifications, obviating the need for transformation and/or tissue culture. This might contribute significantly to applications for plant functional genomics and to expedite CRISPR/Cas technology for trait discovery and development for crop improvement.

6.7 CONCLUSION

Targeting of pathogenicity-related genes, modification of host susceptibility genes to interrupt compatible host-pathogen interactions, and transcriptional regulation of factors involved in resistance or susceptibility responses are some of the strategies that can be facilitated by genome editing techniques for conferring virus resistance. The success and applicability of the first two strategies are evidenced by the studies discussed above, and the third one is yet to be evaluated. The evolution of cleavage-resistant viruses triggered by CRISPR/Cas editing of the viral genome is considered a serious concern, particularly when we try to engineer resistance at the field level. Insertion of resistance-related genes through homology-directed repair mechanisms is also supposed to be effective in developing virus resistance. Identification of more genes that can function as potential targets for editing, especially in the host, robust gRNA designing, exploiting versatile editing opportunities offered by various CRISPR systems and modified CRISPR/Cas, etc. would facilitate efficient resistance building in crops against pathogenic viruses. It is crucial to ensure that the host's health and development are not impaired while host genes are targeted. Since editing host factors enables the elimination of transgenes through segregation once mutations are induced, it is a promising strategy to create non-transgenic crops. The use of virus vectors as tools for transformation-free genome editing of plants offers a promising alternative that can also help to enlarge the range of plant species that can be edited. Precise gene knockouts by CRISPR/Cas are useful for the elucidation of regulatory and metabolic networks underlying defense and susceptibility responses. A significant amount of sequence information regarding several crop species is now available, and this should accelerate genomic manipulation efforts through editing techniques for improving different traits, including disease resistance.

Additional applications of CRISPR-based approaches for crop improvement require optimization of delivery mechanisms of CRISPR/Cas constructs, transformation and regeneration protocols for edited lines, improvements in specificity, and reduced non-target effects. Devising CRISPR editing cassettes based on recent advancements such as base editing and prime editing would further enhance the specificity and efficiency of editing and thereby help to provide effective and durable virus resistance. Previous studies and recent developments assure the potential of the multifaceted editing platforms provided by CRISPR/Cas in conferring resistance to viruses in crop plants, especially when synergistically deployed with other existing control measures.

ACKNOWLEDGMENTS

Hanu Pappu acknowledges the receipt of a fellowship from the OECD Co-operative Research Programme: Sustainable Agricultural and Food Systems in 2022.

REFERENCES

Abudayyeh, O.O., Gootenberg, J.S., Essletzbichler, P. RNA targeting with CRISPR-Cas13. *Nature.* 2017, 550, 280–284.

Abudayyeh, O.O., Gootenberg, J.S., Konermann, S. C2c2 is a single-component programmable RNA-guided RNA-targeting CRISPR effector. *Science.* 2016, 353, aaf5573.

Ali, Z., Abul-faraj, A., Idris, A., et al. CRISPR/Cas9-mediated viral interference in plants. *Genome Biol.* 2015, 16(1), 238.

Ali, Z., Abul-faraj, A., Li, L., et al. Efficient virus-mediated genome editing in plants using the CRISPR/Cas9 system. *Mol Plant.* 2015, 8, 1288–1291.

Ali, Z., Ali, S., Tashkandi, M., et al. CRISPR/Cas9-mediated immunity to geminiviruses: differential interference and evasion. *Sci Rep.* 2016, 6, 26912.

Alon, D.M., Hak, H., Bornstein, M., Pines, G., Spiegelman, Z. Differential detection of the tobamoviruses tomato mosaic virus (ToMV) and tomato brown rugose fruit virus (ToBRFV) using CRISPR-Cas12a. *Plants.* 2021, 10, 1256.

Aman, R., Ali, Z., Butt, H., et al. RNA virus interference via CRISPR/Cas13a system in plants. *Genome Biol.* 2018, 19(1), 1–9.

Aman, R., Mahas, A., Marsic, T., Hassan, N., Mahfouz, M.M. Efficient, rapid, and sensitive detection of plant RNA viruses with one-Pot RT-RPA-CRISPR/Cas12a assay. *Front Microbiol.* 2020, 11, 610872.

Ariga, H., Toki, S., Ishibashi, K. Potato virus X vector mediated DNA-free genome editing in plants. *Plant Cell Physiol.* 2020, 61, 1946–1953.

Baltes, N.J., Hummel, A.W., Konecna, E., et al. Conferring resistance to geminiviruses with the CRISPR-Cas prokaryotic immune system. *Nat Plants.* 2015, 1, 15145.

Bisht, D.S., Bhatia, V., Bhattacharya, R. Improving plant-resistance to insect-pests and pathogens: the new opportunities through targeted genome editing. *Semin Cell Dev Biol.* 2019, 96, 65–76.

Boch, J., Scholze, H., Schornack, S. Breaking the code of DNA binding specificity of TAL-type III effectors. *Science.* 2009, 326, 1509–1512.

Chandrasekaran, J., Brumin, M., Wolf, D., et al. Development of broad virus resistance in non-transgenic cucumber using CRISPR/Cas9 technology. *Mol Plant Pathol.* 2016, 17(7), 1140–1153.

Chen, J.S., Ma, E., Harrington, L.B., et al. CRISPR-Cas12a target binding unleashes indiscriminate single-stranded DNase activity. *Science.* 2018, 360, 436–439.

Cody, W.B., Scholthof, H.B., Mirkov, T.E. Multiplexed gene editing and protein overexpression using a Tobacco mosaic virus viral vector. *Plant Physiol.* 2017, 175, 23–35.

Dong, Z.Q., Chen, T.T., Zhang, J., et al. Establishment of a highly efficient virus-inducible CRISPR/Cas9 system in insect cells. *Antiviral Res.* 2016, 130, 50–57.

Ellison, E.E., Nagalakshmi, U., Gamo, M.E., Huang, P.-J., Dinesh-Kumar, S., Voytas, D.F. Multiplexed heritable gene editing using RNA viruses and mobile single guide RNAs. *Nat Plants.* 2020, 6, 620–624.

Gaj, T., Gersbach, C.A., Barbas, C.F. ZFN, TALEN, and CRISPR/Cas-based methods for genome engineering. *Trends Biotechnol.* 2013, 31(7), 397–405.

Gomez, M.A., Lin, Z.D., Moll, T., et al. Simultaneous CRISPR/Cas9-mediated editing of cassava eIF 4E isoforms nCBP-1 and nCBP-2 reduces cassava brown streak disease symptom severity and incidence. *Plant Biotechnol J.* 2019, 17(2), 421–434.

Gootenberg, J.S., Abudayyeh, O.O., Lee, J.W., et al. Nucleic acid detection with CRISPR-Cas13a/C2c2. *Science.* 2017, 356, 438–442.

Gratz, S.J., Cummings, A.M., Nguyen, J.N., et al. Genome engineering of Drosophila with the CRISPR RNA-guided Cas9 nuclease. *Genetics.* 2013, 194, 1029–1035.

Harrington, L.B., Burstein, D., Chen, J.S., et al. Programmed DNA destruction by miniature CRISPR-Cas14 enzymes. *Science.* 2018, 362, 839–842.

Heu, C.C., McCullough, F.M., Luan, J., Rasgon, J.L. CRISPR-Cas9-based genome editing in the silverleaf whitefly (bemisiatabaci). *CRISPR J.* 2020, 3, 89–96.

Iqbal, Z., Sattar, M.N., Shafiq, M. CRISPR/Cas9: a tool to circumscribe cotton leaf curl disease. *Front Plant Sci.* 2016, 7, 475.

Ji, X., Zhang, H., Zhang, Y., Wang, Y., Gao, C. Establishing a CRISPR-Cas-like immune system conferring DNA virus resistance in plants. *Nat Plants.* 2015, 1(10), 15144.

Jiao, J., Kong, K., Han, J., et al. Field detection of multiple RNA viruses/viroids in apple using a CRISPR/Cas12a-based visual assay. *Plant Biotechnol J.* 2021, 19, 394–405.

Jinek, M., Chylinski, K., Fonfara, I., et al. A programmable dual-RNA-guided DNA endonuclease in adaptive bacterial immunity. *Science.* 2012, 337, 816–821.

Kellner, M.J., Koob, J.G., Gootenberg, J.S., Abudayyeh, O.O., Zhang, F. SHERLOCK: nucleic acid detection with CRISPR nucleases. *Nat Protoc.* 2019, 14, 2986–3012.

Khan, Z., Khan, S.H., Ahmad, A., Aslam, S., Mubarik, M.S., Khan, S. CRISPR/dCas9-mediated inhibition of replication of begomoviruses. *Int J Agric Biol.* 2019, 21(4), 711–718.

Kim, H., Kim, S.T., Ryu, J., et al. CRISPR/Cpf1-mediated DNA-free plant genome editing. *Nat Commun.* 2017, 8, 14406.

Kis, A., Hamar, É., Tholt, G., Bán, R., Havelda, Z. Creating highly efficient resistance against wheat dwarf virus in barley by employing CRISPR/Cas9 system. *Plant Biotechnol J.* 2019, 17(6), 1004.

Ma, X., Zhang, X., Liu, H., Li, Z. Highly efficient DNA-free plant genome editing using virally delivered CRISPR-Cas9. *Nat Plants.* 2020, 6, 773–779.

Macovei, A., Sevilla, N.R., Cantos, C., et al. Novel alleles of rice eIF4G generated by CRISPR/Cas9-targeted mutagenesis confer resistance to Rice tungro spherical virus. *Plant Biotechnol J.* 2018, 16(11), 1918–1927.

Mahas, A., Aman, R., Mahfouz, M. CRISPR-Cas13d mediates robust RNA virus interference in plants. *Genome Biol.* 2019, 20(1), 1–16.

Mahas, A., Hassan, N., Aman, R., et al. LAMP-coupled CRISPR-Cas12a module for rapid and sensitive detection of plant DNA viruses. *Viruses.* 2021, 13, 466.

Makarova, K.S., Wolf, Y.I., Iranzo, J., et al. Evolutionary classification of CRISPR-Cas systems: a burst of class 2 and derived variants. *Nat Rev Microbiol.* 2020, 18, 67–83.

Makhotenko, A.V., Khromov, A.V., Snigir, E.A., et al. Functional analysis of coilin in virus resistance and stress tolerance of potato *Solanum tuberosum* using CRISPR-Cas9 editing. *DoklBiochemBiophys.* 2019, 484(1), 88–91.

Mehta, D., Stürchler, A., Anjanappa, R.B., et al. Linking CRISPR-Cas9 interference in cassava to the evolution of editing-resistant geminiviruses. *Genome Biol.* 2019, 20(1), 80.

Moradpour, M., Abdulah, S.N.A. CRISPR/dCas9 platforms in plants: strategies and applications beyond genome editing. *Plant Biotechnol J.* 2020, 18(1), 32–44.

Mushtaq, M., Sakina, A., Wani, S.H., et al. Harnessing genome editing techniques to engineer disease resistance in plants. *Front Plant Sci.* 2019, 10, 550.

Pabo, C.O., Peisach, E., Grant, R.A. Design and selection of novel Cys2His2 zinc finger proteins. *Annu Rev Biochem.* 2001, 70, 313–340.

Pardee, K., Green, A.A., Takahashi, M.K., et al. Rapid, low-cost detection of zika virus using programmable biomolecular components. *Cell.* 2016, 165, 1255–1266.

Pyott, D.E., Sheehan, E., Molnar, A. Engineering of CRISPR/Cas9-mediated potyvirus resistance in transgene-free *Arabidopsis* plants. *Mol Plant Pathol.* 2016, 17(8), 1276–1288.

Ramachandran, V., Weiland, J.J., Bolton, M.D. CRISPR-based isothermal next-generation diagnostic method for virus detection in sugarbeet. *Front Microbiol.* 2021, 12, 679994.

Sharma, S.K., Gupta, O.P., Pathaw, N., et al. CRISPR-Cas-led revolution in diagnosis and management of emerging plant viruses: new avenues toward food and nutritional security. *Front Nutr.* 2021, 8, 751512.

Taning, C.N.T., Van Eynde, B., Yu, N., Ma, S., Smagghe, G. CRISPR/Cas9 in insects: applications, best practices and biosafety concerns. *J Insect Physiol.* 2007, 98, 245–257.

Tashkandi, M., Ali, Z., Aljedaani, F., Shami, A., Mahfouz, M.M. Engineering resistance against Tomato yellow leaf curl virus via the CRISPR/Cas9 system in tomato. *Plant Signal Behav.* 2018, 13(10), e1525996.

Tripathi, J.N., Ntui, V.O., Ron, M., Muiruri, S.K., Britt, A., Tripathi, L. CRISPR/Cas9 editing of endogenous banana streak virus in the B genome of *Musa* spp. overcomes a major challenge in banana breeding. *Commun Biol.* 2019, 2(1), 46.

Uranga, M., Aragonés, V., Selma, S., Vázquez-Vilar, M., Orzáez, D., Daròs, J.-A. Efficient Cas9 multiplex editing using unspaced sgRNA arrays engineering in a Potato virus X vector. *Plant J.* 2021a, 106, 555–565.

Uranga, M., Vazquez-Vilar, M., Orzáez, D., Daròs, J.A. CRISPR-Cas12a genome editing at the whole-plant level using two compatible RNA virus vectors. *CRISPR J.* 2021b, 4, 761–769.

Wang, A. and Krishnaswamy, S. Eukaryotic translation initiation factor 4E-mediated recessive resistance to plant viruses and its utility in crop improvement. *Mol Plant Pathol.* 2012, 13(7), 795–803.

Xue, Z., Wu, M., Wen, K., et al. CRISPR/Cas9 mediates efficient conditional mutagenesis in Drosophila. *G3 (Bethesda).* 2014, 4, 2167–2173.

Yin, K., Han, T., Xie, K., Zhao, J., Song, J., Liu, Y. Engineer complete resistance to Cotton Leaf Curl Multan virus by the CRISPR/Cas9 system in *Nicotiana benthamiana*. Phytopathol Res. 2019, 1(1), 1–9.

Yoon, Y.J., Venkatesh, J., Lee, J.H., et al. Genome editing of eIF4E1 in tomato confers resistance to pepper mottle virus. *Front Plant Sci.* 2020, 11, 1098.

Yu, Y., Pan, Z., Wang, X. Targeting of SPCSV-RNase3 via CRISPR-Cas13 confers resistance against sweet potato virus disease. *Mol Plant Pathol.* 2022, 23, 104–117.

Yu, Z., Ren, M., Wang, Z., et al. Highly efficient genome modifications mediated by CRISPR/Cas9 in Drosophila. *Genetics.* 2013, 195, 289–291.

Zhan, X., Zhang, F., Zhong, Z., et al. Generation of virus-resistant potato plants by RNA genome targeting. *Plant Biotechnol J.* 2019, 17(9), 1814–1822.

Zhang, P., Du, H., Wang, J., et al. Multiplex CRISPR/Cas9-mediated metabolic engineering increases soya bean isoflavone content and resistance to soya bean mosaic virus. *Plant Biotechnol J.* 2020, 18(6), 1384–1395.

Zhang, T., Zhao, Y., Ye, J., et al. Establishing CRISPR/Cas13a immune system conferring RNA virus resistance in both dicot and monocot plants. *Plant Biotechnol J.* 2019, 17(7), 1185–1187.

Zhang, Y., Qian, L., Wei, W., et al. Paired design of dCas9 as a systematic platform for the detection of featured nucleic acid sequences in pathogenic strains. *ACS Synth Biol.* 2017, 6, 211–216.

Zhao, Y., Yang, X., Zhou, G., et al. Engineering plant virus resistance: from RNA silencing to genome editing strategies. *Plant Biotechnol J.* 2020, 18, 328–336.

Zhu, Z., Li, R., Zhang, H., et al. PAM-free loop-mediated isothermal amplification coupled with CRISPR/Cas12a cleavage (Cas-PfLAMP) for rapid detection of rice pathogens. *Biosens Bioelectron.* 2022, 204, 114076.

7 Genome Editing for Multiple Disease Resistance in Crop Plants

Madhav M. S., Aparna E., Barbadikar K. M.
ICAR-Indian Institute of Rice Research

Devanna B. N.
ICAR-National Rice Research Institute

7.1 INTRODUCTION

Plants are vulnerable to attack by a wide range of pathogens, such as bacterial, fungal, and viral pathogens, causing various diseases that result in a significant loss to crop yields and quality (Law et al., 2017). The global yield losses due to pests and disease estimated for the period from 2010 to 2014 were 30.0% in rice, 21.5% in wheat, and 22.5% in maize (Savary et al., 2019). In the case of resistance, a multilayer defense system is activated during a pathogen attack to combat the invading pathogen. Basically, the plant defense system is made up of a two-tier immune sensing system that includes molecular receptors that detect potential infections both outside and inside the plant cell (Jones et al., 2016), pattern recognition receptors (PRRs) on the cell surface to detect conserved microbial patterns, and pathogen proteins produced in the apoplast to trigger PRR-triggered immunity (PTI). In the cytoplasm, disease resistance genes (*R* genes) encode intracellular nucleotide-binding domain and leucine-rich repeat–containing (NBS-LRR) or NOD-like receptors (NLRs), which recognize pathogen effectors, which otherwise are important for pathogenesis in host plants (Dodds and Rathjen, 2010; Jones et al., 2016). This recognition triggers an NLR-mediated response, also known as NLR- or effector-triggered immunity (NTI/ETI), which activates the plant immune system, thereby effectively stopping pathogen growth. Among the other approaches, applications of fungicides are largely used to combat diseases where there is no host immunity (Heinrichs and Muniappan, 2017). However, this approach causes environmental deterioration and has adverse effects on the metabolic pathway of the plants (Aktar et al., 2009). Plants that are naturally resistant to diseases can help to reduce pesticide use and increase yield. On the contrary, constitutively activated disease defense frequently results in a growth trade-off, resulting in diminished growth or even a compromise in yield (Gao et al., 2020; van Butselaar and van den Ackerveken, 2020). The *R* gene-mediated genetic resistance has been extensively used by introgression of these genes into elite cultivars either by genetic crosses or via transfer of NLRs between plant species for engineering broad-spectrum resistance using genetic

DOI: 10.1201/9781003403142-7

modification (Dangl et al., 2013). In conventional and transgenic-mediated breeding, several single or stacked *R* genes have been introduced into crops with the goal of creating long-term resistance to invading pathogens (Kumari et al., 2017, 2018). Although effective against a diversity of pathogens and pests, NLRs tend to have a narrow spectrum of pathogen recognition. Therefore, researchers are adopting novel plant breeding techniques such as genome-editing technology (GET) that are more flexible, reliable, and sustainable to increase production in a way that does not negatively impact natural resources (Chakraborty et al., 2010).

7.2 GENOME-EDITING TECHNOLOGY

Genome-editing technology (GET) includes oligo-directed mutagenesis (ODM) and sequence-specific nucleases (SSNs), such as zinc finger nucleases (ZFN), transcription-activator-like effector nucleases (TALEN), or RNA-guided nuclease Cas9 (CRISPR-Cas9 system) CRISPR/Cpf1, base editing (BE), and prime editing (PE), which thereby greatly reduce the background of unwanted off-target mutations, which can be applied to develop disease resistance in crop plants. The ZFNs are endonucleases that contain multiple zinc finger DNA-binding domains and several zinc fingers as a tandem array, recognizing a 3-bp module. The hybrid FokI restriction enzyme fused to the zinc finger DNA-binding domain is a common ZFN example, with specificity directed by the combination of zinc fingers. TALENs are similar in principle to ZFNs, although the endonuclease domain is now paired with multiple transcription activator-like (TAL) effector domains that recognize single base pairs. As a result, the number of TAL effector domains present in the nuclease determines the specificity of the TALEN target site. TALEN, for example, may very precisely target a given genomic location; however, programming them at the amino acid level is time-consuming, and delivering them all into the crop nucleus is difficult. Since it became possible to use the bacterial CRISPR/Cas system in eukaryotic cells, which is easy to train, relatively cheap, and multiplex compatible, TALEN has been quickly replaced by CRISPR/Cas (Thompson et al., 2018) as the most popular GET. CRISPR/Cas9 is a natural bacterial adaptive immune defense system against invasive DNA sequences, consisting of only two components: guide RNA (gRNA) aligns with the target sequence, directing the double-strand break induced by the endonuclease Cas9. Synthetic gRNA can be designed to target a specific plant or pest/pathogen gene to induce deletions, point mutations, or insertions (Ghogare et al., 2021). Targeted genome editing has become the preferred genetic tool for resistance incorporation in plants against various phytopathogen-mediated diseases. These techniques have proved to be powerful tools for the successful modification of genome sequences in a simple and precise manner by generating targeted DNA double-strand breaks, which are then repaired mainly by either error-prone non-homologous end joining (NHEJ) or high-fidelity homologous recombination (Fiaz et al., 2021).

7.2.1 Plant Genome Editing to Enhance Immunity

Several traditional methods, such as the pure line method, pedigree method, back-cross method, recurrent selection, and mutation breeding, have been successfully developed for breeding disease resistance in crop plants (Allen, 1921; Miah et al., 2013; Guimaraes et al., 2000; Sreewongchai et al., 2010; Sharma et al., 2012). Traditional resistance breeding procedures, on the other hand, are time-consuming, and resistance alleles are occasionally connected to genes that affect important agronomic traits. Plant genomes may be targeted and rearranged with high precision using genome-editing technology (Puchta, 2017). Hence, genome editing, as compared to natural breeding, has the potential to speed up crop development and food production.

Any successful genome-editing approach requires known host genome sequences and molecular information about the target gene. An increasing number of genome sequences are available for varied plant species, and intensive genetic and molecular studies result in finer molecular details of plant innate immunity, providing increasing numbers of targets, particularly the negative regulators of plant defense (e.g., host *S* genes). Therefore, *S* gene knockout is a straightforward way to develop plant disease resistance against a specific pathogen (Zaidi et al., 2018). However, it is important to note that *S* genes are frequently involved in plant growth and development, and mutations in them might have a potential undesirable effect, thereby compromising this approach (Figure 7.1). Genome editing can also greatly facilitate gene expression modulation via promoter disruption and changes in cis-regulatory elements, as it is likely to benefit disease resistance with less detrimental pleiotropic effects (Pandiarajan and Grover, 2018). Alternatively, knocking in negative regulators of plant defense genes/*R* genes, as well as precise editing via targeted replacement or base editing, can result in allelic variants or novel alleles for disease resistance. Modulation of amino acids in receptor proteins in the plant may prevent the evasion of secreted pathogen effectors, and these receptors could be targeted without complete abolition of their natural function. Modification of master regulators involved in salicylic acid (SA)-mediated systemic acquired resistance (SAR) of plant defense responses, such as *NPR1* and *NPR3*, can also reduce the pathogen infection levels (Kuai et al., 2015). Currently, genome-editing strategies are being used to target the

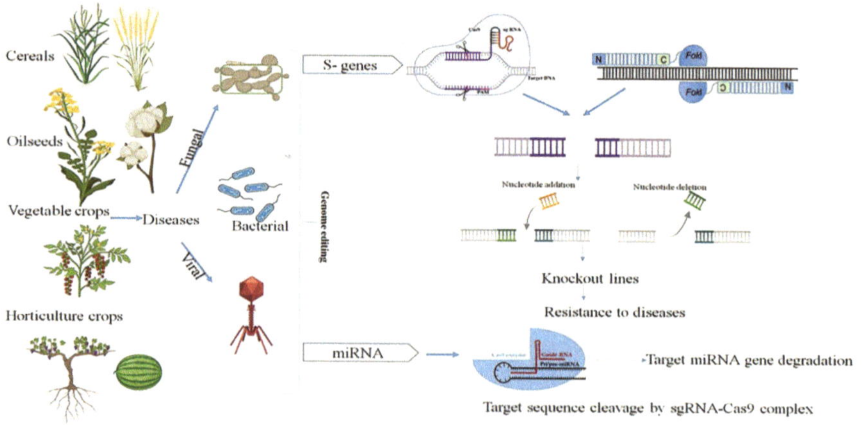

FIGURE 7.1 Pipeline for gene editing using TALEN and CRISPR/Cas9 for resistance against fungal, bacterial, and viral diseases.

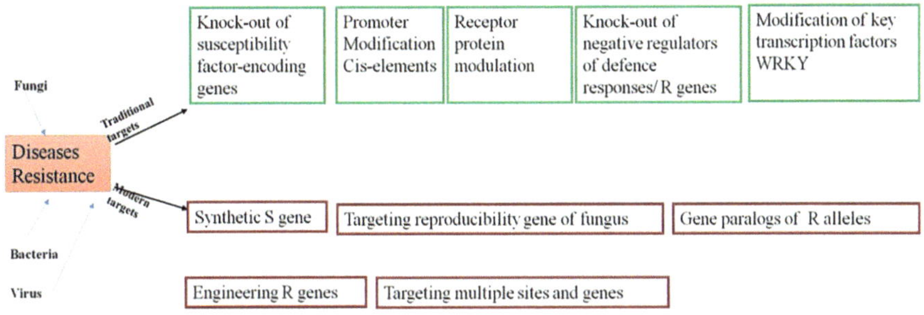

FIGURE 7.2 Targets of GE for imparting disease resistance using CRISPR/Cas.

reproducibility of pathogen genes in order to disrupt their function, thereby establishing a novel defense mechanism in the plant (Gil et al., 2021) (Figure 7.2).

7.2.1.1 Application of Genome-Editing Technologies in Crop Plants for Enhancing Resistance to Fungal Diseases

Fungal pathogens are the primary cause of all of the world's most devastating plant diseases, resulting in significant crop yield and quality losses and associated massive economic losses. Fungi are considered to be responsible for around 30% of new plant diseases (Giraud et al., 2010), demanding the development of new management strategies. Most harmful plant pathogens are fungi and oomycetes, which are taxonomically distinct but have similar filamentous development and host-infecting structures (Dong et al., 2015). Now, in the era of genome editing and system biology, it's vital to look at the efficacy of these technologies in understanding the biology of plant-pathogen interaction and establishing broad-spectrum resistance. Different genome editing-based methodologies can be used for this, which can be divided into two categories: (1) modification of pathogen targets in crops and (2) regulation of the host immune response beginning with recognition (Schenke and Cai, 2020). Primarily, alteration of the pathogen's targets could be possible through functional knockout of the host susceptibility (*S*) genes and engineering the cis-elements in the promoter region of *S* genes. On the other hand, modification of the coding sequences of the central regulators of plant defense response and knockdown of non-coding RNAs, such as miRNAs and siRNAs, which are involved in immune responses by genome-editing tools, can possibly help to develop resistance against fungal pathogens (Figure 7.1).

7.2.1.1.1 Cereals

Fungal infections are a major yield-limiting factor in cereals, with estimated yield losses of 15%–20%; when they are widespread, losses might exceed 50% (Jaczewska-Kalicka, 2002; Korbas, 2009). *Magnaporthe oryzae*, *Blumeria graminis*, *Puccinia recondita*, *Puccinia graminis*, *Puccina striiformis*, *Septoria tritici*, *Septoria nodorum*, and *Fusarium* are the most important fungi causing diseases in cereal crops.

7.2.1.1.1.1 Rice It is one of the major cereal crops being affected by multiple fungal pathogens, of which rice blast (*Magnaporthe oryzae*) is a serious concern as it is responsible for approximately 30% of crop losses, worldwide. Thus far, various strategies are being adopted to generate blast resistance in rice, and the latest gene-editing technique is the most reliable technique for enhancing resistance. Initial reports on targeting host genes such as *mitogen-activated protein kinase 5* (*OsMPK5*) through CRISPR/Cas9 resulted in broad-spectrum disease resistance in Nipponbare (Xie and Yang, 2013). Wang et al. (2016) reported a significant decrease in the infection rate (blast lesions) compared to wild-type by inducing mutagenesis in *ethylene-responsive factor 922* (*OsERF922*) through CRISPR/Cas9. Furthermore, Ma et al. (2018) demonstrated the blast resistance through the disruption of *OsSEC3A* (which is a subunit of the exocyst complex) through the induction of mutation at multiple sites (3rd and 10th exons of the gene). A similar strategy was also applied to mutate the *Pi21*, which encodes a cytoplasmic proline-rich protein that acts as a susceptibility factor (Fukuoka et al., 2009). Another susceptibility factor, the *BSR-D1* gene, was also targeted for CRISPR/Cas9-mediated mutation to promote blast resistance in rice; the engineering of the promoter cis-element by single nucleotide substitution increases the binding of *MYBS1* transcriptional factor, which suppresses the expression of peroxidase to enhance H_2O_2 production (Li et al., 2017). Recently, enhanced resistance to the rice blast fungus and bacterial blight was achieved by multiplex GE in three genes, i.e., *OsTMS5* (a protein involved in thermosensitive male sterility), *Pyricularia oryzae resistance 21* (*OsPi21*, an S gene that slows plant defense against *M. oryzae*), and *OsXa13* genes, and developed enhanced resistance to infections of *M. grisea* and *Xoo* strain PXO99 (Li et al., 2019). A novel peroxidase synthesizing gene (*Perox3*) of rice is also knocked out by using CRISPR/Cas9 to assess its functional significance in blast disease resistance (Zhu et al., 2020).

Other than rice blast disease, sheath blight is another serious problem in rice cultivation. To confer broad-spectrum defense against sheath blight disease, the *Phytochrome and Flowering Time 1* (*OsPFT1*) gene of *Indica* rice (var. ASD16) was successfully mutated by insertion of a one-base in exons 1 and 3 through CRISPR/Cas9 (Shah et al., 2019), but the functional assessment of the mutant plants has not yet been reported.

7.2.1.1.1.2 Wheat Wheat quality and production are severely affected by fungal diseases such as rust, powdery mildew, wheat blast, and *Fusarium* head blight (FHB). CRISPR/Cas9 has been used in several research centers to improve wheat disease resistance. To incorporate the resistance against downy mildew, targeted mutation in the *TaMLO-A1* allele (one of three homoeo alleles that encode Mildew-Resistance Locus) were carried out, and the resistance was achieved (Wang et al., 2014). Similarly, Zhang et al. (2017) reported the editing of the three *TaEDR1* (a negative role in the defense response against powdery mildew) homologs and showed resistance to powdery mildew. Recently, CRISPR-derived line having the random deletion of a sequence containing the start codon of a *TaHRC* gene (encodes a putative histidine-rich calcium-binding protein) conferred resistance to FHB. Another gene, *TaNFXL*, transcription factor (TF) which is induced by the deoxynivalenol (mycotoxin virulence factor that promotes the growth of the *Fusarium graminearum* fungus), was also targeted by CRISPR/Cas9 to create mutations. The resulting mutant lines showed enhanced resistance to Fusarium head blight (Brauer et al., 2020).

7.2.1.1.1.3 Barley *Fusarium* Head Blight (FHB) and powdery mildew are devastating and dangerous diseases that cause yield loss, low test weights, low seed germination, and contamination of grain with mycotoxins. The knockout mutants of one of the MORC family genes, *HvMORC6a*, which acts as a potential interactor in regulating plant immunity, showed the strongest resistant phenotype to biotrophic (*Blumeria graminis*) and necrotrophic (*Fusarium graminearum*) plant pathogenic fungi (Galli et al., 2022). Another attempt by Kumar et al. (2018) *to create knockout mutants* in *HvMORC1 was also showed resistance to Blumeria graminis* (Bgh) and *Fusarium graminearum.*

7.2.1.1.2 Oil Seed Crops

Oil seed crops, such as canola, cotton, soybean, rapeseed, sunflower, and oil palm, have gained ever-increasing importance to world agriculture over the past decades, providing valuable resources for human society (Vollmann and Rajcan, 2009).

7.2.1.1.2.1 Canola Among fungal diseases, *Sclerotinia* stem rot is a key disease that can lead to poor oil quality and cause grain yield losses ranging from 10% to 80%. Two genes, *BnWRKY11* and *BnWRKY70*, are reported to be differentially expressed after inoculation with *Sclerotinia sclerotiorum* (Wu et al., 2016). Knockout of *BnWRKY70* by CRISPR/Cas9 exhibited enhanced resistance to *Sclerotinia* in *Brassica napus*, since this transcription factor functions as a negative regulator of stem rot resistance. *Brassica napus* is also highly susceptible to wilt disease caused by *Verticillium longisporum.* The loss-of-function mutation in *BnCRT1a* using CRISPR/Cas, and the edited plants showed decreased susceptibility to wilt. In the edited lines, activation of the ethylene signaling pathway was observed, which might be contributing to reduced susceptibility.

7.2.1.1.2.2 Cotton Cotton (*Gossypium spp.* L.) is the largest source of natural fibers in the world. Among the most important diseases affecting cotton are those caused by fungal pathogens such as *Fusarium* wilt (caused by *Fusarium oxysporum* f.sp. *vasinfectum*), *Verticillium* wilt (*Verticillium dahliae*), anthracnose (*Colletotrichum gossypii*), and root rots (*Sclerotium rolfsii* and *Rhizoctonia solani*). Zhang et al. (2018) reported the editing of *Gh14-3-3d,* which is *a* negative regulator for the brassinosteroid signaling *pathway.* The resultant edited lines showed high resistance to *Verticillium* wilt.

7.2.1.1.3 Horticultural Crops

Fruit and vegetable crops are rich in cellulose, vitamins, trace elements, minerals, and other important nutrients, which are essential components of the human diet. However, climate and environmental changes could potentially threaten the production and supply of fruits and vegetables. Many studies have since been published on the possible applications of *CRISPR* technology in protecting plants against biotic and abiotic stresses and improving fruit quality, plant architecture, and shelf life. Currently, genome editing is still in the exploratory stage for combating the fungal diseases of several fruit and vegetable crops.

7.2.1.1.3.1 Tomato In tomatoes, downy and powdery mildews cause serious economic losses. The powdery mildew disease-sensitivity gene *Mlo1* (*Mildew resistant locus 1*) encodes a membrane-associated protein, making it possible to edit *Mlo1*. The mutants obtained via gene editing exhibited resistance to the powdery mildew *Oidium neolycopersici* with no apparent pleotropic effect on plant morphology and fruit yield (Nekrasov et al., 2017). Another *S* gene, *SlPMR4*, the closest tomato ortholog of *Arabidopsis PMR4* (*SlPMR4*), was also explored for CRISPR/Cas9-induced mutation, and the mutant tomato plants showed reduced infection with powdery mildew (Santillan Martınez et al., 2020). Gray mold is another widespread fungal disease caused by the necrotrophic fungus *Botrytis cinerea*, which causes serious post-harvest losses in field and greenhouse-grown tomatoes (Williamson et al., 2007). Its initial symptoms are not obvious, and the lack of effective pesticides makes its prevention and control difficult. CRISPR/Cas9-induced mutations in the tomato acetylenase-encoding genes *ACET1a* and *ACET1b* have recently been found to confer enhanced resistance to *Botrytis cinerea* (Jeon et al., 2020). Another gene, *MAPK3* (*mitogen-activated protein kinase 3*), was also used to induce resistance to *Botrytis cinerea* using CRISPR-Cas9 (Shujuan et al., 2018). In addition to these genes, *SlNPR1* (*non-expressor of pathogenesis-related gene 1NPR1*), a critical component of the plant defense mechanisms in tomato, was mutated using CRISPR/Cas9 and generated a homozygous mutant *slnpr1-1* (1bp insertion). The resultant edited lines displayed enhanced resistance against *Botrytis cinerea* with smaller lesion sizes, higher activities of defense enzymes such as glutathione, and upregulated expression of defense genes involved in JA signaling pathways, which likely contributed to the resistance against *B. cinerea* (Li et al., 2020b). *Solyc08g075770*-knockout via CRISPR-Cas9 resulted in resistance to *Fusarium* wilt in tomatoes (Cahya et al., 2018).

Late blight, caused by *Phytophthora infestans*, is severely damaging the global tomato industry. Micro-RNAs (miRNAs) have been widely demonstrated to play vital roles in plant resistance by repressing their target genes. Two miRNAs, viz., *miR482b* and *miR482c* were targeted and edited by CRISPR/Cas9 to develop pathogen-resistant tomatoes (Hong et al., 2021). Knocking out *SlMYBS2* through CRISPR (*SlMYBS2* gene knockout, *slmybs2-c*) in tomato resulted in a reduction of resistance to *P. infestans*, accompanied by increases in the number of necrotic cells, lesion sizes, and disease index, and concluded that *SlMYBS2* acts as a positive regulator of tomato resistance to *P. infestans* infection by regulating the ROS level and the expression level of *PR* genes (Liu et al., 2021).

7.2.1.1.3.2 Potato Late blight is the most serious disease of potato crops worldwide. However, increased resistance obtained via *R* genes is hampered in potato breeding because these genes are often specific for a pathogen race and can be quickly overcome by the evolution of the pathogen. In parallel, *S* genes are important for pathogenesis, and loss of *S* gene function confers increased resistance in several crops. In order to protect potatoes from late blight, mutations of host-susceptible genes such as *StDND1*, *StCHL1*, and *StDMR6*-1 using CRISPR/Cas9 technology were reported. The knockout mutant showed increased resistance against late blight infection (Kieu et al., 2021).

7.2.1.1.3.3 Watermelon Watermelon (*Citrullus lanatus*), an economically important fruit crop, also suffers from several diseases. Among the diseases, wilt, caused by *Fusarium oxysporum* f.sp. *niveum* (FON), is one of the most devastating diseases affecting watermelon. Earlier studies reported that phytosulfokine (PSK) signaling attenuates plant immune responses using this knowledge. Knockout of the *Clpsk1* gene, encoding the PSK precursor, conferred enhanced resistance to FON (Zhang et al., 2020).

7.2.1.1.3.4 Grapevine Grapevine is one of the most extensively cultivated and economically valuable horticultural crops in the world. Unfortunately, its production is limited by various diseases, including powdery mildew caused by the fungus *Erysiphe necator,* downy mildew caused by *Plasmopara viticola*, and bunch rot caused by *Botrytis cinerea*. Studies suggest that the pathogenicity of adapted powdery mildew species relies on functional host *MLO* genes. Malnoy et al. (2016) used *MLO-7,* a susceptible gene, to increase the resistance to powdery mildew in grapes using CRISPR/Cas9 ribonucleoproteins (RNPs) in the protoplast of the grape cultivar Chardonnay to generate transgene-free resistant powdery mildew grape cultivars. Another *MLO* gene, *VvMLO3*, was targeted, and *VvMLO3*-edited lines showed enhanced resistance to powdery mildew, which was associated with host cell death, cell wall apposition (CWA), and H_2O_2 accumulation (Wan et al., 2020). Mutagenesis in the *VvWRKY52* transcription factor resulted in a biallelic mutation, which enhanced the resistance to gray mold disease in grapevine (Wang et al., 2018). Whereas engineered loss-of-function mutations in the *VvPR4b* (*pathogenesis-related 4-PR4*) gene in the "Thompson Seedless" variety using the CRISPR/Cas9 system resulted in increased susceptibility to *P. viticola*, indicating its active role in the defense of grapevine against downy mildew (Li et al., 2020a).

7.2.1.1.3.5 Cacao Severe *Phytophthora* spp. outbreaks can destroy all cacao fruit on a single farm, resulting in reduced revenue for millions of farmers in underdeveloped nations (Wood and Lass, 2008; Gutiérrez et al., 2016). In cacao, CRISPR/Cas9 technology has been used to target the non-expressor of pathogenesis-related 3 (*TcNPR3*) genes, a suppressor of the defensive response, as a preliminary proof of concept. The edited plants exhibited increased resistance to infection with the cacao pathogen *Phytophthora tropicalis* and elevated expression of downstream defense genes (Fister et al., 2018). Different genes that have been targeted through genome-editing technology to achieve fungal disease resistance in cereals, oilseeds, and horticulture crops are listed in Table 7.1.

7.2.1.2 Application of Genome-Editing Technologies in Crop Plants for Enhancing Resistance to Bacterial Diseases

A few hundred species of bacteria are responsible for crop damage, which frequently exposes manifold disease symptoms. It is important to understand host-bacterial interactions. Management of bacterial diseases is challenging primarily due to asymptomatic infections, lack of effective agrochemicals, large diversity, quick multiplication rate, and easy spread among hosts and geographical regions. Consequently, the most efficient strategies to control bacterial pathogens are mainly

TABLE 7.1
Targeted Genome Editing for Resistance against Fungal Diseases in Cereals, Oilseeds and Horticultural Crops

S. No.	Crop	Target	Function of Gene	Resistance Imparted to Pathogen	Gene Mutation	Reference
			A. Cereal crops			
1	Rice (*Oryza sativa* L.)	OsERF922 (transcription factor)	Negative regulator of blast resistance in rice	Magnaporthe oryzae	CRISPR/Cas9-indel-exon	Wang et al. (2016)
2	Rice (*Oryza sativa* L.)	*OsSEC3A* (exocyst subunit)	Subunit of the exocyst complex, engaged in multiple functions	Enhanced resistance to blast disease (*Magnaporthe oryzae*)	CRISPR/Cas9-indel-coding region	Ma et al. (2018)
3	Rice (*Oryza sativa* L.)	*OsBSR-D1* (transcription factor)	Transcription factor	Broad-spectrum resistance to rice blast (*Magnaporthe oryzae* strains)	CRISPR/Cas9-insertion-coding region	Zhu et al. (2020)
4	Rice (*Oryza sativa* L.)	A single base change (SNP33-G) in the promoter of *Bsr-d1*	Promotes peroxidase expression, suppressing immunity to *M. oryzae*	Enhances binding of *MYBS1* to the promoter and suppresses *bsr-d1* expression to confer resistance	CRISPR/ Cas9-indel-promoter	Liu et al. (2016)
5	Rice (*Oryza sativa* L.)	OsMPK15	A negative regulator of defense in rice	Induced defense against blast pathogen	CRISPR/Cas9-insertion-coding sequence	Hong et al. (2019)
6	Rice (*Oryza sativa* L.)	OsPFT1	Involved in disease susceptibility	Confer resistance to rice sheath blight	CRISPR/ Cas9-insertion-exon	Shah et al. (2019)
7	Wheat (*Triticum aestivum* L.)	TaMLO-A1 allele	Repress immunity against powdery mildew.	Powdery mildew *Blumeria graminis* f.sp. *tritica* (Bgt)	TALENS-indel-exon	Wang et al. (2014)
8	Wheat (*Triticum aestivum* L.)	*TaEdr1* (three homologs)	Negative role in the defense response against powdery mildew	Powdery mildew	CRISPR/Cas9-indel-exon	Zhang et al. (2017)

(*Continued*)

TABLE 7.1 (*Continued*)
Targeted Genome Editing for Resistance against Fungal Diseases in Cereals, Oilseeds and Horticultural Crops

S. No.	Crop	Target	Function of Gene	Resistance Imparted to Pathogen	Gene Mutation	Reference
9	Wheat (*Triticum aestivum* L.)	*TaNFXL1* (transcription factor)	Promote the growth of *Fusarium graminearum*	Enhanced resistance to Fusarium head blight (*Fusarium graminearum*)	CRISPR/Cas9-deletion-coding region	Brauer et al. (2020)
10	Wheat (*Triticum aestivum* L.)	TaHRC	Key determinant of *Fhb1*-mediated resistance	Fhb resistance	CRISPR/ Cas9-deletion-exon and intronic region	Su et al. (2019)
11	Barley (Hordeumv ulgare)	HvMORC6a	Regulating plant immunity	Resistant to FHB Powdery mildew	CRISPR/Cas9-indels-gene	Galli et al. (2022)
12	Barley(Hordeum vulgare)	HvMORC1	Regulating plant immunity	Resistant to FHB Powdery mildew	CRISPR/Cas9-indels-gene region	Kumar et al. (2018)
			B. Oil seed crops			
13	Canola (Brassica napus)	BnWRKY70	Confers Susceptibility to *Sclerotinia*	Resistance to *Sclerotinia*	CRISPR/Cas9 -deletion-exon	Sun et al. (2018)
14	Canola (Brassica napus)	BnCRT1a	Confers Susceptibility to *Verticillium longisporum*	Verticillium longisporum	CRISPR/Cas9-indel-exon	Probsting et al. (2020)
15	Cotton (Gossypium hirsutum)	*Gh14-3-3D* (negative regulator)	Susceptibility to Verticillium dahliae	Resistance to *verticillium* wilt	CRISPR/ Cas9-indel-coding region	Zhang et al. (2018a,b)
			C. Horticultural crops			
16	Solanum lycopersicum (Tomato)	*SlMLO1* (calmodulin binding)	Negative regulator	Resistance to powdery mildew (*Oidium neolycopersici*)	CRISPR/ Cas9-Deletion-exon	Nekrasov et al. (2017)
17	Solanum lycopersicum (Tomato)	*SlPMR4* (callose synthase)	Negative regulator	Enhanced resistance to powdery mildew (*Oidium neolycopersici*)	CRISPR/ Cas9-indel-exon	Santillan Martınez et al. (2020)

(*Continued*)

TABLE 7.1 (*Continued*)
Targeted Genome Editing for Resistance against Fungal Diseases in Cereals, Oilseeds and Horticultural Crops

S. No.	Crop	Target	Function of Gene	Resistance Imparted to Pathogen	Gene Mutation	Reference
18	Solanum lycopersicum (Tomato)	ACET1a and ACET1b	Susceptibility to *Botrytis cinerea*	Resistance to Botrytis cinerea	CRISPR/Cas9-deletion-gene	Jeon et al. (2020)
19	Solanumlycopersicum (Tomato)	miR482b and miR482c	Phytophthora infestans	Resistance to Phytophthora infestans	CRISPR/Cas9-indel	Hong et al. (2021)
20	Solanum lycopersicum (Tomato)	SlNPR1	Susceptibility to *Botrytis cinerea*	Resistance to Botrytis cinerea	CRISPR/Cas-indel-exon	Li et al. (2020b)
21	Solanum lycopersicum (Tomato)	SlMYBS2	Negative regulator for disease susceptibility	Reduced resistance to *P. infestans*	CRISPR/Cas9-indel-gene	Liu et al. (2021)
22	Grapes (Vitis vinifera)	VvMLO-7	Confers Susceptibility to Powdery Mildew	Resistance to Erysiphe necator	CRISPR/Cas9-indel-exon	Malnoy et al. (2016)
23	Grapes (Vitis vinifera)	VvWRKY52	Botrytis bunch rot	Botrytis cinerea	CRISPR/Cas9-indel-exon	Wang et al. (2018)
24	Grapes (Vitis vinifera)	VvMLO3	Confers Susceptibility to Powdery mildew	Erysiphe necator	CRISPR/Cas9-indel-coding region	Wan et al. (2020)
25	Watermelons (Citrullus lanatus)	Clpsk1	Involved in PSK signaling that attenuates plant immune response	Resistance to *Fusarium oxysporum* f.sp. *niveum*	CRISPR/ Cas9-deletion-coding	Zhang et al. (2020)
26	Cacao (Theobroma cacao)	TcNPR3	A suppressor of the defense response	Increased resistance to infection with the cacao pathogen *Phytophthora tropicalis*	CRISPR/ Cas9-deletion-exon	Fister et al. (2018)

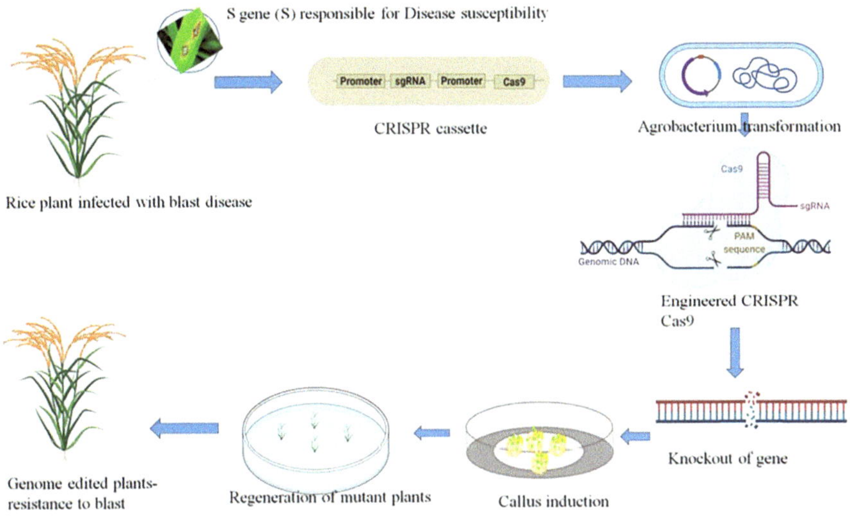

FIGURE 7.3 Illustration of development of disease-resistant lines (ex: rice) by targeting susceptibility (*S*) gene(s) via CRISPR/Cas9 system against bacterial, viral, or fungal pathogens.

based on plant genetic resistance, good agronomic practices, and biocontrol agents. However, due to the marked effects of elucidating the molecular mechanisms of host-bacterial pathogen interactions, many host plant genes, including some *S* genes, that participate in these complex processes have been identified. Bacterial pathogens produce effectors that can bind to the promoter region of *S* genes, promoting their expression and consequent effector-triggered susceptibility (ETS). As *S* genes may be more durable in the field, they are becoming popular targets for breeding crops that are resistant to bacterial diseases via genome editing. Several studies have been published on the application of GE strategies to counteract various bacterial diseases and are depicted in Figure 7.3.

7.2.1.2.1 Rice

Bacterial blight is one of the major vascular bundle diseases caused by *Xanthomonas oryzae* pv. *oryzae* (*Xoo*), has high epidemic potential, and causes 10%–20% losses of yield. The TAL effectors of *Xoo* contribute to pathogen virulence by binding to effector-binding elements in the promoter region of a specific rice *S* gene, including the *SUGAR WILL EVENTUALLY BE EXPORTED TRANSPORTER* (*SWEET*) and the *LATERAL ORGAN BOUNDARIES* (*LOB*), in a sequence-specific manner, resulting in its transcriptional activation (Antony et al., 2010). Many TALE proteins target *S* genes and activate their expression to promote successful infection (Boch and Bonas, 2010; Doyle et al., 2013). Earlier, TALEN was used to mutate predicted TAL effector-binding sites within the *OsSWEET14* (*Os11N3*) promoter region and exhibited heritable broad-spectrum resistance to *X. oryzae* pv. *Oryzae* strains due to lack of *OsSWEET14* induction by the bacterial pathogen (Li et al., 2013). Later, these results were confirmed by using TALEN (Blanvillain-Baufume et al., 2017) and CRISPR/Cas9 technologies (Zafar et al., 2020; Zeng et al., 2020). This approach was also used to knockout *OsSWEET14* by targeting its coding region, and the resulting mutant plants showed broad-spectrum resistance to Asian strains and African *Xoo* strains. Additionally, increased plant height has no detrimental effects on yield (Oliva et al., 2019). Similar results were obtained by Zhou et al. (2015). They generated rice plants harboring mutations in the coding region of *OsSWEET13* (*Xa25*) using CRISPR/Cas9, and the plants exhibited resistance to bacterial blight. Another gene, *OsSWEET11*, regulates rice bacterial blight resistance and participates in pollen development. The mutation caused changes in important agronomic traits and led to a sterile phenotype. So, to minimize any effect on pollen development, a CRISPR/Cas9-specific deletion was engineered in the promoter region of *OsSWEET11* that contains a pathogen-induced element (Yuan et al., 2011). Kim et al. (2019) mutated the promoter region of *OsSWEET11* using CRISPR/Cas9, and homozygous mutant *OsSWEET11* plants displayed significantly enhanced bacterial blight resistance and no significant effect on spikelet fertility.

Xu et al. (2019) also used *CRISPR*/Cas9 to engineer broad-spectrum bacterial blight resistance in rice by disrupting the TAL effector-binding sites of two *OsSWEET11* and *OsSWEET14* in rice cv. Kitaake, which already had a recessive resistance allele of *OsSWEET13*, and the triple mutant line exhibited broad-spectrum resistance to most *Xoo* strains. Oliva et al. (2019) also obtained strong, broad-spectrum resistance when *OsSWEET11*, *OsSWEET13*, and *OsSWEET14* were simultaneously edited by CRISPR/Cas9 in all promoters. No alterations in agronomic traits in triple mutant lines were recorded. Recently, CRISPR/Cas9-mediated genome editing in the promoter region effector-binding elements (EBEs) of the *OsSWEET14* gene has been demonstrated to confer resistance in Super Basmati rice lines against the *Xoo* strain carrying *AvrXa7* TALE (Zafar et al., 2020).

The TAL effectors can also indirectly regulate gene expression and gene transcription at the mRNA level. *BROAD-SPECTRUM RESISTANCE KITAAKE-1* (*BSR-K1*) encodes a tetratricopeptide repeat–containing protein that binds to mRNAs of multiple defense-related *OsPAL* genes, which are required for the biosynthesis of secondary metabolites, including lignin and SA, and *OsPAL1* overexpression confers resistance to fungal and bacterial diseases. So, Zhou et al. (2018) reported that CRISPR/Cas9-derived knockout of *OsBSR-K1* led to *OsPAL* gene expression with accumulation of secondary metabolites and enhanced broad-spectrum resistance to both *M. oryzae* and *X. oryzae* pv. *Oryzae* without major yield penalty.

7.2.1.2.2 Tomato

Two diseases, bacterial speck caused by *Pseudomonas syringae* pv. *tomato* and bacterial spot caused by *Xanthomonas* spp., can cause significant reductions in field crop yields of tomato (*Solanum lycopersicum*). Genome editing has been used to increase tomato resistance against these diseases. One of the genes, downy mildew resistance 6 (*DMR6*), was widely conserved and encodes a salicylate-5-hydroxylase, and loss of function presumably increases the local SA concentration at the infection site. This knowledge was used to engineer a CRISPR/ Cas9-mediated mutation in *S* gene *DMR6*, resulting in improved resistance to a variety of pathogens, including the oomycete pathogen *Phytophthora capsici* and the bacterial pathogen *Pseudomonas syringae* pv. *Tomato* DC3000 (Pto DC3000) (de Toledo Thomazella et al., 2016). In an attempt to impart resistance to bacterial speck disease, the CRISPR/Cas9-mediated mutation of *SlJAZ2,* the closest tomato ortholog of Arabidopsis JAZ (*JASMONATE ZIM DOMAIN*) repressor *AtJAZ2*, displayed bacterial speck resistance to *Pseudomonas syringae* pv. *tomato* in *Sljaz2* mutant plants (Ortigosa et al., 2019).

7.2.1.2.3 Citrus

Citrus canker, caused by *Xanthomonas citri* subsp. *citri* (*Xcc*), is one of the most destructive diseases, causing severe yield losses in all citrus-producing regions (Gottwald et al., 2002). *Citrus sinensis Lateral Organ Boundary 1* (*CsLOB1*), the disease susceptibility gene for citrus bacterial canker disease, encodes a transcription factor that controls plant growth and development. The main transcription activator-like (TAL) effectors of *Xcc* specifically bind to the effector-binding element in the *CsLOB1* promoter to activate expression of *CsLOB1* for favoring citrus canker development (Hu et al., 2014). In orange (*Citrus sinensis*), CRISPR-Cas deletion of the effector-binding region inside the promoter of the susceptibility gene *CsLOB1* gave resistance to the bacterial canker pathogen *Xanthomonas citri spp. Citri* (Xcc) (Peng et al., 2017). Similarly, in grapefruit (*Citrus paradise*), altering the coding area of *CsLOB1* with CRISPR-Cas resulted in increased resistance to *Xcc* (Jia et al., 2017). Another gene, *CsWRKY22*, is involved in the plant immunity response to citrus canker in orange (*Citrus sinensis* (L.) Osbeck). CRISPR/Cas9-mediated editing of *CsWRKY22* to improve orange resistance to citrus canker leads to lines that are less susceptible to citrus canker.

7.2.1.2.4 Apple

Erwinia amylovora, which causes fire blight, is a major threat to the apple and a variety of other commercial and ornamental plants. The bacterium triggers its infection through the DspA/E effector (a 198 kDa effector protein homologous to the type III effector AvrE of *Pseudomonas syringae* pv. tomato), which interacts with the apple susceptibility protein *MdDIPM4* (DspA/E-interacting proteins of *M.* × *domestica*). Pompili et al. (2020) reported *MdDIPM4* knockout in susceptible cultivars using the CRISPR/Cas9 system delivered via *Agrobacterium tumefaciens*. Edited lines with a loss-of-function mutation showed a highly significant reduction in susceptibility to *Erwinia amylovora*.

7.2.1.2.5 Banana

In bananas, *Xanthomonas* wilt disease caused by *Xanthomonas campestris* pv. musacearum is a major constraint on global banana production. This pathogen is a highly transmissible bacterium that spreads quickly through infected plant material, cutting tools, long-distance trade, and vectors such as insects, birds, and bats. By modifying the Musa *DMR6* gene, which has an ortholog in Musa species, we engineered a resistant banana against *Xanthomonas* wilt (BXW). Finally, the CRISPR tool provides resistance to a variety of devastating bacterial diseases in plants. Different genes have been edited through genome-editing technology to achieve bacterial resistance in crop plants (Table 7.2).

TABLE 7.2
Gene/s Engineered Using Genome-Editing Technology to Develop Bacterial Disease Resistance in Crop Plants

S. No	Crop	Target	Function of Gene	Resistance Imparted to Pathogen	Gene Mutation	References
			A. Cereals			
1	Rice (*Oryza sativa* L.)	OsSWEET13	Negative regulator of bacterial blight resistance	*Xanthomonas oryzae* pv. *oryzae* (Xoo)	CRISPR/Cas9- deletion in exonic region	Zhou et al. (2015)
2	Rice (*Oryza sativa* L.)	OsSWEET11	Negative regulator of bacterial blight resistance	*Xanthomonas oryzae* pv. *oryzae* (Xoo)	CRISPR/ Cas9-indel-exon	Kim et al. (2019)
3	Rice (*Oryza sativa* L.)	TALE-binding elements (EBEs) in *OsSWEET13* promoter, *OsSWEET11*, and *OsSWEET14* genes	Sucrose transporter. Negative regulator of bacterial blight resistance	*Xanthomonas oryzae* pv. *oryzae* (Xoo)	CRISPR/Cas9-gene and promoter disruption	Xu et al. (2019)
4	Rice (*Oryza sativa* L.)	OsSWEET11, OsSWEET13, and OsSWEET14	Negative regulator of bacterial blight resistance	*Xanthomonas oryzae* pv. *oryzae* (Xoo)	CRISPR/Cas9- promoter disruption	Oliva et al. (2019)
5	Rice (*Oryza sativa* L.)	OsSWEET14	Negative regulator of bacterial blight resistance	Resistance in Super Basmati rice lines against Xoo	CRISPR/ Cas9-deletion-gene	Zafar et al. (2020)
6	Rice (*Oryza sativa* L.)	OsBSR-K1	Tetratricopeptide repeats	*X. oryzae* pv. *oryzae* and *M. oryzae*	CRISPR/Cas9	Zhou et al. (2018)

(*Continued*)

TABLE 7.2 (*Continued*)
Gene/s Engineered Using Genome-Editing Technology to Develop Bacterial Disease Resistance in Crop Plants

S. No	Crop	Target	Function of Gene	Resistance Imparted to Pathogen	Gene Mutation	References
			B. horticultural crops			
7	Tomato	SlDMR6	Negative regulator	Resistant to Phytophthora capsici, and the bacterial pathogens Pseudomonas syringae pv. Tomato	CRISPR/Cas9-deletions in exon	de Toledo Thomazella et al. (2016)
8	Orange (*Citrus sinensis*)	CsLOB1	Negative regulator of *Xanthomonas citri* subsp. *citri* resistance	Resistance to the bacterial canker	CRISPR/Cas9-deletion in promoter	Peng et al. (2017)
9	Grapefruit (*Citrus paradise*)	CsLOB1	Negative regulator of *Xanthomonas citri* subsp. *citri* resistance	Resistance to the bacterial canker	CRISPR/Cas9-indel in exon	Jia et al. (2017)
10	Orange *Citrus sinensis* (L.)	CsWRKY22	Negative regulator of *Xanthomonas citri* subsp. *citri* resistance	Resistance to citrus canker	CRISPR/Cas9-deletion-promoter region	Long et al. (2021)
11	Apple	MdDIPM4	Susceptibility to *Erwinia amylovora*	Resistance to Erwinia amylovora	CRISPR/Cas9-deletion-coding region	Pompili et al. (2020)
12	Banana	DMR6	Susceptibility to Xanthomonas Wilt	Resistance to Xanthomonas Wilt	CRISPR/Cas9-indel-exon region	Tripathi et al. (2021)

7.2.1.3 Application of Genome-Editing Technologies in Crop Plants for Enhancing Resistance to Viral Diseases

7.2.1.3.1 Cereals

Rice tungro disease (RTD) is a serious viral disease caused by an interaction between rice tungro spherical virus (RTSV) and rice tungro bacilliform virus (RTBV). The translational initiation factor, eIF4G, controls and regulates RTSV. Resistance to Rice Tungro Streak Spherical Viral Disease was conferred by a mutation in native eIF4G via CRISPR/Cas9 in the susceptible IR64 rice variety. The induced frame shift mutations were then passed down to future generations without regard for any agronomic parameters (Macovei et al., 2018).

Wheat dwarf virus (WDV) is an economically important, phloem-limited, insect-transmitted virus belonging to the Gemini viridae family that causes severe yield losses, and the natural resistance resources against this are limited (Nygren et al., 2015). The wheat dwarf virus was attenuated by simultaneous editing of coat protein and movement protein using the CRISPR editing system, and the resulting edited plants showed enhanced resistance to the WDV.

7.2.1.3.2 Horticultural Crops

7.2.1.3.2.1 Tomato Tomato (yellow) leaf curl disease is a very destructive disease of tomatoes caused by begomo viruses transmitted by whiteflies. The CRISPR/Cas9 RNPs have been employed for inducing mutations in virus genes encoding coat protein (CP) or replicase (Rep), and the edited tomato plants showed significantly increased resistance to the infection of tomato yellow leaf curl virus (TYLCV) (Tashkandi et al., 2018). Another susceptibility factor gene, *Ty-5/SlPelo* (involved in ribosome recycling during protein synthesis), was edited by a single nucleotide insertion and led to frameshifts and premature stop codons, which correlated with resistance to TYLCV as well as reduced viral accumulation (Pramanik et al., 2021). *Pepper mottle virus* (*PepMoV*) from the potyvirus group was also a destructive pathogen that infected tomatoes (Melzer et al., 2012). To develop resistance against *PepMoV*, the CRISPR/Cas9 RNPs have been employed for inducing mutations in eIF4E1 and developing resistance to *PepMoV* in tomatoes.

7.2.1.3.2.2 Potato *Potato virus Y* (*PVY*) is a major potato pathogen that is spread by aphids and causes significant annual crop losses worldwide. By delivering Cas9 and sgRNAs into potato apical meristematic tissues via biolistic bombardment/vacuum infiltration, CRISPR/Cas9 RNPs were used to induce mutations in the susceptible protein *coilin*. The *coilin* gene mutations in the transgene-free potato plants conferred resistance to the RNA-based potato virus Y. (Makhotenko et al., 2019).

7.2.1.3.2.3 Cassava Cassava is an important tropical subsistence crop affected by cassava brown streak disease caused by *Cassava brown streak virus* (CBSV), a whitefly-transmitted member of the genus Ipomo virus and a member of the Potyviridae family. Cassava brown streak virus was attenuated by simultaneous editing of *eIF4E*(iso) and two novel cap-binding proteins, *nCBP1* and *nCBP2*, to develop a vaccine against viral disease. The resulting plants were resistant to the severity of cassava brown streak disease and had less root necrosis (Gomez et al., 2019).

7.2.1.3.2.4 Cucumber and Papaya Members of Potyviridae cause significant crop losses in a variety of crops. This family of viruses has an RNA genome that encodes a polyprotein that is cleaved by viral proteases. Viral protein is involved in polyprotein translation as well as other functions during the virus's life cycle. The silencing of the eIF4E gene resulted in broad RNA virus resistance. Natural mutations in the *eIF4E* and *eIF(iso)4E* genes have been linked to potyvirus resistance in fruit crops. CRISPR-Cas9 was used in cucumber to inactivate the eIF4E gene, and the resulting homozygotic mutant plants were immune to cucumber vein yellowing virus and resistant to potyviruses, zucchini yellow mosaic virus, and papaya ring spot mosaic virus (Chandrasekaran et al., 2016). Different genes edited through the CRISPR/cas tool to achieve viral resistance in plants are listed in Table 7.3.

To assess the possibility of widespread practical application of genome editing in crop plants, we summarized data on a number of genes edited for disease resistance for each crop (Figure 7.3). The majority of genome-edited reports were found in rice, followed by tomato. Several crops had only one reported edited event (apple, cucumber, cotton, cassava, grapefruit, potato, watermelon, and cocoa), and a few crops (orange, canola, and banana) reported two events modified for pathogen resistance (Figure 7.4).

7.2.2 Editing of Biocontrol Agents (BCAS)

CRISPR/Cas9 has been used to investigate genes involved in BCA virulence. CRISPR/Cas9 coupled with blastospore-mediated transformation was first used to edit the single and multiple genes in the fungus (Chen et al., 2017). Improving expression through the CRISPR/Cas9 system in *lcsL* gene that codes for a putative bZIP transcription factor, a member of *lcs* cluster, increases the yield of leucinostatins, which are the most prominent secondary metabolites in *Purpureocillium lilacinum* and a promising commercial agent for controlling plant-parasitic nematodes and plant pathogens (Jiao et al., 2019). Similarly, the biocontrol activity of *B. subtilis* HS3 and *B. mycoides* EC18 against two fungal pathogens, *Rhizoctonia solani* and *Fusarium culmorum*, was also improved by the disruption of *4'-phosphopantetheinyl transferase* (sfp), which is required for the production of several lipopeptide antibiotics (Yi et al., 2018).

7.3 CONCLUSIONS AND PROSPECTS FOR THE FUTURE

Given the remarkable success of genome-editing technology in various fields, it is reasonable to expect that plant pathology research will continue to benefit from the ever-improving toolbox being validated and made available for a broader range of plant species and pathogens. The development of new tools to improve plant protection is critical in the context of modern agricultural, environmental, and ecological challenges, particularly in light of global climate change and a burgeoning human population. GE is a novel tool for improving food safety and security in a sustainable manner, as well as mitigating the challenges posed by the world's growing population and climate change.

TABLE 7.3
Gene/s Engineered Using Genome-Editing Technology to Develop Viral Disease Resistance in Crop Plants

S. No	Crop	Target	Function of Gene	Resistance Imparted to Pathogen	Transformation method	References
			A. Cereals			
1	Rice	Translation initiation factor 4 gamma gene eIF4G	Essential roles in the translation of viral RNA genomes	Rice tungro spherical virus (RTSV)	CRISPR- CAS9-indel-coding region	Macovei et al. (2018)
2	Barley	sgRNAs-MP, CP, Rep/Rep, IR/WDV	Viral growth and development	Wheat dwarf virus (WDV)	CRISPR-CAS9-indel-coding sequence	Kis et al. (2019)
			B. Horticultural crops			
3	Tomato	Coat protein (CP) or replicase (Rep)	Viral growth and development	Tomato yellow leaf curl virus	CRISPR-Cas9-indels-CP and Rep regions	Tashkandi et al. (2018)
4	Tomato	Ty-5/SlPelo	Susceptible gene	Tomato yellow leaf curl virus	CRISPR-Cas9-deletion-target region	Pramanik et al. (2021)
5	Tomato	eIF4E1	Viral growth and development	Pepper mottle virus	CRISPR-Cas9-deletion-target gene	Yoon et al. (2020)
6	Potato	Susceptible protein, coilin	Metabolism of RNA and assembly of various ribonucleoprotein (RNP) complexes	Potato virus Y	CRISPR-Cas9-deletion-target gene	Makhotenko et al. (2019)
7	Cassava	eIF4E1	Viral growth and development	Cassava brown streak virus	CRISPR- Cas9-indel-exon	Gomeze et al. (2019)
8	Cucumber	eIF4E	Essential roles in the translation of viral RNA genomes	Immune to Cucumber vein yellowing and resistant to the potyviruses, *Zucchini yellow mosaic virus* and *Papaya ring spot mosaic virus*	CRISPR/Cas9-deletion-exon	Chandrasekaran et al. (2016)

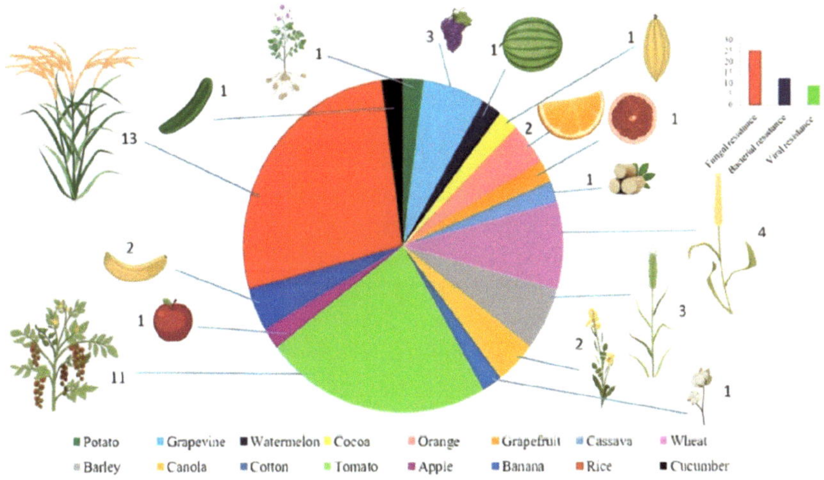

FIGURE 7.4 Number of gene(s) modified using genome-editing technology in cereals, oilseeds, and horticultural crops for disease resistance against pathogens. Bar graph represents the number of genes modified for disease resistance in fungal, bacterial, and viral pathogens.

Resistance breeding with genome editing does not require genetic crosses and segregant selection, thus driving rapid advancements. Indeed, a number of GE efforts resulted in plants with potentially useful traits that could be easily implemented in the field. Conclusively, the studies discussed here show that the *S* gene has the potential to develop resistance against a wide range of pathogens. These *S* genes are generally conserved in nature, and the discovery of one gene can lead to the discovery of more genes among or within plant species. Resistance acquired through susceptibility modification provides enormous opportunities for crop safety because a specific allele that confers susceptibility to one pathogen may also confer resistance to another pathogen (Lorang et al., 2012; McGrann et al., 2014). Collectively, these studies demonstrated the superiority of host *S* genes over *R* genes in the development of pathogen resistance, emphasizing the importance of discovering new S genes that can be exploited in the future. S genes are generally reported to be involved in plant growth and development, and null mutations in these genes sometimes may cause adverse effects on the development of plants. This phenomenon may well hamper the use of these genes in editing for disease resistance. The *R* genes, such as NB-LRRs, are valuable resources for transferring resistance from wild species to elite varieties. The critical differences between *R* genes from wild species and elite varieties are sometimes limited to a single nucleotide variant (Lindner et al., 2020). In this case, newly developed base editors (Anzalone et al., 2020) could be used to generate specific base changes. Furthermore, genes can be inserted anywhere in the genome; for example, introducing new resistance genes next to already existing resistance loci could provide the breeder with more flexibility. Precisely adding and removing genes, and gene stacks could be created and updated. This kind of staking enables us to improve downstream genetic segregation, which leads to a significant advance in plant breeding. Information generated for one NLR could be exploited to improve NLR in other plant species, as was evident from the work on synthetic immune receptors (Giannakopoulou et al., 2015), which highlighted a new direction for breeding crops resistant to phylogenetically divergent pathogens. Targeting regulatory mechanisms such as protein post-translational modification (PTMs) could also be used to fine-tune resistance responses.

Broad-spectrum resistance can also be achieved by targeting multiple resistance genes at one time using CRISPR/Cas. Exploring variable TALE-binding elements of multiple susceptibility genes is a viable option. This strategy was utilized by Tao et al. (2021), wherein a triple mutant of three genes, viz., *Pi21*, *Bsr-d1*, *and Xa5*, and three *SWEET* genes, displayed elevated resistance to blight and bacterial blight in rice. The dual RNA-sequencing can also be deployed for understanding the host and fungal/bacterial/viral interactions. The effectors or invading pathogen/viral transcript information is known, and a holistic picture of interaction can be captured. The dual transcriptome at varying stages also sheds light on the progress of disease and defense from the host side. Such information can lead to the identification of new genes or targets for genome editing. Involvement of long non-coding RNAs (lncRNAs) in several diseases was reported recently, such as powdery mildew and stripe rust in wheat, white mold in canola, and yellow leaf curl virus in tomato. The literature on lncRNA as a potential regulator of disease resistance is continuously upscaling. In a recent study, more than 500 lncRNAs were found to be involved in the rice-*Xanthomonas oryzae* pv. *oryzae* (Xoo) infection (Yu et al., 2020). Interestingly, expression of lncRNA, *ALEX1*, was highly inducible upon infection and activated the jasmonic acid pathway, imparting resistance to bacterial blight. It is possible that such functionally characterized lncRNAs can be overexpressed using the CRISPR/Cas systems for inducing broad-spectrum resistance in plants.

Technically, there is still a lot of room for GE advancement using the CRISPR/Cas system. Comparative analyses of the various methods available may enable us to identify the best and most robust approaches suited to the specific needs, which may streamline the processes. As more Cas9 structural data becomes available, protein engineering is being used to improve Cas9 efficiency for each species. Eventually, the discovery of different CRISPR systems and components from other bacteria will most likely lead to the identification of additional Cas enzymes. Moreover, recently developed CRISPR activation (CRISPRa) or CRISPR interference (CRISPRi) approaches may offer to specifically regulate target genes and enable multiplex gene regulation; they can be used to modulate genes for breeding crops that have broad-spectrum resistance in the future. Three new genome-editing systems have recently been established, which include mitochondria-targeted transcription activator-like effector nucleases (mitoTALENs) (Kazama et al., 2019), APOBEC-Cas9 fusion-induced deletion systems (AFIDs) (Wang et al., 2020), and prime editing systems, which are likely to further revolutionize basic research and precision breeding in plants.

Despite the undeniable value of GE resources in general and CRISPR/Cas in particular, scientists are still constrained by regulatory issues as well as public opinions and perceptions, implying that these technologies are still a long way from commercialization. It is critical to have governance mechanisms in place to ensure that this technology is developed and used in an ethical and safe manner. Gene-edited products can be considered as non-transgenic plants as recent methods of CRISPR/Cas make the plants without any remnants of the CRISPR/Cas system. In addition, the use of efficient biocontrol agents that carry the CRISPR/Cas system can also increase resistance in a non-transgenic way. The legislative and public perception barriers are gradually eroding. The general public may begin to trust these new technologies and scientists more as they realize how critical it is to respond quickly, easily, and safely in the extreme circumstances of potentially serious plant disease outbreaks that could endanger global food security. If this is the case, it may pave the way for precise breeding to become more common in response to climate change and food scarcity. The potential impact of GE, and particularly CRISPR technology, in synthetic plant biology to breed resistance against diseases is enormous and will have a direct impact on agricultural sustainability on a never-before-seen scale.

REFERENCES

Aktar, M.W.; Sengupta, D.; Chowdhury, A. Impact of pesticides use in agriculture: their benefits and hazards. *Interdiscipl Toxicol.* 2009; *2*(1), pp 1.

Allen, R.F. Resistance to stem rust in Kanred wheat. *Science.* 1921, *53*, pp 575–576.

Antony, G.; Zhou, J.; Huang, S.; et al. Rice *xa13* recessive resistance to bacterial blight is defeated by induction of the disease susceptibility gene *Os-11N3*. *Plant Cell.* 2010; *22*(11), pp 3864–3876.

Anzalone, A.V.; Koblan, L.W.; Liu, D.R. Genome editing with CRISPR-Cas nucleases, base editors, transposases and prime editors. *Nat Biotechnol.* 2020; *38*, pp 824–844.

Blanvillain-Baufumé, S.; Reschke, M.; Solé, M.; et al. Targeted promoter editing for rice resistance to *Xanthomonas oryzaepv. oryzae* reveals differential activities for *SWEET14*-inducing TAL effectors. *Plant Biotechnol J.* 2017; *15*, pp 306–317.

Boch, J.; Bonas, U. Xanthomonas AvrBs3 family-type III effectors: discovery and function. *Annu Rev Phytopathol.* 2010; *48*, pp 419–436.

Brauer, E.K.; Balcerzak, M.; Rocheleau, H.; et al. Genome editing of a deoxynivalenol-induced transcription factor confers resistance to *Fusarium graminearum* in wheat. Mol Plant-Microbe Interact. 2020; *33*, pp 553–560.

Cahya, P.; Barbetti, M.J.; Barker, S.J. A novel tomato fusarium wilt tolerance gene. *Front. Microbiol.* 2018; *9*, pp 1226.

Chakraborty, S.; Chakraborty, N.; Agrawal, L.; et al. Next-generation protein-rich potato expressing the seed protein gene *AmA1* is a result of proteome rebalancing in transgenic tuber. *Proc Nat Acad Sci USA.* 2010; *107*, pp 17533–17538.

Chandrasekaran, J.; Brumin, M.; Wolf, D.; et al. Development of broad virus resistance in non-transgenic cucumber using CRISPR/Cas9 technology. *Mol Plant Pathol.* 2016; *17*(7), pp 1140–1153.

Chen, J.; Lai, Y.; Wang, L.; et al. CRISPR/Cas9-mediated efficient genome editing via blastospore-based transformation in entomopathogenic fungus *Beauveria bassiana*. *Sci Rep.* 2017; 7, pp 45763.

Dangl, J.L.; Horvath, D.M.; Staskawicz, B.J. Pivoting the plant immune system from dissection to deployment. *Science.* 2013; *341*, pp 746–751.

Dodds, P.N.; Rathjen, J.P. Plant immunity: towards an integrated view of plant-pathogen interactions. *Nat Rev Genet.* 2010; *11*, pp 539–548.

Dong, S.; Raffaele, S.; Kamoun, S. The two-speed genomes of filamentous pathogens: waltz with plants. *Curr Opin Genet Dev.* 2015; *35*, pp 57–65.

Doyle, E.L.; Stoddard, B.L.; Voytas, D.F.; Bogdanove, A J. TAL effectors: highly adaptable phytobacterial virulence factors and readily engineered DNA-targeting proteins. *Trends Cell Biol.* 2013; *23*, pp 390–398.

Fiaz, S.; Ahmar, S.; Saeed, S.; Riaz, A.; Mora-Poblete, F.; Jung, K.H. Evolution and application of genome editing techniques for achieving food and nutritional security. *Int J Mol Sci.* 2021; *22*, pp 5585.

Fister, A.S.; Landherr, L.; Maximova, S.N; Guiltinan, M.J. Transient expression of CRISPR/Cas9 machinery targeting *TcNPR3* enhances defense response in *Theobroma cacao*. Front Plant Sci. 2018; *9*, pp 268.

Fukuoka, S.; Saka, N.; Koga, H.; et al. Loss of function of a proline-containing protein confers durable disease resistance in rice. *Science*. 2009; *325*, pp 998–1001.

Galli, M.; Martiny, E.; Imani, J.; et al. CRISPR/Sp Cas9-mediated double knockout of barley *Microrchidia MORC1* and *MORC6a* reveals their strong involvement in plant immunity, transcriptional gene silencing and plant growth. *Plant Biotechnol J*. 2022; *20*(1), pp 89–102.

Gao, Z.; Liu, Q.; Zhang, Y.; et al. *OsCUL3a*-associated molecular switches have functions in cell metabolism, cell death, and disease resistance. *J Agricultural Food Chem*. 2020; *68*(19), pp 5471–5482.

Ghogare, R.; Ludwig, Y.; Bueno, G.M.; Slamet-Loedin, I.H.; Dhingra, A. Genome editing reagent delivery in plants. *Transgen Res*. 2021; *30*(4), pp 321–335.

Giannakopoulou, A.; Steele, J.F.; Segretin, M.E.; et al. Tomato I2 immune receptor can be engineered to confer partial resistance to the oomycete *Phytophthora infestans* in addition to the fungus *Fusarium oxysporum*. *Mol Plant Microbe Interact*. 2015; *28*(12), pp 1316–1329.

Gil Lee, H.; Kim, D.H.; Choi, Y.-R.; et al. Enhancing plant immunity by expression of pathogen targeted CRISPR-Cas9 in plants. *Gene Genome Ed*. 2021; *1*, pp 100001.

Giraud, T.; Gladieux, P.; Gavrilets, S. Linking the emergence of fungal plant diseases with ecological speciation. *Trends Ecol Evol*. 2010; *25*(7), pp 387–395.

Gomez, M.A.; Lin, Z.D.; Moll, T.; et al. Simultaneous CRISPR/Cas9-mediated editing of cassava *eIF 4E* isoforms *nCBP-1* and *nCBP-2* reduces cassava brown streak disease symptom severity and incidence. *Plant Biotechnol J*. 2019; *17*(2), pp 421–434.

Gottwald, T.R.; Graham, J.H.; Schubert, T.S. Citrus canker: the pathogen and its impact. *Plant Health Prog*. 2002; *10*, pp 1–34.

Guimaraes, E.P.; Correa-Victoria F. Use of recurrent selection for develop resistance *Pyricularria grisea Sacc*. on rice. In *Advances in Rice Population Improvement* (ed. EP Guimaraes), Santo Antonio de Goias, Brazil: Embrapa Rice and Beans. 2000; pp 165–175.

Gutiérrez, O.A.; Campbell, A.S.; Phillips-Mora, W. Breeding for disease resistance in cacao. In B.A. Bailey and L.W. Meinhardt (Eds.), *Cacao Diseases*. Springer, International Publishing Switzerland; 2016; pp 567–609.

Heinrichs, E.A.; Muniappan, R. IPM for tropical crops: rice. *CAB Rev*. 2017; *12* (1), pp 1–30.

Hong, Y.; Liu, Q.; Cao, Y.; et al. The *OsMPK15* negatively regulates Magnaporthe oryza and Xoo disease resistance via SA and JA signaling pathway in rice. *Front Plant Sci*. 2019; *10*, pp 752.

Hong, Y.; Meng, J.; He, X.; Zhang, Y.; Liu, Y.; Zhang, C.; Qi, H.; Luan, Y. Editing *miR482b* and *miR482c* simultaneously by CRISPR/Cas9 enhanced tomato resistance to *Phytophthora infestans*. *Phytopathology(r)*. 2021; *111*(6), pp 1008–1016.

Hu, Y.; Zhang, J.L.; Jia, H.G.; et al. *Lateral organ boundaries 1* is a disease susceptibility gene for citrus bacterial canker disease. *PNAS*. 2014; *111*, pp E521–E529.

Jaczewska-Kalicka, A. Occurrence and harmfulness of the most important diseases of winter wheat in Central Poland. *J Plant Prot Res*. 2002; *42*, pp 93–101.

Jeon, J.E.; Kim, J.G.; Fischer, C.R.; et al. A pathogen-responsive gene cluster for highly modified fatty acids in tomato. *Cell*. 2020; *180*(1), pp 176–187

Jia, H.; Zhang, Y.; Orbović, V.; et al. Genome editing of the disease susceptibility gene *Cs LOB 1* in citrus confers resistance to citrus canker. *Plant Biotechnol J*. 2017; *15*(7), pp 817–823.

Jiao, Y.; Li, Y.; Li, Y.; et al. Functional genetic analysis of the leucinostatin biosynthesis transcription regulator *lcsL* in *Purpureocillium lilacinum* using CRISPR-Cas9 technology. *Appl Microbiol Biotechnol*. 2019; *103*, pp 6187–6194.

Jones, J.D.; Vance, R.E.; Dangl, J.L. Intracellular innate immune surveillance devices in plants and animals. *Science*. 2016; *354*, pp 6395.

Kazama, T.; Okuno, M.; Watari, Y.; et al. Curing cytoplasmic male sterility via TALEN-mediated mitochondrial genome editing. *Nature Plants*. 2019; *5*(7), pp 722–730.

Kieu, N.P.; Lenman, M.; Wang, E.S.; Petersen, B.L.; Andreasson, E. Mutations introduced in susceptibility genes through CRISPR/Cas9 genome editing confer increased late blight resistance in potatoes. *Sci Rep*. 2021; *11*(1), pp 1–12.

Kim, YA.; Moon, H.; Park, C.J. CRISPR/Cas9-targeted mutagenesis of *Os8N3* in rice to confer resistance to *Xanthomonasoryzae* pv. *oryzae*. *Rice*. 2019; *12*, pp 67.

Kis, A.; Hamar, É.; Tholt, G.; Bán, R.; Havelda, Z. Creating highly efficient resistance against wheat dwarf virus in barley by employing CRISPR/Cas9 system. *Plant Biotechnol J*. 2019; *17*(6), pp 1004.

Korbas, M. Disease-causing fungi in wheat. *Agro-Serwis*. 2009; *4*, pp 71–77.

Kuai, X.; MacLeod, B.J.; Després, C. Integrating data on the Arabidopsis NPR1/NPR3/NPR4 salicylic acid receptors; a differentiating argument. *Front Plant Sci*. 2015; *6*, pp 235.

Kumar, N.; Galli, M.; Ordon, J.; Stuttmann, J.; Kogel, K.H.; Imani, J. Further analysis of barley *MORC1* using a highly efficient RNA-guided Cas9 gene-editing system. *Plant Biotechnol J*. 2018; *16*(11), pp 1892–1903.

Kumari, M.; Devanna, B.N.; Singh, P.K.; Rajashekara, H.; Sharma, V.; Sharma, T.R. Stacking of blast resistance orthologue genes in susceptible indica rice line improves resistance against *Magnaporthe oryzae*. *3 Biotech*. 2018; *8*(1), pp 1–7.

Kumari, M.; Rai, A.K.; Devanna, B.N.; et al. Co-transformation mediated stacking of blast resistance genes *Pi54* and *Pi54rh* in rice provides broad spectrum resistance against *Magnaporthe oryzae*. Plant Cell Rep. 2017; *36*(11), pp 1747–1755.

Law, J.; Ser, H.; Khan, T.; et al. The potential of Streptomyces as biocontrol agents against the rice blast fungus, *Magnaporthe oryzae* (*Pyricularia oryzae*). *Front Microbiol*. 2017; *8*, pp 3.

Li, J.F.; Norville, J.E.; Aach, J.; et al. Multiplex and homologous recombination-mediated genome editing in Arabidopsis and *Nicotiana benthamiana* using guide RNA and Cas9. *Nat Biotechnol*. 2013; *31*, pp 688–691.

Li, M.Y.; Jiao, Y.T.; Wang, Y.T.; et al. CRISPR/Cas9-mediated *VvPR4b* editing decreases downy mildew resistance in grapevine (*Vitis vinifera* L.). *Hortic Res*. 2020a; *7*, pp 149.

Li, R.; Wang, L.; Li, Y.; et al. Knockout of *SlNPR1* enhances tomato plants resistance against *Botrytis cinerea* by modulating ROS homeostasis and JA/ET signaling pathways. *Physiol Plant*. 2020b; *170*, pp 569–579.

Li, S.; Shen, L.; Hu, P.; et al. Developing disease-resistant thermosensitive male sterile rice by multiplex gene editing. *J Integr Plant Biol*. 2019; *61*(12), pp 1201–1205.

Li, W.; Zhu, Z.; Chern, M.; et al. A natural allele of a transcription factor in rice confers broad spectrum blast resistance. *Cell*. 2017; *170*, pp 114–126.e15.

Lindner, S.; Keller, B.; Pal Singh, S.; et al. Single residues in the LRR domain of the wheat *PM3A* immune receptor can control the strength and the spectrum of the immune response. *Plant J*. 2020; 104, pp 200–214.

Liu, C.; Zhang, Y.; Tan, Y.; et al. CRISPR/Cas9-mediated SlMYBS2 mutagenesis reduces tomato resistance to *Phytophthora infestans*. *Int J Mol Sci*. 2021; *22*(21), pp 11423.

Liu, J.; Chen, X.; Liang, X.; et al. Alternative splicing of rice *WRKY62* and *WRKY76* transcription factor genes in pathogen defense. *Plant Physiol*. 2016; *171*(2), pp 1427–1442.

Long, Q.; Du, M.; Long, J.; et al. Transcription factor *WRKY22* regulates canker susceptibility in sweet orange (*Citrus sinensis* Osbeck) by enhancing cell enlargement and *CsLOB1* expression. *Hortic Res*. 2021; 8, pp 50.

Lorang, J.; Kidarsa, T.; Bradford, C.S.; et al. Tricking the guard: exploiting plant defense for disease susceptibility. *Science*. 2012; *338*(6107), pp 659–662.

Ma, J.; Chen, J.; Wang, M.; et al. Disruption of *OsSEC3A* increases the content of salicylic acid and induces plant defense responses in rice. *J Exp Bot*. 2018; 69, pp 1051–1064.

Macovei, A.; Sevilla, N.R.; Cantos, C.; et al. Novel alleles of rice eIF4G generated by CRISPR/Cas9-targeted mutagenesis confer resistance to *rice tungro spherical virus*. *Plant Biotechnol J*. 2018; *16*(11), pp 1918–1927.

Makhotenko, A.V.; Khromov, A.V.; Snigir, E. A.; et al. Functional analysis of *coilin* in virus resistance and stress tolerance of potato *Solanum tuberosum* using CRISPR-Cas9 editing. In *Doklady Biochemistry and Biophysics.* 2019; *484*, 1: pp 88–91.

Malnoy, M.; Viola, R.; Jung, M.H.; et al. DNA free genetically edited grapevine and apple protoplast using CRISPR/Cas9 ribonucleoproteins. *Front Plant Sci.* 2016; 7, pp 1904.

McGrann, G.R.; Stavrinides, A.; Russell, J.; et al. A trade off between *mlo* resistance to powdery mildew and increased susceptibility of barley to a newly important disease, *Ramularia* leaf spot. *J Exp Bot.* 2014; *65*(4), pp 1025–1037.

Melzer, M.J.; Sugano, J.S.; Cabanas, D.; et al. First report of *pepper mottle virus* infecting tomato in Hawaii. *Plant Dis.* 2012; *96*, pp 917–917.

Miah, G.; Rafii, M.Y.; Ismail, M.R.; et al. Blast resistance in rice: a review of conventional breeding to molecular approaches. *Mol Biol.* 2013; *40*, pp 2369–2388.

Nekrasov, V.; Wang, C.; Win, J; et al. Rapid generation of a transgene-free powdery mildew resistant tomato by genome deletion. *Sci Rep.* 2017; *7*, pp 482.

Nygren, J.; Shad, N.; Kvarnheden, A.; Westerbergh, A. Variation in susceptibility to wheat dwarf virus among wild and domesticated wheat. *PloS One.* 2015; *10*(4), pp e0121580.

Oliva, R.; Ji, C.; Atienza-Grande, G.; et al. Broad-spectrum resistance to bacterial blight in rice using genome editing. *Nat Biotechnol.* 2019; *37*(11), pp 1344–1350.

Ortigosa, A.; Gimenez-Ibanez, S.; Leonhardt, N.; et al. Design of a bacterial speck resistant tomato by CRISPR/Cas9-mediated editing of *SlJAZ2*. *Plant Biotechnol J.* 2019; *17*, pp 665–673.

Pandiarajan, R.; Grover, A. In vivo promoter engineering in plants: are we ready? *Plant Sci.* 2018; *277*, pp 132–138.

Peng, A.; Chen, S.; Lei, T.; et al. Engineering canker-resistant plants through CRISPR/Cas9-targeted editing of the susceptibility gene *Cs LOB 1* promoter in citrus. *Plant Biotechnol J.* 2017; *15*(12), pp 1509–1519.

Pompili, V.; Dalla Costa, L.; Piazza, S.; Pindo, M.; Malnoy, M. Reduced fire blight susceptibility in apple cultivars using a high-efficiency CRISPR/Cas9-FLP/FRT-based gene editing system. *Plant Biotechnol J.* 2020; *18*(3), pp 845–858.

Pramanik, D.; Shelake, R.M.; Park, J.; et al. CRISPR/Cas9-mediated generation of pathogen-resistant tomato against tomato yellow leaf curl virus and powdery mildew. *Internat J Mol Sci.* 2021; *22*, pp 1878.

Puchta, H. Applying CRISPR/Cas for genome engineering in plants: the best is yet to come. *Curr Opin Plant Biol.* 2017; *36*, pp 1–8.

Santillan Martinez, M.I.; Bracuto, V.; Koseoglou, E.; et al. CRISPR/Cas9-targeted mutagenesis of the tomato susceptibility gene *PMR4* for resistance against powdery mildew. *BMC Plant Biol.* 2020; *20*, pp 284.

Savary, S.; Willocquet, L.; Pethybridge, et al. The global burden of pathogens and pests on major food crops. *Nat Ecol Evol.* 2019; *3*(3), pp.430–439.

Schenke, D.; Cai, D. Applications of CRISPR/Cas to improve crop disease resistance: beyond inactivation of susceptibility factors. *iScience.* 2020; *23*(9), pp 101478.

Shah, P.R.; Varanavasiappan, S.; Kokiladevi, E.; et al. Genome editing of rice *PFT1* gene to study its role in rice sheath blight disease resistance. *Int J Curr Microbiol App Sci.* 2019; *8*, pp 2356–2364.

Sharma, T.R.; Rai, A.K.; Gupta, S.K.; Vijayan, J.; Devanna, B.N.; Ray, S. Rice blast management through host-plant resistance: retrospect and prospects. *Agr Res.* 2012; *1*(1), pp 37–52.

Shujuan, Z.; Liu, W.; Ruirui, Z.; et al. Knockout of *SlMAPK3* reduced disease resistance to *Botrytis cinerea* in tomato plants. *J Agric Food Chem.* 2018; *66*, pp 8949–8956.

Sreewongchai, T.; Toojinda, T.; Thanintorn, N.; et al. Development of elite indica rice lines with wide spectrum of resistance to Thai blast isolates by pyramiding multiple resistance QTLs. *Plant Breed.* 2010; *129*, pp 176–180.

Su, Z.; Bernardo, A.; Tian, B.; et al. A deletion mutation in *TaHRC* confers Fhb1 resistance to Fusarium head blight in wheat. *Nat Genet.* 2019; *51*(7), pp 1099–1105.

Tao, H.; Shi, X.; He, F.; et al. Engineering broad-spectrum disease-resistant rice by editing multiple susceptibility genes. *J Integr Plant Biol.* 2021; *63*(9), pp 1639–1648.

Tashkandi, M.; Ali, Z.; Aljedaani, F.; Shami, A.; Mahfouz, M.M. Engineering resistance against *Tomato yellow leaf curl virus* via the CRISPR/Cas9 system in tomato. *Plant Signal Behav*. 2018; *13*(10), pp e1525996.

Thomazella, D.P.; Brail, Q.; Dahlbeck, D.; Staskawicz, B.; CRISPR-Cas9 mediated mutagenesis of a *DMR6* ortholog in tomato confers broad-spectrum disease resistance. *BioRxiv*. 2016; pp 064824.

Thompson, D.B.; Aboulhouda, S.; Hysolli, H.; et al. The future of multiplexed eukaryotic genome engineering. *ACS Chem Biol.* 2018; 13, pp 313–325.

Van Butselaar, T.; Van den Ackerveken, G. Salicylic acid steers the growth-immunity tradeoff. *Trends Plant Sci.* 2020; *25*(6), pp 566–576.

Vollmann, J.; Rajcan, I. Oil crop breeding and genetics. In Johann Vollmann and Istvan Rajcan (Eds.), *Oil Crops*. 2009; pp 1–30. Springer, New York.

Wan, D.Y.; Guo, Y.; Cheng, Y.; et al. CRISPR/Cas9-mediated mutagenesis of *VvMLO3* results in enhanced resistance to powdery mildew in grapevine (*Vitis vinifera*). *Hortic Research.* 2020; 7, pp 116.

Wang, F; Wang, C.; Liu, P.; et al. Enhanced rice blast resistance by CRISPR/Cas9-targeted mutagenesis of the ERF transcription factor gene OsERF922. *PLoS One*. 2016; *11*(4), pp e0154027.

Wang, S.; Zong, Y.; Lin, Q.; et al. Precise, predictable multi-nucleotide deletions in rice and wheat using APOBEC-Cas9. *Nat Biotechnol.* 2020; *38*(12), pp 1460–1465.

Wang, X.; Tu, M.; Wang, D.; et al. CRISPR/Cas9-mediated efficient targeted mutagenesis in grape in the first generation. *Plant Biotechnol. J.* 2018, *16*, pp 844–855.

Wang, Y.; Cheng, X.; Shan, Q.; et al. Simultaneous editing of three homoeoalleles in hexaploid bread wheat confers heritable resistance to powdery mildew. *Nat Biotechnol.* 2014; *32*(9), pp 947–951.

Williamson, B.; Tudzynski, B.; Tudzynski, P.; Van Kan, J.A. *Botrytis cinerea*: the cause of grey mould disease. *Mol Plant Pathol.* 2007; *8*(5), pp 561–580.

Wood, G.A.R; Lass, R. Cocoa: Longman Scientific & *Technical*. 2008. New York: John Wiley & Sons

Wu, J.; Zhao, Q.; Yang, Q.; et al. Comparative transcriptomic analysis uncovers the complex genetic network for resistance to *Sclerotinia sclerotiorum* in Brassica napus. *Scientific reports*. 2016; *6*(1), pp 1–16.

Xie, K.; Yang, Y. RNA-guided genome editing in plants using a CRISPR-Cas system. *Mol. Plant.* 2013; *6*, pp 1975–1983.

Xu, Z.; Xu, X.; Gong, Q.; et al. *Engineering Broad-Spectrum Bacterial Blight Resistance by Simultaneously Disrupting Variable TALE-binding Elements of Multiple Susceptibility Genes in Rice. Molecular Plant.* 2019; S1674205219302898.

Yi, Y.; Li, Z.; Song, C.; Kuipers, OP. Exploring plant-microbe interactions of the rhizobacteria *Bacillus subtilis* and *Bacillus mycoides* by use of the CRISPR-Cas9 system. *Environ Microbiol.* 2018; *20*, pp 4245–4260.

Yu, Y.; Zhou, Y.F.; Feng, Y.Z.; et al. Transcriptional landscape of pathogen-responsive lncRNAs in rice unveils the role of *ALEX1* in jasmonate pathway and disease resistance. *Plant Biotechnol J.* 2020, *18*(3), pp 679–690.

Yuan, T.; Li, X.; Xiao, J.; Wang, S. Characterization of *Xanthomonas oryzae*-responsive cis-acting element in the promoter of rice race-specific susceptibility gene Xa13. *Molecular Plant*. 2011; *4*(2), pp 300–309.

Zafar, K.; Khan, M.Z.; Amin, I.; et al. Precise CRISPR-Cas9 mediated genome editing in Super Basmati rice for resistance against bacterial blight by targeting the major susceptibility gene. *Front Plant Sci.* 2020; *11*, pp 575.

Zaidi, S.S.A.; Mukhtar, M.S.; Mansoor, S. Genome editing: targeting susceptibility genes for plant disease resistance. *Trends Biotechnol.* 2018; *36*, pp 898–906.

Zeng, X.; Luo, Y.; Vu, N.T.Q.; et al. CRISPR/Cas9-mediated mutation of OsSWEET14 in rice cv. Zhonghua11 confers resistance to *Xanthomonas oryzae pv. oryzae* without yield penalty. BMC Plant Biol. 2020; *20*(1), pp 1–11.

Zhang, H.; Si, X.; Ji, X.; et al. Genome editing of upstream open reading frames enables translational control in plants. *Nat Biotechnol.* 2018a; *36*, pp 894–898.

Zhang, M.; Liu, Q.; Yang, X.; et al. CRISPR/Cas9-mediated mutagenesis of *Clpsk1* in watermelon to confer resistance to *Fusarium oxysporumf.sp. niveum. Plant Cell Rep.* 2020; *39*, pp 589–595.

Zhang, Y.; Bai, Y.; Wu, G.; et al. Simultaneous modification of three homoeologs of *TaEDR 1* by genome editing enhances powdery mildew resistance in wheat. *Plant J.* 2017; *91*(4), pp 714–724.

Zhang, Z.; Ge, X.; Luo, X.; et al. Simultaneous editing of two copies of *Gh14-3-3d* confers enhanced transgene-clean plant defense against *Verticillium dahliae* in allotetraploid upland cotton. Front Plant Sci. 2018b; 9, pp 842.

Zhou, J.; Peng, Z.; Long, J.; et al. Gene targeting by the TAL effector *PthXo2* reveals cryptic resistance gene for bacterial blight of rice. *Plant J.* 2015; *82*(4), pp 632–643.

Zhou, X.; Haicheng, L.; Mawsheng, C.; et al. Loss of function of a rice TPR-domain RNA-binding protein confers broad-spectrum disease resistance. *Proceed Natl Acad Sci U S A.* 2018; *115*(120), pp 3174–3179.

Zhu, Z.; Yin, J.; Chern, M.; et al. New insights into bsr-d1-mediated broad-spectrum resistance to rice blast. *Mol Plant Pathol.* 2020; *21*, pp 951–960.

8 Combating Plant Diseases through CRISPR-Based Genome-Editing Approaches

Dipankar Barman, Pragya Yadav, Jyoti Priya, Divya Bharathi R, Fatima Nazish Khan, Shivani Nagar, Pranjal Yadava, Archana Watts, Soham Ray, and Viswanathan Chinnusamy
ICAR-Indian Agricultural Research Institute

8.1 INTRODUCTION

Plant diseases are one of the most severe threats to improving crop productivity and achieving food security worldwide. Besides, rising global temperatures and climate change also play a major role in the alarming rise in disease incidence. Plant disease is an abnormality in the structure and/or function of the host plant cells and/or tissue as a result of a continuous irritation caused by a pathogenic agent or an environmental factor (Cooper and Gardener, 2007). A disease is not static in nature; it induces a series of physicochemical changes in the plant that ultimately benefit the pathogen but hamper the plant's natural growth and development. All plants, to some extent, are subject to disease, but most of the plants are resistant to most of the diseases most of the time. Hence, it is said that disease is an exception, not a rule. However, serious disease conditions may affect the vital functions of crops, including photosynthesis, assimilate partitioning, etc., leading to low productivity or complete crop loss.

8.1.1 Factors Affecting Plant Disease Development and Progression

Successful establishment of any disease in a plant requires intersection of three components at a given point of time – susceptible host (a host that cannot resist pathogen invasion), virulent pathogen (a pathogen that is capable of causing disease) and congenial environment (that supports disease development and progression). Together, these three constitute a disease triangle, and the absence of any of these three hinders the process of disease development (Figure 8.1). The biotic agents (pathogens) causing plant diseases include viruses, bacteria, fungi, protozoa, nematodes, as well

DOI: 10.1201/9781003403142-8

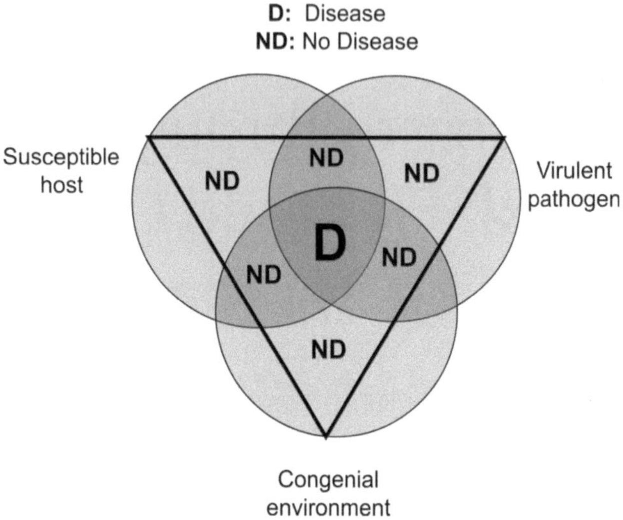

FIGURE 8.1 Pictorial depiction of disease triangle. Successful disease development requires intersection of susceptible host, virulent pathogen and congenial environment. In all other cases, disease does not develop.

as parasitic plants, but the first three are most common. In general, plant pathogens infect specifically particular plant species or groups, and sometimes the pathogen may strike a large crop, causing devastating effects to the entire area and they can be transmitted between plants. Plant pathogens are able to move with respect to time and space and possess tremendous genomic plasticity, thus evolving and adapting themselves to their surrounding harsh environment. This capability allows them to overcome the surveillance mechanism of resistant genotypes and the breakdown of resistance. Hence, it is required to understand clearly how pathogens spread among crop plants, how they evolve and what are the weak links in plant defence machinery that the pathogen capitalizes upon. Presently, with the advent of genome-editing technologies, finding such weak links in the plant defence system and rectifying those errors is an active area of research.

8.1.2 Significant Phytopathogens

Viruses are non-living, obligate intracellular parasites that replicate only inside living organisms. They are comprised of a nucleic acid and protein coat called a capsid, which is often made up of proteins and lipids with an overall size range of around 30 nm (Gergerich and Dolja, 2006). Plant viruses can be single-stranded RNA viruses, double-stranded RNA viruses, or single-stranded DNA viruses. About 700 plant viruses are known to cause devastating diseases, and they have wide host ranges, from bacteria and archaea to plants and animals (Koonin et al., 2006). According to Czosnek and Laterrot (1997), the viral infection can cause up to 98% crop loss in most of the tropical and sub-tropical regions. Shkalikov et al. (2010) categorized the symptoms of viral diseases into five types, which include growth suppression, discoloration, deformation, necrosis and impaired reproduction.

Another type of infectious agent are viroids, taxonomically belonging to the viruses' group, but they lack a capsid protein coat and possess a single-stranded RNA molecule 50–80 times smaller than the viral genome. They replicate via a rolling circle mechanism and utilize host enzyme machinery for their replication (Nohales et al., 2012). Stunted growth of the plant, discoloration and deformed organs are the main symptoms of viroid diseases (Shkalikov et al., 2010).

Bacteria are omnipresent prokaryotic cellular organisms with chromosomal DNA as their genetic element. It also possesses extra chromosomal circular DNA called the plasmid, which carries important antibiotic resistance marker genes as well as virulence factors. Bacteria belonging to most of the bacterial genera are devastating plant pathogens. For example, *Xanthomonas* bacteria cause at least 350 different plant diseases (Leyns et al., 1984). There is a twofold increase in the prevalence of bacterial diseases with an increase in the average daily temperature of 3°C–4°C annually, while the prevalence of plant infection grows by 30%–50% (Ignatov et al., 2015). The main symptoms of bacterial diseases include wilting, necrosis, chlorosis, rot, overgrowth (galls) and scab.

Fungi, unlike bacteria, are eukaryotic microorganisms with complex cell structures and chitin-rich cell walls, whose genome size is smaller among almost all eukaryotes but larger than that of prokaryotes. Generally, fungi grow by forming a thread-like thinly branched hyphae called mycelium that may be septate or aseptate (Zeilinger et al., 2016). Fungi, as phytopathogens, can infect plants at all stages, and it is more difficult to get rid of the infection once it gets established in the host plants. Phytopathogenic fungi are the most dangerous plant pathogens to cause harm at all stages of crop production. About 10,000 fungal species are identified to cause disease in plants (Hussain and Usman, 2019), and undoubtfully, fungus causes more harm to crop plants than any other phytopathogen. Shkalikov et al. (2010) listed the symptoms of fungal infection in the plants as wilting, spotting, formation of moulds and pustules, rot, overgrowth, deformation and mummification.

8.1.3 Crop Losses due to Diseases

Savary et al. (2019) documented the worldwide yield losses associated with 137 pathogens and pests at the global level and in food security hotspots for five major crops (wheat, rice, maize, potato and soybean). The range of yield loss at the global level is estimated as 21.5%, 30.0%, 22.5%, 17.2% and 21.4% for wheat, rice, maize, potato and soybean, respectively. Similarly, the range of yield loss in food security hotspots is 10.1%–28.1%, 24.6%–40.9%, 19.5%–41.1%, 8.1%–21.0% and 11.0%–32.4% for wheat, rice, maize, potato and soybean, respectively. From these data, it is clearly known that the highest losses are associated with areas where pests and diseases are frequently emerging and re-emerging, as well as areas with a higher population. For pathogen incidence and multiplication inside the host cell, it requires a favourable environment, and unfortunately, in recent years, the prevailing climate change and changing agricultural practices have coincided with the environment that is conducive to the growth of pathogens. For example, the advent of modern technologies such as monoculture practices helped to produce more food on less land, but at the same time, the crop became more susceptible to infectious pathogens, and due to this, about 30% of losses were recorded during storage, transport and distribution. Moreover, due to the use of non-selective pesticides and fungicides, the beneficial organisms get killed along with the pathogens, which severely upsets the ecological balance prevailing in nature.

Thus, to meet the food requirements of the world's growing population, the use of disease-resistant crop varieties can be a sustainable and environmentally friendly responsible approach. However, the conventional plant breeding methods used traditionally are labour-intensive and time-consuming. Besides, the resistance imparted by the introduction of the resistance (*R*) gene tends to break down after a few years (boom-and-bust cycle). With the arrival of genome-editing technologies, especially the CRISPR-Cas9 (Clustered Regularly Interspaced Short Palindromic Repeats-Cas9) system, one can precisely edit the plant genome to introduce improved crop traits rapidly. Furthermore, with these CRISPR methodologies, one can target and successfully knock out the susceptibility genes (*S*), thereby generating disease resistance in high-yielding crop plants. Unlike the phenomenon of resistance frequent breakdown as in the case of *R* gene-mediated resistance, the resistance imparted by the correction of *S* genes tends to impart long-term protection. This chapter summarizes the mechanism of disease resistance by the host plants and how the susceptibility genes are targeted for generating disease-resistance cultivars against bacterial, fungal and viral diseases.

8.2 GENETIC COMPONENTS OF RESISTANCE AND SUSCEPTIBILITY

Plants are continuously exposed to numerous pathogens but are generally capable of resisting most of the pathogen attacks most of the time. So, resistance can be considered as a rule, whereas disease is an exception. Unlike animals, plants lack central antibody-directed disease-resistance mechanisms. The pioneering work on understanding disease-resistance strategies in plants was laid down by the study of "inheritance of reaction to rust in flax" by Flor (1947). Subsequently, Flor's gene-for-gene hypothesis established the foundation for elucidating the genetic basis of plant disease resistance. Presently, we know that pants possess two distinct and genetically unlinked but partially overlapping defence strategies against pathogens. The first one is tolerance, which is achieved by a sequence of events that make the interior of the host cell inhospitable to the pathogen without inducing a hypersensitive response (HR). The second is resistance, which is mediated by a very specific interaction between resistance protein (R) encoded by the host and avirulence protein (Avr) encoded by the pathogen, which initiates a cascade of reactions inside the host cell that culminates in HR-mediated programmed cell death (PCD) of the infected host cell. It is further important to remember that "tolerance" is basically a host-driven phenomenon that does not impart complete protection against the infection but maintains the infection at a manageable level so that the normal physiology of the plant is not disturbed too much. "Tolerance" imparts protection of relatively lower amplitude, but the protection is of a broad spectrum and also limits pathogenic variant development. Resistance, on the other hand, is a phenomenon involving genetic factors from both the host (*R* gene) and the pathogen (*Avr* gene). It is capable of terminating the infection (pathogen spread in the host) and providing a high level of protection. However, the protection is fragile and is frequently broken down by mutant pathogenic variants.

The high level of resistance in the plant is achieved by a multi-layered defence system (Figure 8.2) and mostly relies on the signal transduction pathways for mounting defence responses against the invading pathogens (Walbot, 1985). The first layer of defence is constituted by preformed and induced barriers of pathogen invasion and establishment. This includes cell wall apposition (CWA), secondary metabolites and pathogenesis-related (PR) proteins. The second layer of defence includes interactions between resistance (*R*) and avirulence (Avr) proteins, which induce HR-mediated

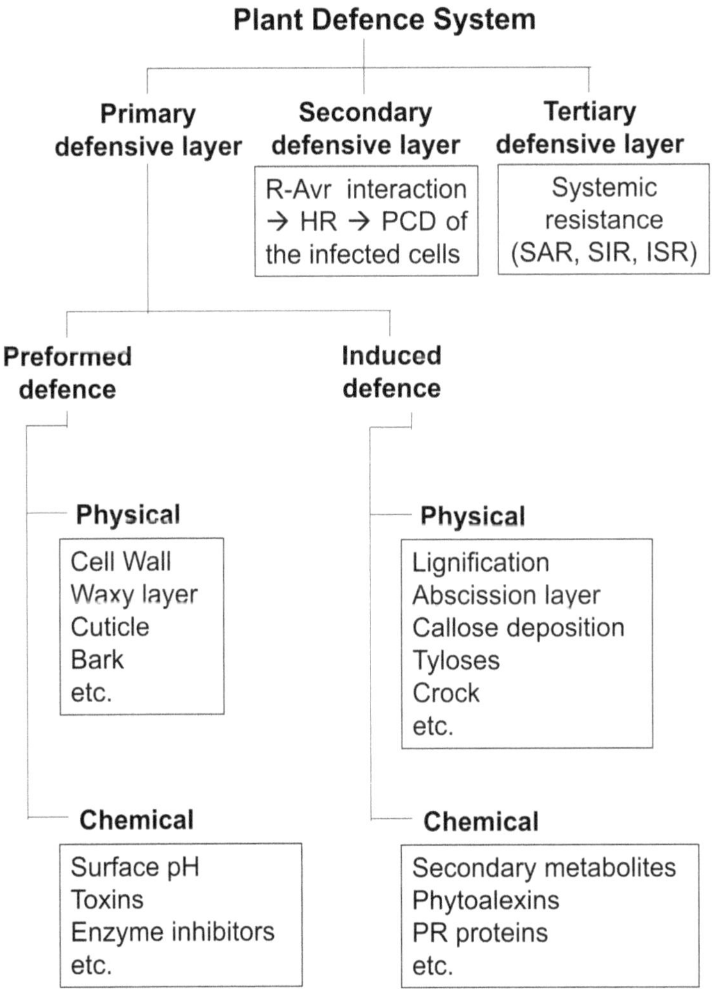

FIGURE 8.2 Multi-layered defence system of plants. It includes three tyre defence mechanisms, among which the first tyre is responsible for "tolerance" and the second tyre is responsible for "resistance".

death of the infected cell. The third layer of defence, which is mediated through different types of acquired resistance that are raised in the whole plant system, is systemic in nature. The latest model of plant defence against biotic stress – the "zig-zag model" – has been proposed by Jones and Dangl (2006) (Figure 8.3). According to this model, pathogens carry specific signature elements, which are called pathogen-associated molecular patterns (PAMPs), such as flagellin protein and any cell wall component. These are recognized by pattern recognition receptors (PPRs) present in the plants, and a primary defence response is developed, which is called PAMP-triggered immunity (PTI). This is not a very high level of defence and can be broken down by the potential pathogens, which secrete some pathogen effector molecules make the plant susceptible. This is called effector-triggered susceptibility (ETS). These effectors are Avr gene products. So the resistant host, which has the corresponding *R* gene, can recognize these effectors by the *R* gene product, and a high level of defence response arises, which is called effector-triggered immunity (ETI). Now the pathogen is under evolutionary pressure to survive on the host plant. So it tries to shade its old effector and develop a new and more potent effector, and if it

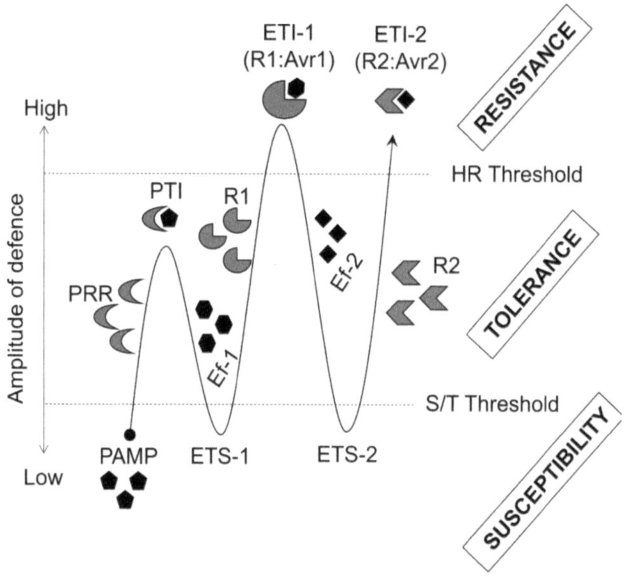

FIGURE 8.3 Zigzac model of plant defence. Pathogens carry specific signature elements called PAMPs. PAMPs are recognized by PRRs present in the plants, and a primary defence response is developed which is called PAMP-triggered immunity (PTI). This is not a very high level of defence (tolerance) and can be overridden by the virulent pathogens, which secrete some pathogen effector (Ef) molecules and make the plant susceptible. This phenomenon is termed as ETS. The effectors can be recognized by specific R proteins. That particular effector is now termed as AVR-protein. The R-AVR interaction imparts high-level resistance to the host, mediated by HR-induced PCD – a defence response termed as ETI. Mutations in AVR disrupt the R-AVR interaction, and ETS reoccurs, while mutations in the *R* gene can re-establish the R-AVR interaction, initiating a new round of ETI. Thus, the co-evolution of R- and AVR-proteins continues.

gets success in doing so, then it again lowers the level of defence below the threshold of susceptibility, and again the plant becomes susceptible to that pathogen attack. Now that the plant senses the evolutionary pressure to survive from the attack of that pathogen and responds to that, it also tries to modify its *R* gene product so that it can recognize the new Avr product. If the plant successfully does so, then the defence level again rises above the threshold of resistance, and again the plant becomes resistant to pathogen attack. Likewise, the evolutionary struggle goes on between plants and their pathogens. According to this scheme, the amplitude of disease resistance or susceptibility is proportional to [PTI − ETS + ETI]. If this value for a particular host–pathogen interaction is positive, then the plant will show a resistance response, whereas if this value is negative, then the outcome will be susceptibility.

So as we have seen so far, disease resistance in plants is mediated by resistance genes (*R*), which trigger ETI and are effective against pathogen invasion and activation of signalling cascades for disease resistance. Cloned R proteins from plants are categorized into nine major classes based on their structural motifs and cellular and subcellular localization (Hammond-Kosack and Parker, 2003). Most common types of R proteins consist of extracellular leucine-rich repeats (LRRs), an intracellular serine/threonine kinase domain and a transmembrane domain (Hammond-Kosack and Kanyuka, 2007). These domains confer specificity to some of the R proteins such that their interaction is limited to the respective effector proteins of specific races of pathogens only, as seen in the race-specific interaction of the Xa27 *R* gene from rice against the effector protein AvrXa27 from *Xanthomonas oryzae* pv. *oryzae* (Gu et al., 2005), while some R proteins can interact in a race-independent manner (Table 8.1). Susceptibility genes (*S*), on the other hand, express proteins that help in host recognition, successful infection, penetration and disease development in host tissue, and mutating the same causes disease resistance. Genes involving basic compatibility of host and pathogens, cell wall and cuticle modification, membrane penetration, stomatal opening, haustoria establishment, vesicular trafficking, salicylic

TABLE 8.1
List of Resistance (R) and Susceptibility (S) Genes Involved in Mechanism of Plant Disease Resistance

Components of Susceptibility (Susceptibility Factors)	Components of Resistance (Resistance Factors)
CESA; cellulose synthase	WRKY; transcription factor
PMR4; powdery mildew resistance 4	MEKK; MAP kinase kinase kinase
SPL11; E3 Ubiquitin ligase	MKK; MAP kinase kinase
PUB22; E3 Ubiquitin ligase	SA; salicylic acid
S3H; salicylic acid-3-hydroxylase	NLR; nucleotide binding leucine-rich repeat (R-protein)
RIN4; RPM1 (R-gene)-interacting 4	BI-1; BAX inhibitor-1
EDR1; enhanced disease resistance 1	MLO; mildew resistance locus O
ACD11; accelerated cell death 11	LFG; lifeguard
BAP1; calcium dependent lipid binding protein	Pattern recognition receptor (PRR)
SR1; calmodulin-binding transcription factor	MPK3/6; MAP kinase 3/6

acid (SA) catabolism, suppressors of PTI, calcium and calmodulin-binding proteins, endoreduplication and cell expansion, sugar transport and virus replication and movement are some of the major susceptibility factors required for overall disease development (Eckardt, 2002) (Table 8.1).

8.3 APPROACHES TO IMPART DISEASE RESISTANCE IN CROPS THAT DO NOT INVOLVE GENOME EDITING

Evolution is an indispensable part of all beings on earth, extending from prokaryotes to eukaryotes. The most evident example can be witnessed in relation to plant pathogens, particularly bacteria, fungi and viruses, which over the course of time have evolved themselves at the genomic level, thereby imparting them resistance to pesticides and trait-specific genetic pyramiding carried out using conventional as well as molecular-assisted breeding approaches. This has made crop species such as members of the poaceae family, horticulture species and several oilseed crops vulnerable to biotic stresses, resulting in a 20%–40% loss of global agricultural productivity. Many strategies are used in the field to combat these diseases.

8.3.1 Conventional Breeding Approach for Disease Resistance

Improving crop disease resistance through plant breeding has always been a cost-effective and environmentally sound way to reduce the damage caused by plant diseases. The main objective of a crop disease-resistance breeding programme is to accumulate favourable gene(s) in elite cultivars to improve disease resistance in a given scenario. To date, many plant disease-resistance breeding programmes have been successful in developing broad-spectrum and sustainable disease resistance.

In conventional plant breeding, dominant or semi-dominant disease-resistance genes (*R* genes) are introduced into elite cultivars by genetic crosses with donor genotype/wild relatives. The introduction of single/multiple *R* genes in numerous combinations helped achieve durable resistance toward invading microbes and pests. Susceptibility genes (*S* genes) are negative regulators of plant immunity or encode for proteins involved in pathogen-host plant interaction. Mutations in *S* genes can suppress immune responses or inhibit pathogen growth. *S*-gene-mediated disease resistance is more stable compared to *R* genes. Introgression of the non-functional susceptibility gene in rice (pi21) and barley (*mlo*) conferred resistance against the blast fungus *Magnaporthe oryzae* (syn. Pyricularia oryzae) and powdery mildew (caused by *Blumeria graminis* f.sp. *hordei*), respectively.

However, it is challenging to breed disease-resistant varieties with effective, stable and broad-spectrum disease resistance. Breeding for long-lasting resistance is critical in many plant-pathogen systems and typically involves combining multiple and diverse resistance genes. The breeding strategies currently used to improve disease resistance are time-, money- and labour-consuming. The biodiversity/race of crop pests and pathogens has been continuously altering faster. Thus, it will be further challenging to breed resistant varieties. Moreover, the breeding programme's success depends on the source of resistance genes (landraces, related species, mutations and wild relatives), which are limited. There are also resistance breakdown problems (vertifolia effect, boom-and-bust cycle) and horizontal resistance (incomplete but stable protection). Linkage drag is another major problem that requires multiple backcrossing and is sometimes associated with undesired effects on yield (e.g., wheat leaf rust gene Lr34 causes a 5% reduction in grain yield). Under the scenario mentioned above, the disease-resistance breeding programme may need to be complemented with new technologies.

8.3.2 Transgenic and RNAi Approaches for Disease Resistance

Genetic engineering technology allows for the interchange of genetic material across species. The plants developed by genetic engineering contain transgenes. They are classified as genetically modified organisms. Different strategies have been adopted for boosting resistance against diseases in transgenic plants, such as detoxification of pathogen virulence factors; overexpressing *R* genes and pathogenesis-related (PR) genes increasing structural barriers and modification of defence-signalling pathways (Liang et al., 2000). However, genetically modified crops developed through transgenic breeding have biosafety concerns, a regulatory burden and very low acceptability among the public. Till now, only a few disease-resistant transgenic plants are commercially released, which includes a squash variety resistant against three different mosaic viruses and the Hawaiian papaya Rainbow and Sun-Up resistant to Papaya ringspot virus (PRSV). In many countries, cultivation of transgenic crops is banned or limited till now.

The transgenic RNA interference (RNAi) approach also has potential for controlling plant diseases. RNAi targets RNA/mRNAs in a sequence-specific manner and then represses their expression either by ribonuclease-directed degradation or by preventing their translation. RNAi has proven to be a practical approach to controlling plant virus diseases for many different viruses in many plant species. Some of the successful examples are transgenic-RNAi virus-resistant papaya (PRSV), Destiny III yellow crookneck squash (*Cucurbita pepo*) having resistance against zucchini yellow mosaic virus (ZYMV), watermelon mosaic virus and cucumber mosaic virus (CMV), cassava (*Manihot esculenta*) having resistance to cassava black streak disease (Patil et al., 2015) and many more. Although RNAi technology offers great potential for controlling plant virus diseases even when no natural resistance sources are known, the technology could not gain commercial importance.

8.4 GENOME-EDITING APPROACH FOR DISEASE RESISTANCE

The emergence of genome-editing technologies such as Zinc Finger Nucleases (ZFNs), Transcription Activator-Like Effector Nucleases (TALENs) (Voytas et al., 2013) and Clustered Regulatory Interspaced Short Palindromic Repeats)/CRISPR-associated (Cas) has revolutionized plant research as it became possible to create site-specific genetic modification to modulate the gene function.

8.4.1 Zinc Finger Nucleases and Transcription Activator-Like Effector Nucleases

ZFNs are the first engineered endonucleases to recognize and cleave chromosomal DNA. Each ZFN comprises a zinc finger protein DNA-binding domain at the N terminus and the FokI nuclease domain at the C terminus. The DNA-binding domain is fused to the catalytic site of a FokI endonuclease, which allows DNA recognition and target cleavage. It is functional in dimeric form, and each of the "zinc finger" domains recognizes the target site in the triplet of nucleotides (Osakabe et al., 2010, Curtin et al., 2011; Persikov et al., 2015). ZFNs have been used to create gene modifications in *Arabidopsis*, rice, tobacco, maize, soybean and canola (Wright et al., 2005; Townsend et al., 2009; Osakabe et al., 2010; Petolino et al., 2010; Zhang et al., 2010; Even-Faitelson et al., 2011; Qi et al., 2013a; Shukla et al., 2009; Curtin et al., 2011; Gupta et al., 2012; Ainley et al., 2013). The artificial zinc finger proteins were used to create modifications at specific sites in the begomovirus DNA (*Tomato yellow leaf curl China virus* (TYLCCNV) and *Tobacco curly shoot virus*) (Chen et al., 2014).

Further investigation led to the development of TALENs. TALEs are engineered from naturally occurring TAL effector proteins from the plant pathogenic bacteria (*Xanthomonas*). They consist of an artificially fused, customized array of TALEs (DNA-binding domain) with FokI nuclease (cleavage domain). DNA-binding domains comprise 13–28 copies of 33–35 amino acid repeat domains, and each domain recognizes a single nucleotide base. Repeat-variable di-residues determine TALE specificity at positions 12 and 13. TALENs have been successfully used to modify endogenous genes in crops such as rice, maize, *Arabidopsis*, soybean and barley. For disease resistance, TALENs were used in rice to produce plants with disease resistance against *Xanthomonas oryzae* pv. *oryzae* by editing its target sequence of TALE effectors in rice (Zhou et al., 2015). In wheat, disease resistance to powdery mildew was conferred by targeting three simultaneous knockouts of three homoallele Ta*MLO* (*A1, B1, D1*). Broad-spectrum resistance against begomoviruses has been evaluated by using TALEN.

However, some shortcomings are associated with ZFN and TALEN, such as complex design, synthesis and validation, high off-targets and high cost. Moreover, it is even more challenging to achieve multiple gene editing as many ZFNs or TALENs would be required simultaneously, which will be a tedious task.

8.4.2 CRISPR/Cas9

The CRISPR/Cas9 system of genome editing has emerged as transformative, as these tools dramatically expanded the ability to manipulate plant genomes and allowed us to understand the genetic causes behind many diseases. Different from site-specific nucleases such as ZFNs, TALENs and meganucleases, the CRISPR/Cas system has emerged as a more efficient, cost-effective and easy tool for inducing targeted genetic alterations. Unlike ZFNs and TALENs, CRISPR/Cas9 is independent of any protein engineering steps and can be retargeted to new DNA sequences by simply changing the sequence of single-guide RNA. The natural CRISPR system provides acquired immunity against invading foreign DNA via RNA-guided DNA cleavage in bacteria. The bacterial CRISPR/Cas system consists of short CRISPR RNA (crRNA), which is annealed to trans-activating crRNAs (tracrRNAs), and direct sequence-specific cleavage and silencing of pathogenic DNA by Cas proteins. The modified CRISPR-Cas9 system consists of a customized small single-guide RNA (sgRNA) that directs Cas9 nuclease at a specific target site in the genome. Cas9 makes a double-stranded break in the target DNA near the adjacent protospacer motif (PAM) sequence. These double-stranded DNA breaks are subsequently repaired by the homologous recombination or non-homologous end-joining (NHEJ) repair mechanism in the cell (Chen et al., 2019). The NHEJ repair is the error-prone pathway that creates random insertions and deletions (indels) and results in frameshift mutations and targeted gene knockouts.

The most widely used SpCas9 (*Streptococcus pyogenes*) is generally limited to performing genome editing at sites with the PAM sequence "NGG". In order to expand the applicability of genome editing in any region of a genome, numerous Cas9 orthologs and engineered variants of Cas9 have been developed with different PAMs, such as *Staphylococcus aureus* Cas9 (SaCas9), *Neisseria meningitides* (NmCas9) and Cas9-VQR (D1135V/R1335Q/T1337R). Furthermore, engineered variants of Cas9 enzymes expanded the scope of targeted genome editing as they could identify different PAMs, i.e., EQR-Cas9 (NGAG PAM), VRER-Cas9 (NGCG PAM), VQR-Cas9 (NGA PAM), xCas9 (NG, GAA and GTA PAM), SpCas9-NG (NG PAM) and SaKKH-Cas9 (NNNRRT PAM) (Kleinstiver et al., 2015). With the availability of all these variants, the CRISPR/Cas system can be modified easily to retarget or cleave virtually any DNA sequence by re-designing the crRNA. Furthermore, the development of CRISPR/Cas9-based multiplex genome-editing systems allows for targeting multiplexed gene editing. To further improve the multiplex genome-editing efficiency, the tRNA-processing system was used to generate multiple mRNAs from a single polycistronic gene using an endogenous tRNA-processing system that precisely cleaves both ends of the tRNA precursor. This system could boost the targeting and multiplex editing capabilities of the CRISPR/Cas9 system.

Depending on the method of genome editing, CRISPR-Cas genome editing can be grouped into three broad categories: site-directed nuclease-1 (SDN1), SDN2 and SDN3 (Figure 8.4). SDN1 involves the unguided repair of a targeted double-strand DNA break (DSB) by the natural endogenous DNA repair mechanism of the host organism, such as NHEJ. The spontaneous repair of this break can lead to a mutation causing gene silencing, gene knockout or a change in the activity of a gene. These mutations can be base substitution, indels or deletions including large deletions or structural changes. These resultant mutations are comparable to those occurring in nature, obtained through conventional mutagenic treatments or natural variation found in primary and secondary gene pools. SDN2 involves a template-guided repair of a targeted DSB using an externally supplied donor sequence. The donor carries one or several small mutations flanked by two sequences matching both ends of the DSB and is thus recognized as a repair template, allowing the introduction of the mutation(s) at the target site. The resultant mutant carries a modified sequence, leading to an altered expression profile of the gene and/or altered activity of the encoded protein/RNA. Thus, the edited version could be regarded as an allelic form comparable to those available in the primary or secondary gene pool. In the SDN3 approach, the repair template provided is a foreign DNA flake in both ends with native DNA for HR. SDN3 mutants are transgenic. The SDN3 approach is useful for targeting the foreign to the safe harbour loci in the host plant genome so that the native gene is not disrupted due to foreign gene insertion and the foreign gene can be targeted to a highly expressed region of the genome.

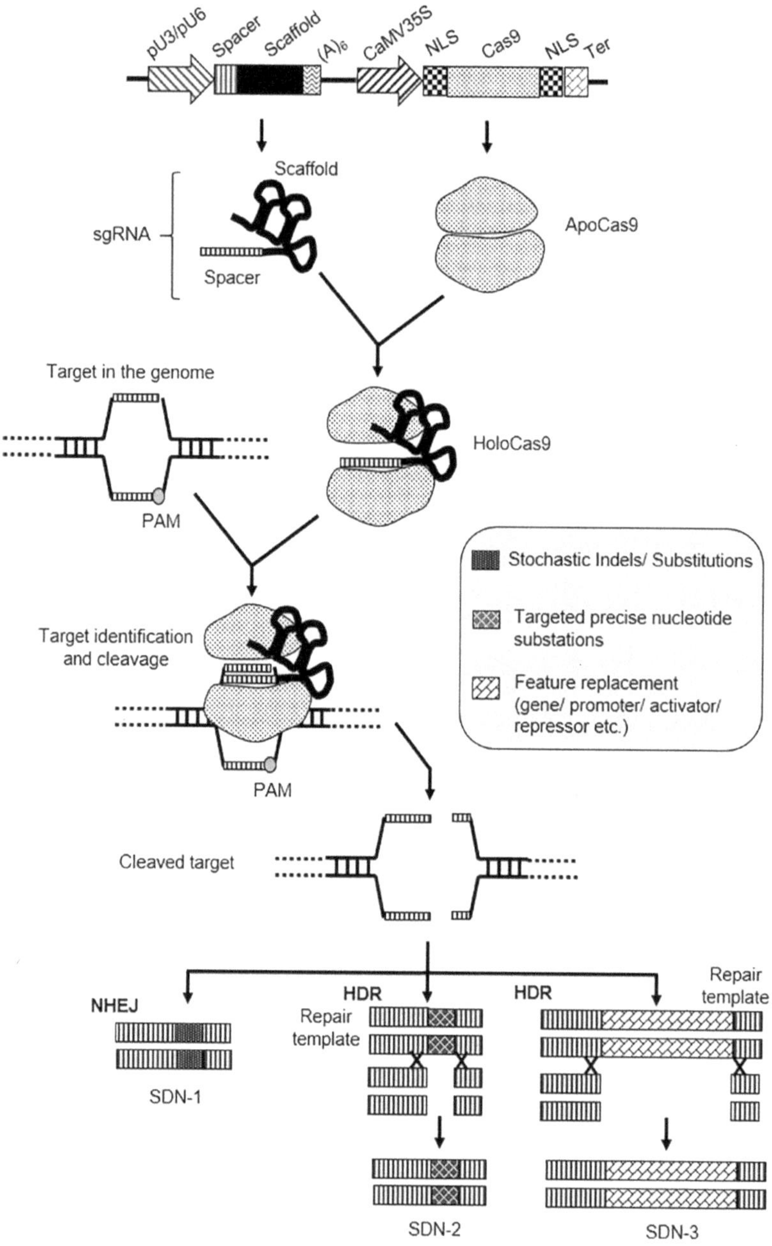

FIGURE 8.4 Mechanism of CRISPR-Cas-mediated genome editing and categories of mutations. The genome-editing reagents, namely (1) the molecular scissors "Cas enzyme" – endonuclease and (2) single-guide RNA (sgRNA), are delivered into the plant cell through agrobacterium-mediated/biolistic genetic transformation/protoplast transformation with RNP. The sgRNA guides Cas enzyme to the target site, and the Cas enzyme makes a double-strand cut at three nucleotides upstream to the PAM (protospacer adjacent motif) site. The double-stranded break is repaired by NHEJ mechanism or homology-depended repair (HDR).

Genome editing has emerged as a New Breeding Technology (NBT) to instantly correct the genetic defects in elite popular cultivars and improve them in a shorter time as compared with that required under conventional breeding and marker-assisted breeding. The CRISPR/Cas9-mediated genome editing and precision mutagenesis system have emerged as new breeding technologies. Globally, several countries are intensively working on the development of products using gene editing. Until 2020, about 231 market-oriented products in agriculture using gene editing have been published. During 2015–2020, transgene-free genome-edited crops have been given non-regulated status in countries, viz. the USA, Australia, Japan, Argentina, Brazil, Colombia, Chile, Guatemala, Honduras, Paraguay, Ecuador, Israel, Kenya, Nigeria, the Philippines and China (Buchholzer and Frommer, 2022). India has also exempted transgene-free genome-edited lines developed by SDN1 and SDN2 approaches from the Rules 1989 vide in its office Memorandum F.No. C-12013/3/2020-CS-III, dated 30th March 2022. The Review Committee on Genetic Manipulation (RCGM), DBT, Government of India has published "Guidelines for the Safety Assessment of Genome-Edited Plants, 2022" vide in its office Memorandum File No. PID-15011/1/2022-PPB-DBT, dated 17 May 2022. Department of Biotechnology, Ministry of Science & Technology, Government of India has also published "Standard Operating Procedures for Regulatory Review of Genome-Edited Plants under SDN-1 and SDN-2 Categories" wide File No. PID-15011/1/22-PPB-DBT dated 4 October 2022.

The USA has given non-regulated status to 168 genome-edited products. Genome-edited canola in the USA and tomatoes and fishes (tiger puffer and red sea bream) in Japan are marketed for consumption.

In plant pathology, CRISPR-Cas9 genome editing is exploited to develop disease resistance against DNA viruses (Ali et al., 2015; Baltes et al., 2015), RNA viruses (Chandrasekaran et al., 2016; Aman et al., 2018), bacteria (Peng et al., 2017) and fungi in crops such as *Arabidopsis*, rice, wheat, tomato, tobacco and grapes. Together, the above-discussed CRISPR-Cas technologies promise to expand our ability to explore and alter any genome and constitute a new and promising paradigm to understand and treat disease. The CRISPR-mediated disease resistance can be achieved either by directly targeting the editing of the pathogen genome with a system (Ali et al., 2015; Aman et al., 2018; Ali et al., 2016) or by targeting host genes to interfere with pathogen vulnerability. Among these two strategies, the targeting host gene allows the development of disease-resistant plants in a transgene-free manner. Among the host genes, targeted editing of susceptibility (S) genes and their promoters using CRISPR offers the opportunity to introduce durable pathogen resistance. Susceptibility genes are required for successful pathogen infection. Thus, they are considered essential for compatible plant-pathogen interactions. Disruption in the function of the *S* gene can confer broad-spectrum disease resistance in many plant species (Figure 8.5). The CRISPR–Cas9 system has been successfully used to confer bacterial blight resistance by a targeted mutation in the promoter regions of *S* genes *OsSWEET14* and *OsSWEET11,* disease resistance against downy mildew by targeting Mildew Resistance Locus (MLO) proteins (Thomazella et al., 2016; Malnoy et al., 2016), Citrus canker resistance by targeting the promoter of *canker susceptibility lateral organ boundaries 1* (*CsLOB1*) gene (Jia et al., 2016) and potyvirus resistance by

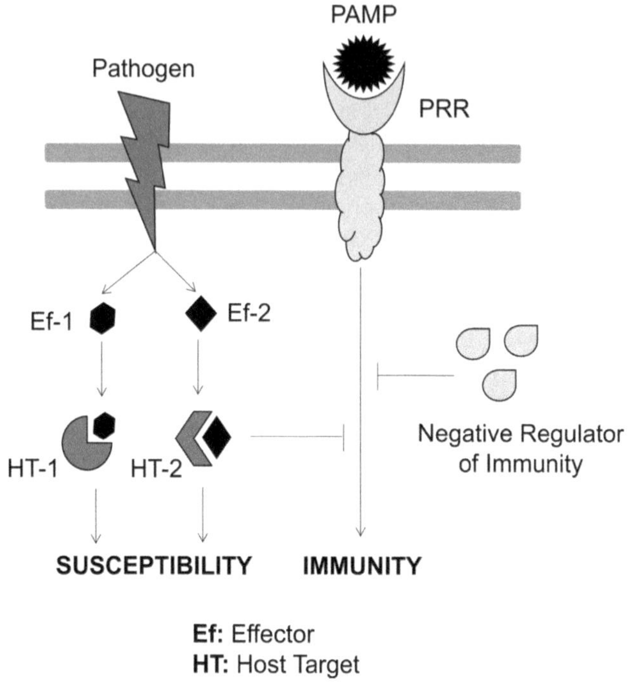

FIGURE 8.5 Functional categorization of *S*-gene-encoded proteins. There are three different groups of S-proteins (encoded by corresponding *S* genes). Firstly (left), the proteins involved in normal growth and development of the plant are the targets of pathogen-secreted effectors to subvert the normal physiology of the plant. Secondly (middle), the inducer/activator of the plant immune system, which are the targets of pathogen-secreted effectors to down-regulate immune response during pathogen infection. Third (right), the suppressors of the immune system present in plants suppress unintended immune reactions to support normal growth and development when there is no threat of infection or disease.

disrupting *eukaryotic translation initiation factor 4E* (*eIF4E*). The transgene-free tomato line resists powdery mildew infections (Nekrasov et al., 2017).

Over the ZFN and TALEN site-directed nucleases, the CRISPR/Cas9 gene-editing technology is easy to implement, cost-effective, confers target-specific editing and allows multiplex editing more efficiently. Using this technique, genome-edited transgene-free plants can be generated. Furthermore, continuous development in CRISPR-Cas-based technologies increases their translational potential as they allow for targeted editing of any region. The generation of transgene-free, edited varieties offers high potential for using resistance lines as donors in plant breeding. Furthermore, continued development of CRISPR-based genome-editing tools such as **CRISPR-**Cas12a/Cpf1, dCAs9, nCas9, base editing, prime editing, gene insertion and deletion might accelerate the pace of crop improvement. Genome-editing technologies are now being successfully employed to develop crop species resistant to pathogen attack. Advancements in genome-editing tools have opened new horizons accomplishing resistance in crops. In recent years, the CRISPR/Cas system has been exploited enormously to respond to several agricultural challenges, including the goal of imparting tolerance to several biotic factors.

8.5 ACHIEVEMENTS OF CRISPR-CAS-MEDIATED GENOME EDITING FOR DISEASE RESISTANCE

In the following subsections, we will be discussing some of the success stories demonstrating the utility of CRISPR-Cas-mediated genome-editing technology in the field of plant disease resistance.

8.5.1 CRISPR-Cas9-Mediated Genome Editing for Combating Bacterial Diseases

Plant diseases caused by bacterial pathogens are major constraints on crop production and productivity and cause significant annual losses on a global scale. Some of the major drawbacks to controlling bacterial diseases are due to constant genetic evolution, absence or limited chemical control measures and late detection of diseases such as blights, mosaics and spots (Vale et al., 2001). Phytopathogenic bacteria such as spots, mosaics and rots impose a significant threat to agricultural production and productivity (Zeng et al., 2010). Bacteria get entry to the plant cell either through natural means such as the stomatal pore and stigma, by secreting some bioactive molecules or by mechanical wounds (Zeng et al., 2010). Under pathogen invasion, the plant activates a defence-related signalling response to eliminate the bacteria, but in some cases, pathogens use complex signalling cascades to produce resistance to plant defence mechanisms, and in other cases, pathogens use host susceptibility (S) genes commonly known as *S* genes to initiate infection (Dong and Ronald, 2019; Dominguez et al., 2016; Das et al., 2019). These *S* genes, which work as negative regulators, can be potential targets for CRISPR-Cas9-mediated editing to achieve plant resistance to bacterial diseases.

Most of the research has focused on the host *S* gene, which serves as a negative regulator for resistance to bacterial infection. Bacterial pathogens secrete effector proteins that can induce plant susceptibility. These effector proteins physically interact and activate the transcription of *S* genes (Hui et al., 2019; Zhang et al., 2019; Oh and Beer, 2007). During pathogen invasion, host plant *S* genes are upregulated by the effector proteins (Kay et al., 2007; Boch and Bonas, 2010). Introgression of natural mutant alleles of *S* genes into elite cultivars or the generation of EMS populations is time-consuming and may introduce unwanted changes into the elite background. Nevertheless, the development of new genetic engineering tools, especially CRISPR-Cas9 genome-editing technology, has been used in the study of multiple *S* gene mutations (Table 8.2).

In rice, studies have been conducted to knock out some important *S* genes, such as SWEET family genes, which play a role in sucrose and glucose efflux under bacterial invasion. Resistance against the bacterial blight pathogen *Xanthomonas oryzae* pv. Oryzae was successfully attained by using CRISPR-Cas9 mutating *OsSWEET13*

TABLE 8.2
CRISPR-Cas9 Genome Editing in Plants to Confer Bacterial Disease Resistance

Plant Species	Bacterial Pathogen	Disease	Target Gene	Target Gene Location	Target Gene Function	References
Oryza sativa	*Xanthomonas oryzae* pv. *Oryzae*	Bacterial blight	*OsSWEET14, OsSWEET11*	Host *S* gene	Sucrose transporter gene	Jiang et al. (2013)
	Xanthomonas oryzae pv. *Oryzae*	Bacterial blight	*SWEET11, SWEET13 and SWEET14*	Host *S* gene	Sucrose transporter gene	Oliva et al. (2019)
	Xanthomonas oryzae pv. *Oryzae*	Bacterial blight	*SWEET13*	Host *S* gene	Sucrose transporter gene	Zhou et al. (2015)
	Burkholderia glumae	Bacterial blight	*OsMPK5*	Host *S* gene	Help in pathogeninfection	Xie and Yang (2013)
Solanum lycopersicum	*Pseudomonas syringae* pv. *tomato*	Bacterial Speck	*SlJAZ2*	Host *S* gene	Coronatine production	Ortigosa et al. (2019)
	Pseudomonas syringae pv. *tomato*	Bacterial blight	*SlDMR6-1*	Host *S* gene	Regulation of PR genes	Thomazella et al. (2021)
	Xanthomonas spp.	Bacterial blight	*SlDMR6-1*	Host *S* gene	Regulation of PR genes	Thomazella et al. (2021)
Citrus spp.	*Xanthomonas citri* subsp. *citri*	Citrus canker	*CsLOB1*	Host *S* gene	Help in pathogeninfection	Peng et al. (2017) and Jia et al. (2016)
Vitis vinifera	*Erwinia amylovora*	Fire blight	*DIPM-1, DIPM-2, DIPM-4*	Host *S* gene	Interact with pathogen effectors	Malnoy et al. (2016)
Malus (Golden delicious)	*Erwinia amylovora*	Fire blight	*DIPM-1, DIPM-2, DIPM-4*	Host *S* gene	Interact with pathogen effectors	Malnoy et al. (2016)

gene, which is involved in sucrose transport during pathogenesis (Zhou et al., 2014). It was reported that the *X. oryzae* effector *PthXo2* protein induces the expression of *OsSWEET13* gene. Knockout mutations of *OsSWEET13* have shown improved resistance against *X. oryzae* (Zhou et al., 2015). Other *S* genes, such as *SWEET11*, *SWEET13* and *SWEET14* in rice, were also edited by the CRISPR-Cas9 approach which generated broad-spectrum resistance to *Xanthomonas oryzae* pv. *Oryzae* (Oliva et al., 2019). The rice gene OsMPK5, which encodes a stress-responsive mitogen-activated protein kinase, a negative regulator of rice defence response, was also edited by the CRISPR-Cas9 mechanism (Xie and Yang, 2013).

It was observed that *AtJAZ2* acts as a co-receptor for coronatine (*COR*), resulting in the stomatal opening, which helps *P. syringae* invade the plant and initiate infection. In tomato, *Jasmonate ZIM-domain-2* ortholog (*SlJAZ2*) acts as a positive regulator for *P. syringae* pv. *Tomato* pathogen invasion. A CRISPR-Cas9-mediated mutation in this gene causes stomatal closure upon bacterial invasion and provides resistance against bacterial speck disease (Ortigosa et al., 2019). Whereas jasmonic acid (JA) defence-signalling system was not changed in *Sljaz21*j as mutant plants, which resulted in resistance to tomato grey mould disease caused by *Botrytis cinerea* (Gimenez-Ibanez et al., 2017). This study gives us ideas to specifically target important genes from the host genome to derive successful extrication of both (JA and SA) defence mechanisms and achieve broad-spectrum pathogen resistance. Similarly, two more tomato genes, viz., *SlDMR6-1* and *SlDMR6-1*, were edited using CRISPR-Cas9, resulting in knockout mutants that were resistant to the tomato bacterial blight disease caused by *Pseudomonas syringae* pv. *Tomato* and *Xanthomonas spp.* Resistance to these pathogens was correlated with an increased level of SA and transcriptional activation of immune responses (Thomazella et al., 2021). In apples and grapes, *DIPM* genes were targeted for CRISPR-Cas9-directed mutations to achieve resistance against fire blight disease. The causal organism of bacterial fire blight disease, *Erwinia amylovora*, produces the *DspE* effector protein, which interacts with the *DspE-interacting proteins* of Malus (*DIPM*) and leads to disease susceptibility (Gimenez-Ibanez et al., 2017). Thus, mutations in this gene caused resistance to bacterial fire blight in grapes and apples.

These results suggest a great possibility of using a CRISPR tool with a broad editing window by using a variety of CRISPR editing vectors that are precise in editing and can also be used to insert desired nucleotide changes by a homology-directed target DNA repair (HDR) mechanism. The CRISPR tool can be used in different types of crops or agriculturally important trees to develop a wide range of biotic stress resistance varieties in a short period of time compared to conventional breeding methods.

8.5.2 CRISPR-Cas9-Mediated Genome Editing for Fungal Disease Resistance

Downy mildew resistance 6 (DMR6) gene of *Arabidopsis* encodes the putative 2-oxoglutarate Fe(II)-dependent oxygenase (2OGO) superfamily responsible for susceptibility to downy mildew disease in *Arabidopsis*. *Fusarium graminearum* (Fg) causes Fusarium head blight (FHB) in *Arabidopsis*. FHB-resistant *Arabidopsis* mutants were developed by knocking out the At2OGO gene (Low et al., 2020). Loss of *At2OGO* function by CRISPR/Cas9 provides resistance to *Fusarium graminearum* by upregulating defence genes that are regulated by phytohormones such as SA, JA and ethylene pathways. Rice blast, the most destructive disease affecting rice globally, is caused by the filamentous fungus *Magnaporthe oryzae*. Plant ethylene responsive factors (ERF), a subfamily of the APETELA2/ethylene response factor (AP2/ERF) transcription factor superfamily in plants, are involved in multiple abiotic and biotic stress tolerances. CRISPR-Cas9 was used to target OsERF922 to provide resistance to *Magnaporthe oryzae* in rice. MILDEW RESISTANT LOCUS O (*MLO*) is *a* conserved gene found in both monocots and dicots that confers susceptibility to powdery mildew diseases. Malnoy et al. (2016) targeted *MLO-7* of the grape cultivar, and Nekrasov et al. (2017) targeted *SlMlo1* (encoding a membrane-associated protein with seven transmembrane domains conferring susceptibility to the fungal pathogen *Oidium neolycopersici*) of the tomato, thereby creating a powdery mildew-resistant grape cultivar and tomato, respectively. Powdery mildew resistant 4 (PMR4) is involved in callose synthesis at specific sites of fungal penetration. CRISPR/Cas-9-targeted tomato pmr4 mutants show reduced susceptibility to *Oidium neolycopersici*, Santillán Martínez et al. (2020). The Solyc08g075770 encodes a small transmembrane protein with putative phosphorylation and glycosylation sites; some putative complementation lines were 76R-like, suggesting that the tomato Solyc08g075770 functions in *Fusarium* wilt tolerance (Prihatna et al., 2018). Schuster et al. (2015) established a type II bacterial CRISPR-Cas9 system for effectively disturbing target genes (*bW2, bE1*) of the fungal pathogen of maize *Ustilago maydis*, which imparts resistance to *U. maydis*. Similarly, by inactivating the *DMR6* (*downy mildew resistance 6*) gene in tomato, the level of SA increased, and it showed resistance to various pathogens, which including *P. syringae*, *P. capsici* and *Xanthomonas* spp. (Thomazella et al. 2016). Wang et al. (2018a) developed a CRISPR-Cas9 RNP transformation system for targeted disruption of URA5- and URA3-generating uracil auxotroph mutants resistant to 5-fluoroorotic acid (5-FOA). Zhang et al. (2018a) found that in *Botrytis cinerea*-infected tomatoes, knockout mutants of *SIMAPK3* (*mitogen-activated protein kinases*) have reduced defence enzyme activity as well as increased ROS production. Hence, they proposed that *SIMAPK3* has a positive role in resistance to *B. cinerea via* regulation of ROS accumulation. In the cacao tree, Fister et al. (2018) targeted deletion of *Non-Expressor of Pathogenesis-Related 3* (*TcNPR3*) gene, which is the repressor of the plant defence response, resulting in increased resistance to infection with *Phytophthora tropicalis* (Table 8.3).

8.5.3 CRISPR-Cas9-Mediated Genome Editing for Managing Viral Diseases

In 2015, the first report of the use of CRISPR-Cas9 technology in developing plants against DNA viruses came into existence. Baltes et al. (2015) and Ji et al. (2015) reported resistance against geminiviruses such as *Beet severe curly top virus*

TABLE 8.3
Plants Edited to Develop Resistance against Fungal Pathogens

S.No.	Plant sp.	Fungal Host/ Pathogen	Disease	Gene Targeted	Approach	Outcome	Reference
1	*Arabidopsis thaliana*	*Fusarium graminearum (Fg)*	Fusarium head blight	*At2OGO*	Knockout	Resistance to FHB	Low et al. (2020)
2	*Brassica napus*	*Leptosphaeria maculans*	Blackleg disease	*HOS1 (Histidine kinase)*	Knockout	Increased resistance against pathogen	Idnurm et al. (2017)
3	*Brassica napus*	*Leptosphaeria maculans*	Blackleg disease	*AvrLM7*	Knockout	Increased virulence of the pathogen towards host plant	Zou et al. (2020)
4	*Glycine max*	*Phytophthora sojae*	Damping off	*Avr4/6*	Gene disruption/ replacement	Showed increased resistance	Fang et al. (2016)
5	*Oryza sativa*	*Magnaporthe oryzae*	Rice blast	*OsERF922*	Knockout	Improved resistance to blast pathogen	Wang et al. (2016)
6	*Oryza sativa*	Magnaporthe oryzae	Rice blast	*ALB1, RSY1*	Site-directed mutagenesis	RNP based CRISPR system to improve the resistance against pathogen	Foster et al. (2018)
7	*Oryza sativa*	Magnaporthe grisea	Fungal blast	*OsMPK5*	Knockout	Improved disease resistance	Xie et al. (2013)
8	*Oryza sativa*	*Ustilaginoidea virens*	False smut	*USTA, UvSLT2*	Gene substitution	Inhibited fungal growth	Liang et al. (2018)
9	*Solanum lycopersicum cv. moneymaker*	*Oidium neolycopersici*	Powdery mildew	*SlMlo1*	Knockout	Improved disease resistance	Nekrasov et al. (2017)
10	*Solanum lycopersicum*	*Fusarium oxysporum*	Wilt	*PKS4*	Site-directed mutagenesis	Improved resistance	Wang et al. (2018a)

(Continued)

TABLE 8.3 (*Continued*)
Plants Edited to Develop Resistance against Fungal Pathogens

S.No.	Plant sp.	Fungal Host/ Pathogen	Disease	Gene Targeted	Approach	Outcome	Reference
11	*Solanum lycopersicum*	*Phytophthora capsici*	Downy mildew	*DMR6*	Site-directed mutagenesis	Showed resistance against fungal pathogens	Thomazella et al. (2016)
12	*Solanum lycopersicum cv. 76R*	*Fusarium*	Wilt	*Solyc08g075770*	Knockout	Showed resistance to Fusarium wilt	Prihatna et al. (2018)
13	*Solanum lycopersicum cv. Ailsa Craig*	*Botrytis cinerea*	Grey mould	*SlMAPK3*	Knockout	Showed improved resistance via ROS accumulation and SA, JA defence pathways	Zhang et al. (2018a)
14	*Solanum lycopersicum cv. Moneymaker*	*Oidium neolycopersici*	Powdery mildew	*Powdery Mildew Resistance 4 (PMR4)*	Knockout	Resistance against powdery mildew	Santillán Martínez et al. (2020)
15	*Theobroma cacao*	*Phytophthora tropicalis*	Black pod disease	*TcNPR3*	Site-directed mutagenesis	Improved disease resistance	Fister et al. (2018)
16	*Triticum aestivum*	*Blumeriag aminis* f.sp. *tritici*	Powdery mildew	*MLO-A1, B1 and D1*	Knockout	Improved disease resistance	Wang et al. (2014)
17	*Triticum aestivum*	*Blumeriag aminis* f.sp. *tritici*	Powdery mildew	*TaEDR1*	Knockout	Improved disease resistance	Zhang et al. (2017)
18	*Vitis vinifera*	(Erysiphe necator)	Powdery mildew	MLO-7	Knockout	Resistance to powdery mildew	Malnoy et al. (2016)
19	*Zea mays*	*Ustilago maydis*	Corn smut	*bW2, bE1*	Site-directed mutagenesis	Investigated the virulence of pathogen	Schuster et al. (2015)

(BSCTV), tomato yellow leaf curl virus (TYLCV) and *bean yellow drawf virus (BeYDV)* in model plants, e.g., *Arabidopsis thaliana* and *Nicotiana benthamiana.* Most ssDNA geminivirus genomes have been targeted by CRISPR-Cas9 technology (Ali et al., 2015). Baltes et al. (2015) employed 11 sgRNAs targeting Rep motifs, Rep-binding sites, hairpins and the intergenic sequence of *Bean yellow dwarf virus* (BeYDV) and reported up to an 87% reduction in the BeYDV viral load in *N. benthamiana* (Table 8.4).

Plant host translational factors are required by RNA viruses for their life cycle, including the eukaryotic translation initiation factors eIF4E, eIF(iso)4E and eIF4G. Pyott et al. (2016) developed mutated *Arabidopsis* plants by introducing specific deleterious point mutations at the eIF(iso)4E locus to provide complete resistance to Turnip mosaic virus (TuMV). Similarly, Chandrasekaran et al. (2016) developed cucumber plants resistant to potyviruses, i.e., cucumber vein yellowing virus (CVYV), ZYMV and papaya ring spot mosaic virus-W (PRSV-W), by mutating independently two different sites of the host susceptibility gene *eIF4E* by CRISPR-Cas9. Only homozygous *eif4e* mutants showed immunity. While heterozygous knockout and wild-type plants did not exhibit any resistance to potyviruses, Macovei et al. (2018) have shown resistance to rice tungro disease caused by Rice tungro spherical virus (RTSV) and Rice tungro bacilliform virus (RTBV) by mutating *eIF4G* alleles in rice plants. Recently, Zhang et al. (2020) used CRISPR-Cas9-mediated multiplex gene-editing technology to target GmF3H1, GmF3H2 and GmFNSII-1 in soy bean to provide tolerance against soya bean mosaic virus (SMV). Isoflavone synthase (IFS) is a key enzyme in isoflavonoid synthesis and shares a common substrate with flavanone-3-hydroxylase (F3H) and flavone synthase II (FNS II). In 2021, Pramanik et al. developed knockout SlPelo mutant plants that showed resistance to TYLCV infection, which causes yellow leaf curl disease in tomatoes.

Protection in response to RNA-based viruses still remains a daunting task for researchers worldwide, owing to the fact that conventional SpCas9 from *Streptococcus pyogenes* specifically recognizes dsDNA. However, the quest for related nucleases led to the inception of enzymes that, upon recognition, can adhere to and cleave RNA, e.g., LwaCas13a from *Leptotrichia wadei* and FnCas9 from *Francisella novicida.* A first report illustrating resistance towards RNA-based viruses expressed FnCas9 and RNA-targeting sgRNAs designated for CMV and tobacco mosaic virus (TMV) in tobacco (*N. Benthamiana*) and *Arabidopsis* plants. These transgenic plants exhibited a 40%–80% lesser accumulation of CMV and TMV compared to non-transformed WT plants (Zhang et al., 2018b). Similarly, Aman et al. (2018) reported exploiting Cas13a, a RNA-guided ribonuclease, to engineer the RNA genome of TuMV, which provides immunity against RNA viruses in Nicotiana benthamiana.

8.6 FUTURE PERSPECTIVES

Great progress has been made in demonstration of the potential of CRISPR-Cas technology in combating plant diseases. Genome editing offers multiple ways of regulating a gene, such as promoter mutations to make a susceptibility gene non-responsive

TABLE 8.4
Plants Edited to Develop Resistance against Viral Pathogens

S. No.	Plant sp.	Viral Host/Pathogen	Disease	Target Gene	Method Used	Outcome	References
1	*Arabidopsis thaliana*	*Turnip mosaic virus (TuMV)*	Mosaic	*eIF4E*	Knockout	Improved resistance without affecting the plant vigour	Pyott et al. (2016)
2	*Arabidopsis thaliana*	*Beet severe curly top virus (BSCTV)*	Curly leaf	IR, CP, Rep	Site-directed mutation	Inhibit virus accumulation and provide resistant to virus infection	Ji et al. (2015)
3	*Arabidopsis thaliana*	*Tobacco mosaic virus (TMV)*	Mosaic	*ORF1a,ORFCP*, 3′-UTR	Site-directed mutation	Improved virus resistance	Zhang et al. (2018b)
4	*Arabidopsis thaliana*	*Clover yellow vein virus (CYVV)*	Necrosis	*eIFE14*	Site-directed mutation	Improved transgene-free genetic resistance in plants	Bastet et al. (2018)
5	*Nicotiana benthamiana*	*Turnip mosaic virus (TuMV)*	Mosaic	TuMV-GFP, HC-Pro, coat protein	Site-directed mutation	Provided RNA-guided immunity against RNA viruses	Aman et al. (2018)
6	*Nicotiana benthamiana*	*Cotton Leaf Curl Kokhran Virus (CLCuKoV), Tomato yellow leaf curl virus (TYLCV), Tomato yellow leaf curl Sardinia virus (TYLCSV)*	Curly leaf, Mosaic	IR, coat protein and Rep	Site-directed mutation	Showed targeting noncoding and intergenic viral sequence can provide resistance	Ali et al. (2016)
7	*Nicotiana benthamiana*	*Tomato yellow leaf curl virus; (TYLCV), Beet curly top virus (BCTV), Merremia mosaic virus (MeMV)*	Curly leaf, Mosaic	Coding, noncoding region	Site-directed mutation	Reduced accumulation of viral DNA	Ali et al. (2015)
8	*Nicotiana benthamiana*	*Bean yellow dwarf virus (BeYDV)*	Dwarfness	Rep and intergenic sequence	Site-directed mutation	Reduced viral load and less infection symptoms	Baltes et al. (2015)
9	*Nicotiana benthamiana*	*Beet severe curly top virus (BSCTV)*	Curly leaf	IR, CP, Rep	Site-directed mutation	Inhibit accumulation of virus	Ji et al. (2015)

(Continued)

TABLE 8.4 (*Continued*)
Plants Edited to Develop Resistance against Viral Pathogens

S. No.	Plant sp.	Viral Host/Pathogen	Disease	Target Gene	Method Used	Outcome	References
10	*Nicotiana benthamiana*	*Tobacco mosaic virus (TMV)*	Mosaic	*ORF1a, ORFCP, 3′- UTR*	Site-directed mutation	Improved virus resistance and inheritable mutation in plants	Zhang et al. (2018b)
11	*Hordeum vulgare*	*Wheat dwarf virus (WDV)*	Dwarfness	Rep, MP, LIR	Site-directed mutation	Improved resistance against virus	Kis et al. (2019)
12	*Glycine max*	*soya bean mosaic virus*	Mosaic	GmF3H1, GmF3H2 and GmFNSII-1(Multiplexing)	Knockout	Resistance to SMV	Zhang et al. (2020)
13	*Solanum lycopersicum*	*Tomato yellow leaf curl virus (TYLCV)*	Yellow leaf curl disease	SlPelo, Cp and Rep of virus	Knockout	Suppressed accumulation of TYLCV	Pramanik et al. (2021)
14	*cv. Ailsa Craig*	Tomato Mosaic virus (ToMV)	Mosaic	DCL2b	Knockout	Antiviral Defence	Wang et al. (2018b)
15	*Musa spp.*	*Endogenous Banana streak virus (eBSV)*	Leaf streak	ORF and IR of BSV (Non-coding, coding regions)	Site-directed mutation	Showed potential of CRISPR system to achieve resistance against virus in banana plants	Tripathi et al. (2019)
16	*Cucumis sativus*	Cucumber vein yellowing virus, Zucchini yellow mosaic virus	mosaic	*eIF4E*	Knock-out	Enhanced resistance against several viral species	Chandrasekaran et al. (2016)
17	*Oryza sativa*	Rice tungro spherical virus (RTSV)	Tungro	*eIF4G*	Knock-out	Improved resistance against virus	Macovei et al. (2018)
18	*Cassava*	*Cassava brown streak virus (CESV), Ugandan cassava brown streak virus (UCBSV)*	*Cassava brown streak disease (CBSD)*	eIF4E, cap-binding protein-1 (nCBP-1)	Knock-out	Tolerance to CBSD	Gomez et al. (2019)

to pathogens, loss of function mutations, gain of function mutations in the coding region, miRNA-resistant mRNAs for enhancing the transcript abundance of the candidate gene, etc. It is also an important functional genomics tool that will be used to fast-track gene function validation for stress resistance. Apart from the potential of the technology, the driving force that can bring these innovations to will be the view taken by the global regulatory agencies and public opinion on the risks and benefits of the technology. The genome-edited disease-resistant varieties shall be under the same kind of pressure for breakdown of resistance as is applicable for conventionally bred mutant products. This may require innovative strategies for pyramiding different edited loci and, possibly, stacking different mechanisms of action. It may also be possible that editing at one site may exhibit only a small allele effect with less than the desired level of resistance. But, in an encouraging regulatory regime, there shall always be scope for pyramiding small effects, adding to big and durable effects. This is unlike transgenics, where, given the enormous regulatory costs, the desired outcome would generally be sought from one or a few genes only. There is a need to harmonize regulations across nations, as agricultural commodities are heavily traded overseas as well. The intellectual property landscape and freedom to operate in the CRISPR-Cas9 area are no better than those of genetically engineered products. Holders of intellectual properties and public institutions should find common grounds for greater sharing of rights and obligations in an equitable manner so that the fruits of innovations are put to work for the benefit of a larger humanity.

ACKNOWLEDGEMENTS

Research work in VC's laboratory is supported by NASF-funded project on "CRISPR Crop Network: Targeted improvement of stress tolerance, nutritional quality and yield of crops by using genome editing" (F.No. NASF/BGAM-9021/2022–23) and BMGF-HHU, Germany funded project on "Transformative strategy for controlling rice disease in developing countries – healthy crops".

REFERENCES

Ali, Z.; Abulfaraj, A.; Idris, A.; Ali, S.; Tashkandi, M.; Mahfouz, M. M. CRISPR/Cas9-mediated viral interference in plants. *Genome Biology*, **2015**, *16*(1), 1–11.

Ali, Z.; Ali, S.; Tashkandi, M.; Zaidi, S. S. E. A.; Mahfouz, M. M. CRISPR/Cas9-mediated immunity to geminiviruses: differential interference and evasion. *Scientific Reports*, **2016**, *6*(1), 1–13.

Aman, R.; Ali, Z.; Butt, H. et al. RNA virus interference via CRISPR/Cas13a system in plants. *Genome Biology*, **2018**, *19*(1), 1–9.

Baltes, N. J.; Hummel, A. W.; Konecna, E. et al. Conferring resistance to geminiviruses with the CRISPR-Cas prokaryotic immune system. *Nature Plants*, **2015**, *1*(10), 1–4.

Bastet, A.; Lederer, B.; Giovinazzo, N. et al. Trans-species synthetic gene design allows resistance pyramiding and broad-spectrum engineering of virus resistance in plants. *Plant Biotechnology Journal*, **2018**, *16*(9), 1569–1581.

Boch, J.; Bonas, U. Xanthomonas AvrBs3 family-type III effectors: discovery and function. *Annual Review of Phytopathology*, **2010**, *48*, 419–436.

Buchholzer, M.; Frommer, W.B. An increasing number of countries regulate genome editing in crops. *New Phytologist*, **2022**, doi:10.1111/nph.18333.

Chandrasekaran, J.; Brumin, M.; Wolf, D. et al. Development of broad virus resistance in non-transgenic cucumber using CRISPR/Cas9 technology. *Molecular Plant Pathology*, **2016**, *17*(7), 1140–1153.

Cooper, J.; Master Gardener, W. S. U. County Extension, SJI edited by Dr. Tom Schultz. Copyright(c) **2007**, Jan, 5.

Czosnek, H.; Laterrot, H. A worldwide survey of tomato yellow leaf curl viruses. *Archives of Virology*, **1997**, *142*(7), 1391–1406.

Das, A.; Sharma, N.; Prasad, M. CRISPR/Cas9: a novel weapon in the arsenal to combat plant diseases. *Frontiers in Plant Science*, **2019**, *9*, 2008.

Dominguez, A. A.; Lim, W. A.; Qi, L. S. Beyond editing: repurposing CRISPR-Cas9 for precision genome regulation and interrogation. *Nature Reviews Molecular Cell Biology*, **2016**, *17*(1), 5–15.

Dong, O. X.; Ronald, P. C. Genetic engineering for disease resistance in plants: recent progress and future perspectives. *Plant Physiology*, **2019**, *180*(1), 26–38.

Eckardt, N. A. Plant disease susceptibility genes? *Plant Cell*, **2002**, *14*(9), 1983–1986.

Fang, Y.; Tyler, B. M. Efficient disruption and replacement of an effector gene in the oomycete Phytophthora sojae using CRISPR/Cas9. *Molecular Plant Pathology*, **2016**, *17*(1), 127–139.

Fister, A. S.; Landherr, L.; Maximova, S. N.; Guiltinan, M. J. Transient expression of CRISPR/Cas9 machinery targeting TcNPR3 enhances defense response in Theobroma cacao. *Frontiers in Plant Science*, **2018**, 9, 268.

Flor, H. H. Inheritance of reaction to rust in flax. *Journal of Agricultural Research*, **1947**, *74*(9), 41.

Foster, A. J.; Martin-Urdiroz, M.; Yan, X.; Wright, H. S.; Soanes, D. M.; Talbot, N. J. CRISPR-Cas9 ribonucleoprotein-mediated co-editing and counterselection in the rice blast fungus. *Scientific Reports*, **2018**, *8*(1), 1–12.

Gergerich, R. C.; Dolja, V. V. Introduction to plant viruses, the invisible foe. *The Plant Health Instructor*, **2006**, 478.

Gimenez-Ibanez, S.; Boter, M.; Ortigosa, A. et al. JAZ 2 controls stomata dynamics during bacterial invasion. *New Phytologist*, **2017**, *213*(3), 1378–1392.

Gomez, M. A.; Lin, Z. D.; Moll, T. et al. Simultaneous CRISPR/Cas9-mediated editing of cassava eIF 4E isoforms nCBP-1 and nCBP-2 reduces cassava brown streak disease symptom severity and incidence. *Plant Biotechnology Journal*, **2019**, *17*(2), 421–434.

Gu, K.; Yang, B.; Tian, D. et al. R gene expression induced by a type-III effector triggers disease resistance in rice. *Nature*, **2005**, *435*(7045), 1122–1125.

Hammond-Kosack, K. E.; Kanyuka, K. Resistance genes (R genes) in plants. eLS, 2007.

Hammond-Kosack, K. E.; Parker, J. E. Deciphering plant-pathogen communication: fresh perspectives for molecular resistance breeding. *Current Opinion in Biotechnology*, **2003**, *14*(2), 177–193.

Hui, S.; Shi, Y.; Tian, J. et al. TALE-carrying bacterial pathogens trap host nuclear import receptors for facilitation of infection of rice. *Molecular Plant Pathology,* **2019**, *20*(4), 519–532.

Hussain, F., and Usman, F. Fungal Biotic Stresses in Plants and Its Control Strategy. In: Abiotic and Biotic Stress in Plants. IntechOpen. 2019. doi:10.5772/intechopen.83406.

Idnurm, A.; Urquhart, A. S.; Vummadi, D. R.; Chang, S.; Van de Wouw, A. P.; López-Ruiz, F. J. Spontaneous and CRISPR/Cas9-induced mutation of the osmosensor histidine kinase of the canola pathogen *Leptosphaeria maculans*. *Fungal Biology and Biotechnology*, **2017**, *4*(1), 1–12.

Ignatov, A. N.; Egorova, M. S.; Khodykina, M. V. Spreading of bacterial and phytoplasma diseases in Russia. *Zashchita i karantin rastenii*, **2015**, *5*, 6–9.

Ji, X.; Zhang, H.; Zhang, Y.; Wang, Y.; Gao, C. Establishing a CRISPR-Cas-like immune system conferring DNA virus resistance in plants. *Nature Plants*, **2015**, *1*(10), 1–4.

Jia, H.; Orbovic, V.; Jones, J. B.; Wang, N. Modification of the PthA4 effector binding elements in Type I Cs LOB 1 promoter using Cas9/sg RNA to produce transgenic Duncan grapefruit alleviating XccΔpthA4: dCs LOB 1.3 infection. *Plant Biotechnology Journal*, **2016**, *14*(5), 1291–1301.

Jiang, W.; Zhou, H.; Bi, H.; Fromm, M.; Yang, B.; Weeks, D. P. Demonstration of CRISPR/Cas9/sgRNA-mediated targeted gene modification in *Arabidopsis*, tobacco, sorghum and rice. *Nucleic Acids Research,* **2013**, *41*(20), e188–e188.

Jones, J. D.; Dangl, J. L. The plant immune system. Nature, **2006**, *444*(7117), 323–329.

Kay, S.; Hahn, S.; Marois, E.; Hause, G.; Bonas, U. A bacterial effector acts as a plant transcription factor and induces a cell size regulator. *Science*, **2007**, *318*(5850), 648–651.

Kis, A.; Hamar, É.; Tholt, G.; Bán, R.; Havelda, Z. Creating highly efficient resistance against wheat dwarf virus in barley by employing CRISPR/Cas9 system. *Plant Biotechnology Journal*, **2019**, *17*(6), 1004.

Koonin, E. V.; Senkevich, T. G.; Dolja, V. V. The ancient Virus World and evolution of cells. *Biology Direct*, **2006**, *1*(1), 1–27.

Leyns, F.; De Cleene, M.; Swings, J. G.; De Ley, J. The host range of the genus Xanthomonas. *The Botanical Review*, **1984**, *50*(3), 308–356.

Liang, Y.; Han, Y.; Wang, C.; Jiang, C.; Xu, J. R. Targeted deletion of the USTA and UvSLT2 genes efficiently in Ustilaginoidea virens with the CRISPR-Cas9 system. *Frontiers in Plant Science*, **2018**, *9*, 699.

Low, Y. C.; Lawton, M. A.; Di, R. Validation of barley 2OGO gene as a functional orthologue of *Arabidopsis* DMR6 gene in Fusarium head blight susceptibility. *Scientific Reports,* **2020**, *10*(1), 1–13.

Macovei, A.; Sevilla, N. R.; Cantos, C. et al. Novel alleles of rice eIF4G generated by CRISPR/Cas9-targeted mutagenesis confer resistance to Rice tungro spherical virus. *Plant Biotechnology Journal,* **2018**, *16*(11), 1918–1927.

Malnoy, M.; Viola, R.; Jung, M. H. et al. DNA-free genetically edited grapevine and apple protoplast using CRISPR/Cas9 ribonucleoproteins. *Frontiers in Plant Science*, **2016**, *7*, 1904.

Nohales, M. Á.; Flores, R.; Daròs, J. A. Viroid RNA redirects host DNA ligase 1 to act as an RNA ligase. *Proceedings of the National Academy of Sciences*, **2012**, *109*(34), 13805–13810.

Oh, C. S.; Beer, S. V. AtHIPM, an ortholog of the apple HrpN-interacting protein, is a negative regulator of plant growth and mediates the growth-enhancing effect of HrpN in *Arabidopsis*. *Plant Physiology*, **2007**, *145*(2), 426–436.

Oliva, R.; Ji, C.; Atienza-Grande, G. et al. Broad-spectrum resistance to bacterial blight in rice using genome editing. *Nature Biotechnology*, **2019**, *37*(11), 1344–1350.

Ortigosa, A.; Gimenez-Ibanez, S.; Leonhardt, N.; Solano, R. Design of a bacterial speck resistant tomato by CRISPR/Cas9-mediated editing of Sl JAZ 2. *Plant Biotechnology Journal*, **2019**, *17*(3), 665–673.

Peng, A.; Chen, S.; Lei, T. et al. Engineering canker-resistant plants through CRISPR/Cas9-targeted editing of the susceptibility gene Cs LOB 1 promoter in citrus. *Plant Biotechnology Journal*, **2017**, *15*(12), 1509–1519.

Pramanik, D.; Shelake, R. M.; Park, J. et al. CRISPR/Cas9-mediated generation of pathogen-resistant tomato against tomato yellow leaf curl virus and powdery mildew. *International Journal of Molecular Sciences*, **2021**, *22*(4), 1878.

Prihatna, C.; Barbetti, M. J.; Barker, S. J. A novel tomato fusarium wilt tolerance gene. *Frontiers in Microbiology*, **2018**, *9*, 1226.

Pyott, D. E.; Sheehan, E.; Molnar, A. Engineering of CRISPR/Cas9-mediated potyvirus resistance in transgene-free *Arabidopsis* plants. *Molecular Plant Pathology*, **2016**, *17*(8), 1276–1288.

Santillán Martínez, M. I.; Bracuto, V.; Koseoglou, E. et al. CRISPR/Cas9-targeted mutagenesis of the tomato susceptibility gene PMR4 for resistance against powdery mildew. *BMC Plant Biology,* **2020**, *20*(1), 1–13.

Savary, S.; Willocquet, L.; Pethybridge, S. J.; Esker, P.; McRoberts, N.; Nelson, A. The global burden of pathogens and pests on major food crops. *Nature Ecology & Evolution*, **2019**, *3*(3), 430–439.

Schuster, M.; Schweizer, G.; Reissmann, S.; Kahmann, R. Genome editing in *Ustilago maydis* using the CRISPR-Cas system. *Fungal Genetics and Biology*, **2015**, *89*, 3–9.

Shkalikov, V. A.; Beloshapkina, O. O.; Bukreev, D. D. et al. Plant protection from disease. *M.: Kolos*, **2010**, 404.

Thomazella, D. P.; Brail, Q.; Dahlbeck, D.; Staskawicz, B. CRISPR-Cas9 mediated mutagenesis of a DMR6 ortholog in tomato confers broad-spectrum disease resistance. *BioRxiv*, **2016**, 064824.

Thomazella, D. P. D. T.; Seong, K.; Mackelprang, R. et al. Loss of function of a DMR6 ortholog in tomato confers broad-spectrum disease resistance. *Proceedings of the National Academy of Sciences*, **2021**, *118*(27), e2026152118.

Tripathi, J. N.; Ntui, V. O.; Ron, M.; Muiruri, S. K.; Britt, A.; Tripathi, L. CRISPR/Cas9 editing of endogenous banana streak virus in the B genome of *Musa spp.* overcomes a major challenge in banana breeding. *Communications Biology*, **2019**, *2*(1), 1–11.

Vale, F. X. R.; Parlevliet, J. E.; Zambolim, L. Concepts in plant disease resistance. *Fitopatologia Brasileira*, **2001**, *26*, 577–589.

Walbot, V. On the life strategies of plants and animals. *Trends in Genetics*, **1985**, *1*, 165–169.

Wang, Q.; Cobine, P. A.; Coleman, J. J. Efficient genome editing in *Fusarium oxysporum* based on CRISPR/Cas9 ribonucleoprotein complexes. *Fungal Genetics and Biology*, **2018a**, *117*, 21–29.

Wang, T.; Deng, Z.; Zhang, X. et al. Tomato DCL2b is required for the biosynthesis of 22 nt small RNAs, the resulting secondary siRNAs, and the host defense against ToMV. *Horticulture Research*, **2018b**, *5*, 62.

Xie, K.; Yang, Y. RNA-guided genome editing in plants using a CRISPR-Cas system. *Molecular Plant*, **2013**, *6*(6), 1975–1983.

Zeilinger, S.; Gupta, V. K.; Dahms, T. E. et al. Friends or foes? Emerging insights from fungal interactions with plants. *FEMS Microbiology Reviews*, **2016**, *40*(2), 182–207.

Zeng, W.; Melotto, M.; He, S. Y. Plant stomata: a checkpoint of host immunity and pathogen virulence. *Current Opinion in Biotechnology*, **2010**, *21*(5), 599–603.

Zhang, L.; Hu, Y.; Li, P.; Wang, X.; Dong, H. Silencing of an aquaporin gene diminishes bacterial blight disease in rice. *Australasian Plant Pathology*, **2019**, *48*(2), 143–158.

Zhang, M.; Kadota, Y.; Prodromou, C.; Shirasu, K.; Pearl, L. H. Structural basis for assembly of Hsp90-Sgt1-CHORD protein complexes: implications for chaperoning of NLR innate immunity receptors. *Molecular Cell*, **2010**, *39*(2), 269–281.

Zhang, P.; Du, H.; Wang, J. et al. Multiplex CRISPR/Cas9-mediated metabolic engineering increases soya bean isoflavone content and resistance to soya bean mosaic virus. *Plant Biotechnology Journal*, **2020**, *18*(6), 1384–1395.

Zhang, S.; Wang, L.; Zhao, R. et al. Knockout of SlMAPK3 reduced disease resistance to Botrytis cinerea in tomato plants. *Journal of Agricultural and Food Chemistry*, **2018a**, *66*(34), 8949–8956.

Zhang, T.; Zheng, Q.; Yi, X. et al. Establishing RNA virus resistance in plants by harnessing CRISPR immune system. *Plant Biotechnology Journal*, **2018b**, *16*(8), 1415–1423.

Zhang, Y.; Bai, Y.; Wu, G. et al. Simultaneous modification of three homoeologs of Ta EDR 1 by genome editing enhances powdery mildew resistance in wheat. *The Plant Journal*, **2017**, *91*(4), 714–724.

Zhou, H.; Liu, B.; Weeks, D. P.; Spalding, M. H.; Yang, B. Large chromosomal deletions and heritable small genetic changes induced by CRISPR/Cas9 in rice. *Nucleic Acids Research*, **2014**, *42*(17), 10903–10914.
Zhou, J.; Peng, Z.; Long, J. et al. Gene targeting by the TAL effector PthXo2 reveals cryptic resistance gene for bacterial blight of rice. *The Plant Journal*, **2015**, *82*(4), 632–643.
Zou, Z.; Liu, F.; Selin, C.; Fernando, W. D. Generation and characterization of a virulent Leptosphaeria maculans isolate carrying a mutated AvrLm7 gene using the CRISPR/Cas9 system. *Frontiers in Microbiology*, **2020**, *11*, 1969.

9 Multi-omics Approaches and Their Applications for Disease Management in Cereal Crops

Devanna B. N.
ICAR-National Rice Research Institute

Pramesh D.
University of Agricultural Science

Prashant S. H.
ICAR-National Rice Research Institute

Rajashekara H.
ICAR-Vivekananda Parvatiya Krishi Anusandhan Sansthan

Kumari M.
ICAR-National Institute for Plant Biotechnology

Raghu S.
ICAR-National Rice Research Institute

Deepak Pawar
ICAR-Directorate of Weed Research (DWR)

Awaji S. M., Parameswaran C., Katara J. L., Samantaray S., Patra B. C., and Madhav M. S.
ICAR-National Rice Research Institute

Sharma T. R.
Indian Council of Agricultural Research

DOI: 10.1201/9781003403142-9

9.1 INTRODUCTION

Globally, cereals are the cornerstone of food and nutritional security, with a contribution of around 32% to the total crop production, and a major chunk of this (>85%) comes from maize, rice, and wheat (CIMMYT, 2019; FAOSTAT, 2021). Besides these three crops, barley, sorghum, bajra, and minor millets contribute towards the gross cereal output, which is pegged to be around 2,793 million tonnes for the year 2021 (FAOSTAT, 2021). The ever-increasing global human population and changing climatic scenarios are exerting more pressure on agricultural production. Besides, cereal production is subjected to various biotic and abiotic stresses. Among the biotic stresses, diseases caused by different pathogens are significantly contributing to losses in yield and quality of cereals (Balint-Kurti and Johal, 2009; Laha et al., 2017; Figueroa et al., 2018; Różewicz et al., 2021). Among other factors, the changing climatic scenarios are also influencing the emergence of diseases in cereals that otherwise minor in nature (Raghu et al., 2021). Therefore, there is a need to understand these cereal diseases through multipronged approaches and elucidate the underlying mechanisms of resistance/susceptibility. The advancements in next-generation sequencing (NGS) technologies have paved the way for the emergence of diverse omics tools such as genomics, transcriptomics, proteomics, metabolomics, and phenomics. These multi-omics approaches have played an important role in understanding the crop response to various biotic and abiotic stresses (Muthamilarasan and Singh, 2019; Yang et al., 2021a).

In the present chapter, we discuss multi-omics approaches (Figure 9.1), applications, foreseen implementations in enhanced understanding and management of cereal diseases, and also briefly deliberate on the emerging omics approaches and future perspectives.

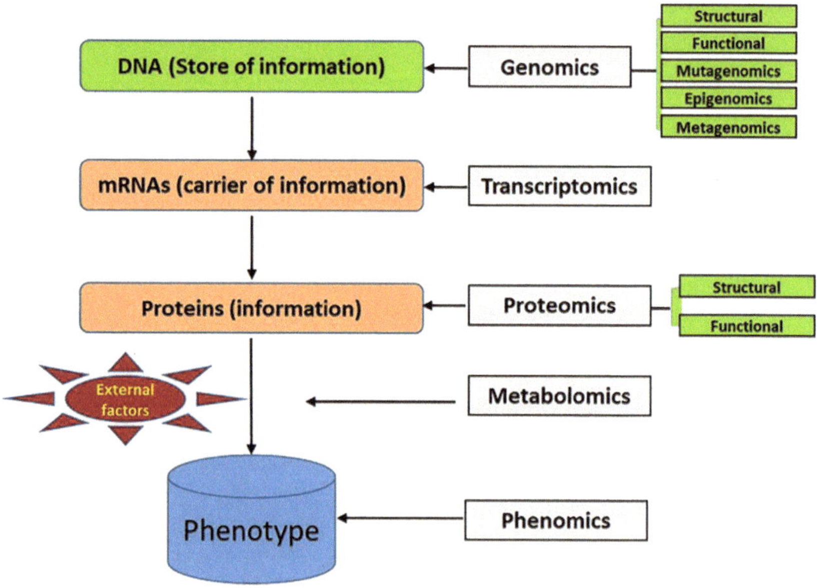

FIGURE 9.1 Schematic diagram depicting the integration of multiomic tools for the development of disease resistance in cereal crops.

9.2 BASIC CONCEPTS OF OMICS APPROACHES

9.2.1 Genomics

Genomics is the study of structural, functional, and evolutionary aspects of the genome of an organism. Plant genomics is an important component in understanding the genetic makeup of the plant, knowing the genomic plasticity under different conditions, and deciphering the various mechanisms governing the plant's development and stress response. Among other organisms, plants are also considered as a significant component in balancing carbon cycles and are being used in making many human medicines, and plants are essential model organisms for understanding the biological system. There are almost 600 plant genome assemblies present in the public repositories. So far, almost 183 plant reference sequences have been published (www.plabipd.de). But very few plant genome assemblies are chromosome-level and are haplotype-phased (Schnable et al., 2009a, b). Plant genomes are complicated to assemble compared to bacteria because their size is higher compared to other vertebrates. The largest plant genome is 28 Gb in size. Plant science studies are facilitated by the availability of the genome and transcriptome sequences of different cereal crops. Genome assemblies have complete information about genome, genic, non-genic, and repetitive elements. The availability of draft and chromosomal-level assemblies is highly useful in all genomics studies. Once the genome is sequenced, gene prediction and functional annotation are the major challenges. In this regard, the transcriptomic data is highly useful for identifying the expressed genes and non-coding transcripts from the whole-genome sequence. For cereal crops for which high-quality genome sequence and annotation are available, genome-wide diversity between different accessions can be studied using resequencing. Whole-genome resequencing will help in identifying single nucleotide variants (SNP), insertions (Ins) and deletions (Del), inversions, and transpositions. Also, several bioinformatics pipelines are available for genome, transcriptome, and resequencing data analysis. Further, these sequence variants are used as markers for linkage and association mapping of several traits. Functional genomics includes the transcriptional level changes of the expressed part of the genome and functional characterization, aiming at the identification of the function of genes. The transcriptome represents a total set of transcribed regions. Different technologies have been traditionally applied to infer and quantify the transcriptome, such as cDNA or expressed sequence tag (EST) library sequencing, serial analysis of gene expression (SAGE), and microarray hybridizations. However, these techniques have low throughput, high cost, low sensitivity, and cloning bias. The most frequently used technique for gene expression profiling, microarray, suffers from high background signals (Miller et al., 2002) and also relies on pre-determined probes designed against known target sequences, hence being unable to identify novel transcribed regions. Also, these probes cover only a portion of the total annotated gene, and hence the complete picture of the transcript is not represented. The development of next-generation high-throughput RNA-sequencing technologies, called (*m*)RNA-Seq, offers novel methods for identifying, mapping, and quantifying transcripts. It has already been demonstrated that deep RNA sequencing is a powerful tool for comparing gene expression and also for discovering the complete 5′ prime and 3′ prime untranslated regions, novel transcripts, novel splice junctions, alternative transcription start sites (TSS), and rare transcripts (Wang et al., 2009). The results of RNA-Seq also show high levels of reproducibility for both technical and biological replicates. RNA transcribed from DNA does not only have coding potential; but it also has noncoding potential. RNA molecules that are not translated into proteins are called noncoding RNA. These noncoding RNAs have attracted attention because of their regulatory role in gene expression. Noncoding RNAs are classified based on their size into small noncoding RNAs (snRNAs) and long noncoding RNAs (lncRNAs), with sizes usually more than 200 nucleotides in length (Chen et al., 2020). lncRNAs are further divided (according to the position of nearby protein-coding genes) into long intergenic noncoding RNAs (lincRNAs), natural antisense transcripts (NATs), and intronic RNAs (incRNAs) (Ma et al., 2013).

9.2.2 Transcriptomics

Plant transcriptomics is a widely used approach for analysing plant responses to various biotic and abiotic stresses. Blumenberg (2019) stated transcriptomics as the study of an organism's total transcriptome (the sum of all RNA transcripts) during a certain developmental stage or physiological state. The transcriptomes of a range of different populations can be used to relate a genotype to a certain trait. Transcriptomes, as measured by microarrays or next-generation sequencing, have resulted in a formerly untold data storm about transcript identities and levels in plant systems (Fernie and Usadel, 2013). Thanks to the NGS, this approach has revolutionized plant biology, including cereals. These have revealed a lot about the genetic networks that control plant stress responses. To unravel the underlying mesh of networks in crop development, massive population-wide transcriptome studies on rice and other agronomic crops such as wheat, maize, barley, oats, rye, and sorghum have been done in terms of genome sequencing, gene regulation, gene differentiation, post-transcriptional modifications, and gene splicing (Groen et al., 2020). The challenges associated with the existing methodologies for mapping and measuring transcriptomes have been resolved due to recent advancements in sequencing technology and their affordability (Sharma et al., 2016). Hybridization (microarray) and sequencing (RNA-Seq and HT/RL SAGE) technologies can be used to measure an organism's transcripts. Sequencing [expressed sequence tags (ESTs) (Parkinson and Blaxter, 2009), RNA-Seq (Wang et al., 2009), and serial analysis of gene expression (SAGE) (Moustafa and Cross, 2016)] are used in most transcriptomic approaches or hybridization [microarray and suppression subtractive hybridization (SSH) (Sahebi et al., 2015)]. Comparative transcriptomics, on the other hand, has generated more information about a plant's response to multiple stresses. As the probes on the chips are built based on annotated data, the expression profiles of unannotated transcripts cannot be examined in microarrays. To convert tag counts to appropriate transcript expression levels, HT-SuperSAGE additionally needs gene structural information. RNA-Seq and high-throughput SuperSAGE (HT-SuperSAGE) are two next-generation sequencing (NGS) methods that give almost the same sensitivity and accuracy of measured expression levels. RNA sequencing is a standard tool for studying gene expression patterns that has been utilized in a variety of host-pathogen interaction experiments (Zhu et al., 2013; Jain et al., 2017; Lu et al., 2019; Sari et al., 2019). However, HT-SuperSAGE is substantially more cost-effective than RNA-Seq (Soanes et al., 2012). In brief, transcriptome investigations enable the detection of mRNA, long noncoding RNAs, and short RNAs, as well as the comprehension of gene organizations and expression patterns (Wang et al., 2009). As a result, transcriptomics, in conjunction with other biotechnological methods, aids in the development of stress tolerance in plants in response to various biotic stresses. Figure 9.2 very concisely outlines the generally followed steps involved in performing a transcriptomic study.

The transcript-assembly techniques have evolved in the last several years. Overall, transcriptome databases provide information on genome-wide expression patterns. Plant expression databases (PlexDB) and EXPath databases (Dash et al., 2012; Chien et al., 2015) are extensively utilized for biotic stress research in rice. The Cereal Small RNA Database is used to access the pyrosequencing-generated small-RNA sequences for rice and maize (Johnson et al., 2009).

9.2.3 Proteomics

The term proteome was first coined to describe the set of proteins encoded by the genome, and the study of the proteome is called proteomics (Wilkins et al., 1996; Tyers and Mann, 2003). The proteome is the total set of proteins produced and/or modified by any organism, including plants. The study of total expressed proteins on

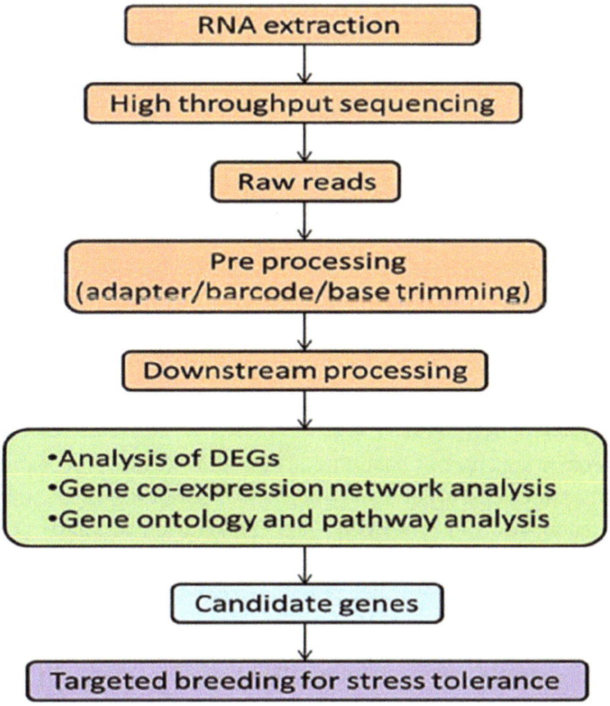

FIGURE 9.2 Steps involved in the transcriptome analysis to identify genes associated with the trait of interest in plants.

a large scale in an organism is termed as proteomics. Unlike the previously discussed transcriptomics, proteomics reveals various post-translational modifications (PTMs) and directly measures the abundance of proteins (Chivasa et al., 2005). For the sake of better analysis and interpretation, proteomics studies have been further divided into different categories, such as sequence, structural, functional, and expression proteomics (Mosa et al., 2017; Aizat and Hassan, 2018). The sequence proteomics part consists of the linear amino acid sequences that are direct manifestations of transcription and post-transcription modifications. Protein sequence is determined using Edman degradation using a protein sequenator and high-performance liquid chromatography (HPLC; Niall, 1973; Twyman, 2013). The structural aspects of proteomics deal with understanding their putative functions based on the structure. Different tools used for protein structure analysis include nuclear magnetic resonance (NMR), electron microscopy, X-ray diffraction of protein crystals, and various computational tools for homology-based modelling (Alberts et al., 2002; Sali et al., 2003; Aslam et al., 2017; Woolfson, 2018). Besides, tools such as matrix-assisted laser desorption ionization time-of-flight mass spectrometry (MALDI-TOF), collision-induced dissociation (CID), and electrospray ionization (ESI) are being used for the identification of molecular weights of proteins (Baggerman et al., 2005; Mehta and Silva, 2015). The actual functional aspect of proteins is studied under functional proteomics, and those functions are examined through various methods, such as yeast hybridization tools, protein microarray profiling, and *in vitro* and *in vivo* studies (Hirano et al., 2004).

Proteomics complements other functional genomics approaches, such as microarray-based expression profiles, phenotypic profiling, systematic genetics, and small-molecule-based arrays (Tong et al., 2001; Gerlai, 2002; Shoemaker and Linsley, 2002; Kuruvilla et al., 2002). Integrating all such data sets through bioinformatics will yield comprehensive information about gene function. The eco-friendliest and cost-effective strategy of plant disease management is through the deployment of resistant cultivars, which are developed through a clear understanding of the host-pathogen interaction (Manikandan et al., 2018; Rampitsch and Bykova, 2012). In recent times, proteomics has become a potential tool for obtaining molecular information about several hosts and pathogen proteins involved in virulence, thus opening up new strategies for plant disease management. This technique allows qualitative and quantitative measurement of different proteins involved during the different infection stages that directly influence cellular biochemistry (Manikandan et al., 2018).

The proteomic approaches have three important stages: preparation of the sample, protein/peptide separation using gel electrophoreses, and identification of proteins using mass spectrometry (MS). The emerging methods of proteomic study have revealed information at the cellular and subcellular level, helping in understanding the unique proteomic behaviour of specific organs by aiding in understanding processes involving biotic stress tolerance (Devanna et al., 2014; Singh et al., 2020).

9.2.3.1 Proteomics Approaches

Broadly, two types of approaches are used in proteomics: gel-based and gel-free proteomics (Figure 9.3).

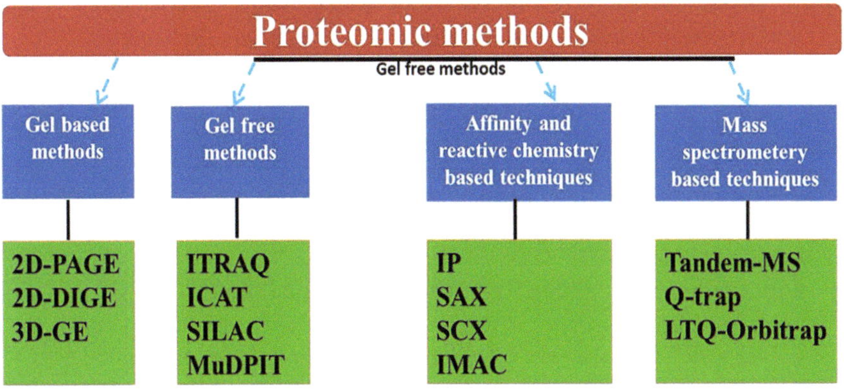

FIGURE 9.3 Classifications of proteomics approaches based on media of separation (2D-PAGE, Two-Dimensional Polyacrylamide Gel Electrophoresis; 2D-DIGE, Fluorescent Two-Dimensional Difference Gel Electrophoresis; 3D-GE, Three-Dimensional Gel Electrophoresis; ITRAQ, Isobaric Tags for Relative and Absolute Quantitation; ICAT, Isotope-Coded Affinity Tags; SILAC, Stable Isotope Labelling by/with Amino acids in Cell culture; MuDPIT, Multidimensional Protein Identification Technology; IP, Immunoprecipitation; SAX, Strong Anion Exchange; SCX, Strong Cation Exchange; IMAC, Immobilized Metal Affinity Chromatography; Tandem-MS, Tandem mass spectrometry; Q-trap, The Quadrupole-Trap Mass Spectrometry; LTQ-Orbitrap, linear trap quadrupole Mass Spectrometry).

a. Two-Dimensional Polyacrylamide Gel Electrophoresis

   Two-dimensional polyacrylamide gel electrophoresis (2D-PAGE) exploits two important properties of proteins (isoelectric point and molecular size) to separate exact proteins from the entire protein mixture (O'Farrell, 1975). Primarily, based on the charges, proteins are separated in one direction, and for the second direction, it depends on the protein size.

b. Fluorescent Two-Dimensional Difference Gel Electrophoresis

   The fluorescent two-dimensional difference gel electrophoresis (2D-DIGE) implies 2D-PAGE to further compare proteins showing differential expression among two conditions. In this technique, the proteins to be compared are treated with different fluorescent dyes (cy2, cy3, and cy5). This blended protein mixture is run using a single 2D gel. These gels are then scanned using software tools (Ünlü et al., 1997; Marouga et al., 2005). To overcome the limitations of gel-based 2D PAGE, such as insensitivity to low abundance proteins and lower reproducibility, gel-free proteomics is considered to be more appropriate.

c. Isotope-Coded Affinity Tags

   Isotope-coded affinity tags (ICAT) is a gel-free proteomics technique that uses chemical labelling reagents to find and quantify the proteins in the populations of two cells (Gygi et al., 1999). Here C13 is labelled with isotopes, and two populations are tagged with different ICAT reagents. The isolation of tagged peptides is done through avidin affinity chromatography, followed by analysis through microcapillary liquid chromatography and electron spray tandem mass spectrometry (Gygi et al., 1999).

d. Multidimensional Protein Identification Technology

Multidimensional protein identification technology (MuDPIT) is particularly used to study mixtures of complex proteins (Issaq et al., 2005). This method is very fast and more sensitive, with a complete list of proteins existing in a specific sample of protein (Rose et al., 2004). This technique combines both SCX and RP-HPLC separation; hence, it is effectively used in an extensive series of proteomics experiments.

e. Isobaric Tags for Relative and Absolute Quantitation

Isobaric tags for relative and absolute quantitation (iTRAQ) another gel-free method used for the quantification of proteins obtained from different sources (Lund et al., 2007). It is also processed by using isotope-labelled tags. This technique depends on covalent labelling by using isotopic tags with different masses at the N-terminus and peptide side chain amines obtained after protein digestion. The peptide separation from the mixture is achieved by nano-chromatography and these peptides are analysed further via tandem mass spectrometry (MS/MS). In order to find out the corresponding proteins of labelled peptides, a database search is performed through fragmentation. The benefit of using iTRAQ technique is that it can quantify many samples with easy analysis including more analytical accuracy and precision. It has the capability to compare four or eight different samples in a single MS-based experiment. Recently, various iTRAQ-based proteomic studies related to both biotic and abiotic stresses have been reported in rice.

9.2.4 Metabolomics

Metabolomics is the scientific study of chemical processes involving metabolites, the small-molecule substrates, intermediates, and products of cell metabolism. Specifically, metabolomics is the "systematic study of the unique chemical fingerprints that specific cellular processes leave behind," the study of their small-molecule metabolite profiles (Daviss, 2005). Metabolomics techniques have been widely used in different research fields of fungal pathogen–plant interactions, such as identifying pathogens, determining infection mechanisms of fungi, and detecting pathogen interactions with the host (Chen et al., 2019). Metabolites play a central role in the plant-microbe interaction; after pathogen attack, the plant's primary as well as secondary metabolic pathways are activated to produce more energy, and secondary metabolites have antimicrobial activity to prevent pathogen invasion and spread. Metabolomics basically involves the chemical analysis of metabolite compounds involved in disease resistance, and this field has progressed very rapidly in the past two decades in multiple spatial and temporal dimensions (Irie et al., 2014; Bartels and Svatoš, 2015; Sumner et al., 2011). Apart from this, pathogens can also interface with the normal metabolism of plants to accomplish their nutritional needs and produce some pathogenic biochemical factors, such as toxin and extracellular polysaccharides (EPS), that have pathogenicity for the host plant.

9.2.5 Phenomics

Plant phenomics encompasses plant architecture and systematic measurement of expression of traits. The concept of phenomics corresponding to genomics was first proposed by Nicholas Schork in 1997 in human disease research (Schork, 1997). Phenomics includes phenotypic analysis of the morphological, physiological, and biochemical variations/characters of an organism using high-throughput technologies. It also includes the phenotypic association with genetic, epigenetic, and environmental variations (Houle et al., 2010; Großkinsky et al., 2015; Zhao et al., 2019). Phenotypic variations are the results of dynamic interactions between plant's genotypes and their surrounding environment. Phenomics helps in the dissection of these complex interactions. Phenomics combines biological science with informatics to solve the problem of rapid and accurate characterization/analysis of plant's phenotypes and phenotypic traits (Kumar and Dasgupta, 2020).

Globally, agriculture is facing major challenges to ensure food security. Among them, biotic stresses caused by various phytopathogens pose a significant threat. Over the last decade and a half, advances in genomics and gene technology have increased confidence in providing solutions to these challenges, resulting in the creation of a massive repository of genomic data. In order to use this information for crop improvement, including cereals, there is a need for novel approaches for identifying quantitative phenotypes and explaining the genetic basis of disease resistance in a high-throughput mode. However, due to rapid developments in genotyping as compared to phenotyping, a genotype-to-phenotype gap has been created. To help bridge this gap, a comprehensive framework for high-throughput phenotyping is the need of the hour (Figure 9.4).

Developments in high-throughput genotyping have enabled the construction of large mapping populations and diversity panels for phenotyping under various biotic stresses. Although molecular breeding strategies have shifted their emphasis to genotypic selection, the following phenotypic data are still required: (1) phenotypes are used for validation of a selection and to train models in genomic selection; (2) a single phenotyping cycle is used to identify markers for subsequent selection through generations using maker-assisted recurrent selection; and (3) phenotyping is critical to identifying promising events in transgenics. Advances in phenotyping are critical for capitalizing on advances in conventional, molecular, and transgenic breeding.

Phenomics tools can also be used to better understand the mechanisms of plant-pathogen interactions and to uncover functional aspects of plant susceptibility

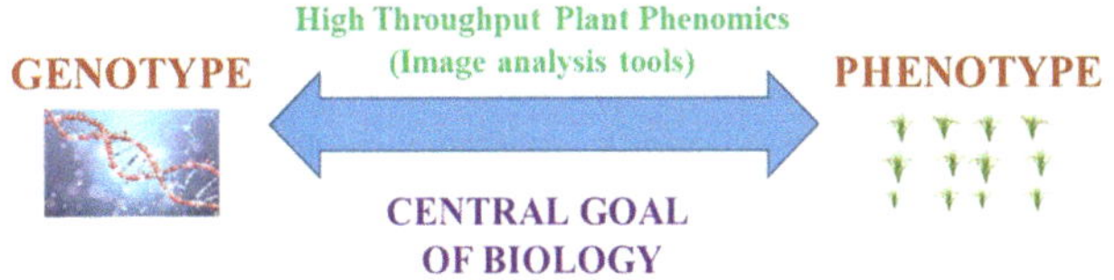

FIGURE 9.4 Schematic presentation of link between genotype and phenotype in crops.

to disease. Precise phenotyping, for example, may identify plant architecture that fosters disease-friendly conditions (e.g., influence of the canopy aerodynamics on creating water pockets). Aside from evaluating plants, phenomic-related tools are useful for quantifying environmental factors such as temperature and humidity, which provide additional information about plant performance. It is critical to use a multidimensional approach when distinguishing between biotic and abiotic stresses. Many disease symptoms, such as wilting and drying of leaf tips, are similar to those associated with water stress. Knowing about environmental factors allows for a more accurate diagnosis of the disease; for example, knowing about sufficient soil water content may rule out potential water stress. Combining different phenomics methods that better explain the aetiology of the disease also aids in identification and management of the disease.

9.3 GENOMICS APPROACH FOR CEREAL DISEASE MANAGEMENT

Plant pathology research and disease management approaches have greatly benefited from the revolution in the field of genomics. This progress has accelerated our understanding of the molecular mechanisms underlying the disease response of hosts, provided new markers for quick pathogen detection and diagnosis, and provided new perspectives into the genetics of pathogen populations on a broader scale. The accessibility to whole genomic sequence resources, along with next-generation sequencing (NGS) techniques, has spurred genomics-based techniques for improving disease resistance in crops. NGS techniques have also accelerated the availability of complete plant and pathogen genomes and facilitated metagenomics research of plant-associated microbial communities. Moreover, NGS techniques can now be used to quickly sequence crop and/or pathogen genomes or develop transcriptomic markers linked with pathogen virulence or plant defence traits, thereby hastening the development of plant disease management (Klosterman et al., 2016; Sharma et al., 2018). Crop and pathogen genomics efforts have yielded immediate, measurable results in several systems. Among others, one successful example wherein the applications of genomics were used for developing effective strategies for disease resistance is the map-based cloning of a major blast resistance gene, *Pi54* from indica rice line Tetep (Rai et al., 2011), further mining for its alleles (Kumari et al., 2013), and cloning of its orthologs, *Pi54rh* (Das et al., 2012) and *Pi54of* (Devanna et al., 2014). Equally, the genomic information of the fungus *Magnaporthe oryzae* led to the cloning and characterization of the fungal avirulence gene *Avr-Pi54* (Ray et al., 2016).

The ecological implementation of genomes (ecogenomics), both at the particular organism and local scale, is increasingly enhancing our comprehension of the intricate link that exists between plants, diseases, the environment, and, eventually, beneficial species (Martin, 2014). Metagenomics and metatranscriptomics could even allow us to better understand the entire microbiota, not just in terms of ecological system but also to identify that tiny proportion of the microflora that, by regulating the activity of plant microbial agents in lieu of the host organism, can be established as biocontrol agents (as a strain or in consortiums). The co-occurrence network analysis has been utilized to characterize microbial communities associated with disease incidence and also to study the diversity of bacterial and fungal pathogens infecting plants (Cobo-Diaz et al., 2019). A screening technique for novel antagonist candidates against pathogenic *Fusarium* spp. could benefit from such an approach.

Using Illumina sequence data, Doucouré et al. (2018) performed genome-wide analysis of *Xantomonas oryzae* pv. *oryzae* for the long-term implementation of broad-spectrum and lasting tolerance to these harmful bacterial pathogens in rice. Broberg et al. (2018) performed a genome-wide association study (GWAS) by resequencing the genomes of 52 isolated strains of the useful fungus *Clonostachys rosea*. They studied the aspects related to conduciveness for biocontrol, with a close emphasis on the cold stress of such fungal species and an overall inference for the control of plant diseases. The genomics tools have largely contributed towards understanding the structural aspects of most of the cereals (Table 9.1) and also the pathogens causing diseases in these crops, leading to an understanding of the structural, functional, and evolutionary aspects of these pathogens (Table 9.2). Recently, Devanna et al. (2021) have very elaborately described the applications of genomics resources for the management of rice blast disease.

TABLE 9.1
List of Cereal Crops with Genomes Sequenced

Crop	Genome Size (Mb)	Genotype	Genes	References
Rice	~389	Nipponbare	37,544	Sasaki (2005)
	~430	93–11	46,022–55,615	Yu et al. (2002)
Wheat	~17,000	Chinese Spring	124,201	IWGSC (2014)
Barley	~5,100	Morex	26,159	Mayer et al. (2012)
Rye	~7.9	Lo7	34,441	Rabanus-Wallace et al. (2021)
Oat	~3.72 (2x)	Cc 7277	51,100	Maughan et al. (2019)
	~4.17	CN 19328	49,105	
Maize	~2,300	B73	32,000	Schnable et al. (2009)
Sorghum	~750	BTx623	27,640	Paterson et al. (2009)
Pearl millet	~17,900	Tift23D2B1-P1-P5	38,579	Varshney et al. (2017)
Finger millet	~1,196	ML-365	85,243	Hittalmani et al. (2017)

TABLE 9.2
Genome Sequence Information of Pathogens Causing Important Diseases in Cereals

Pathogen	Host Plant	Genome Size (Mb)	Strain/ Race	Genes	References
Magnaporthe oryzae	Rice	38.8	70–15	11,109	Dean et al. (2005)
M. oryzae	Rice	42.2 44.85	RML-29, RP-2421	13,297, 13,623	Singh et al. (2019)
Rhizoctonia solani	Rice	36.94	AG1 IA	6,156	Zheng et al. (2013a)
Ustilaginoidea virens	Rice	33.6	IPU010	6,451	Kumagai et al. (2016)
Puccinia striiformis f.sp. *tritici*	Wheat	64.8	130	22,185	Cantu et al. (2011)
P. striiformis	Wheat	110	CYR32	25,288	Zheng et al. (2013b)
P. striiformis	Wheat	70.24, 66.26, 69.77	46S 119, 31, K	18,375, 17,103, 17,569	Kiran et al. (2017)
Puccinia triticina	Wheat	100.62 106.75	77, 106	27,678 26,384	Kiran et al. (2016)
Ustilago maydis	Maize	20.5	521	6,902	Kamper et al. (2006)
Curvularia lunata	Maize	35.5	CX-3	11,234	Gao et al. (2014)
Xanthomonas translucens pv. *translucens*	Barley	4.7	DSM 18974T	3,736	Jaenicke et al. (2016)
Colletotrichum sublineola	Sorghum	46.75	TX430BB	12,699	Baroncelli et al. (2014)

9.4 TRANSCRIPTOMICS APPROACH FOR CEREAL DISEASE MANAGEMENT

9.4.1 Transcriptomic Studies in Rice

In terms of coverage and resolution, RNA-Seq is regarded as the best methodology in contrast to other technologies. The various approaches associated extensively with the understanding of complex mechanisms for stress management in rice include ESTs (Zhou et al., 2003), SAGE (Bao et al., 2005), microarray (Wang et al., 2016; Vijayan et al., 2013), SSH (Chang et al., 2019), and RNA-Seq (Jain et al., 2017). Researchers have employed microarrays and deep sequencing to perform GWAS to find stress-regulated miRNAs (Baldrich and San Segundo, 2016; Arora et al., 2021).

Jain et al. (2017) have used the RNA-Seq technique to understand the *Pi9*-mediated resistance response in rice near-isogenic lines (NILs) PB1. They reported an almost similar transcriptomic profile between PB1 and its NILs. In a similar analysis of CBB23, a rice line with the *Xa23* gene revealed 1,235 DEGs (*p*-value 0.05) post-infection with natural PXo99A and its mutant, P99M2. The DEGs were classified into 11 distinct pathways using Kyoto Encyclopedia of Genes and Genomes (KEGG) pathway analysis, with the ribosome being the most prevalent, followed by phenylpropanoids biosynthesis. The clusters of transcription factors (TFs) that may be implicated in PXO99A resistance were discovered using gene co-expression network analysis. Also, 67 differentially expressed transcription factors (TFs) and 26 peroxidase-sensitive genes were found that might be implicated in disease resistance mechanisms (Wang et al., 2019).

Transcriptional reprogramming provides a significant resource for functional research on rice immune signalling components involved in blast resistance.. Among the differentially expressed genes (DEGs) at 12 hpi, both cultivars had a significant enrichment of ribosome- and protein translation-related genes, demonstrating that protein translation machinery is controlled during the activation of rice immunity. In addition, they found that the metallothionein genes *OsMT1a* and *OsMT1b* positively controlled rice resistance to *M. oryzae*, but the peroxidase gene, *Perox4*, adversely affected resistance to blast pathogens. A study using RNA-Seq technology to dissect out the defensive response genes between *Pish* and *Pik* genes revealed that their lineages had 1,248 and 858 unique DEGs, respectively. The *Pish* line had the wall-associated kinase (*WAK*) gene, while the *Pik* line had the zinc-finger-containing protein gene. *PR-4* and *PR-10*, which are associated with pathogenicity, were detected in both lines. In addition, DEGs from the jasmonic acid, brassinosteriod, and salicylic acid pathways were found in both of these two lines. These distinct and common genes in the Pish and *Pik* lines of rice involved in the blast defence responses will aid in deciphering plant defence processes and making rice blast breeding projects easier. Another study found that miRNA169 inhibits the expression of its target nuclear factor Y-A genes in the blast fungus *M. oryzae*. As a result, rice immunity to the virus was reduced (Li et al., 2017). A transcriptome study of blast-resistant and susceptible cultivars Tetep and HP2216, respectively, revealed 3,553 DEGs in HP2216 and 2,491 DEGs in Tetep, indicating their role in differential disease response (Kumar et al., 2021). Recently, Arora et al. (2021) used an NGS-based miRNAome and reported that the *Pi54* gene-mediated rice blast resistance is affected by several microRNAs, which are also participants in pathogen-triggered immunity (PTI) as well as effector-triggered immunity (ETI). Campo et al. (2013) used high-throughput RNA sequencing in rice seedlings infected with rice dwarf virus (RDV) and rice stripe virus, which were studied using next-generation sequencing (NGS) to look at small RNA expression patterns (RSV). They discovered a new *osa-miR7695* gene that inhibits an alternatively spliced transcript of the *OsNRAMP6* gene (natural resistance-associated macrophage protein 6) while increasing resistance to blast disease when overexpressed.

Besides blast, transcriptomic studies are also reported for other rice diseases. Xu et al. (2014), through microarrays, investigated miRNA expression levels in rice infected with the black-streaked dwarf virus (SRBSDV) and discovered 56 miRNAs and their 24 target genes associated with disease response. Cheng et al. (2020) studied the early resistance against *F. fujikuroi* in two cultivars: Zerawchanica karatals (susceptible) and Tainung 67 (resistant) by analysing the stem transcriptomes of pathogen-inoculated and mock-inoculated line plants using Illumina sequencing, which revealed that the resistant ($n = 169$) line had more DEGs than the susceptible ($n = 118$) line, and gene ontology concepts linked to transcription factor activity and phosphorylation were over-represented in the resistant line DEGs. Zhang et al. (2017) analysed the gene expression patterns generated by the AG1 IA strain of *R. solani* in rice using an RNA-Seq technique. They found a variable transcriptional response under control and infection circumstances by comparing the transcriptomes of a moderately resistant cultivar, TeQing and a susceptible cultivar, Lemont, and 4,802 DEGs were discovered. De Yuan et al. (2020) performed an RNA-Seq study in rice infected with *R. solani* to better understand the rice defence mechanism against sheath blight. When compared to the control group, 3,417 and 2,532 genes were up- and downregulated (> twice or 1/2), respectively. In response to *R. solani* AG1-IA inoculation, the expression levels of 359 transcription factors, including members of the WRKY, MYB, and NAC families, as well as 239 transporter genes, including *ABC*, *MFS*, and *SWEET*, were dramatically altered. Furthermore, through a gain of function analysis for *OsWRKY53* and a loss of function analysis for *OsAKT1*, *OsWRKY53* and *OsAKT1* both adversely affect the rice defensive response against *R. solani*.

A comparative whole-genome transcriptome analysis involving healthy and tungro-affected rice plants provided molecular insights into the global gene expression alterations in rice during infection. Around 959 transcripts belonging to various physiological pathways associated with stress response and hormonal balance were found to be differentially expressed. In summary, the work covers the entire differential gene expression landscape and gives molecular insights into the rice-tungro virus system's infection dynamics (Kumar et al., 2020).

9.4.2 Transcriptomic Studies in Wheat

Advances in NGS technology gave the necessary throughput to dissect large and complicated genomes, such as hexaploid wheat. These tools and platforms provide a strong basis for wheat research, paving the way for a new age of wheat functional genomics that bridges the genotype-phenotype gap.

Wheat transcriptomes have long been studied using cDNA microarrays as the principal method. The relationship between DEGs and yellow rust infection was discovered by various approaches, such as expression profiles, gene ontology, KEGG pathway, and enrichment studies. To understand the mechanism underlying adult plant resistance (APR) to stripe rust (yellow rust), Hao et al. (2016) performed transcriptomic sequencing in elite wheat germplasm, Xingzi 9104 (XZ), which possesses adult plant resistance (APR) to stripe rust (yellow rust). A total of 157,689 unigenes were obtained as a reference. Based on a comparison of *Puccinia striformis* (Pst)- and mock-infected plants, a total of 2,666 and 2,587 differentially expressed genes (DEGs) were discovered to be up- or downregulated following infection. This work added to our knowledge of the transcriptional alterations that occur throughout the mature stage of the Pst-wheat interaction and will aid in our understanding of the precise processes behind APR and Pst infection.

A comparative-transcriptome study enabled us to find a positive regulator of wheat rust susceptibility that alters amino acid metabolism (Corredor-Moreno et al., 2021). The authors focused on mitochondrial branched-chain amino acid (BCAA) aminotransferase 1 (*TaBCAT1*) as a key gene whose expression levels could be tightly linked to disease progression using co-expression cluster analysis. *TaBCAT1* disruption reduced the plant's susceptibility to both yellow and stem rust (*Puccinia graminis* f.sp. *tritici*) infections.

Poretti et al. (2021) compared the transcriptome profiles of susceptible and resistant wheat lines using RNA-Seq to discover genes and pathways affected by pathogenic obligate biotrophic fungi. Transcriptomic analysis of the NILs; Thatcher and Thatcher-*Lr34* revealed that 753 genes were specifically downregulated in the susceptible line after pathogen infection. The inhibition of six main metabolic pathways, including alternative splicing, DNA damage response, nuclear transport, phosphoinositol signalling, ubiquitin-mediated proteolysis, and photosynthesis, was discovered using KEGG enrichment analysis. Powdery mildew and leaf rust elude the wheat defence system by suppressing programmed cell death (PCD) and cellular damage responses, according to our findings.

Using the same approach, Pan et al. (2018) evaluated transcriptomics data from four wheat genotypes (HC374, Nyubai, Wuhan 1, and Shaw) at 2 and 4 days' post-inoculation (dpi) with the fungal *Fusarium graminearum*, the causal agent of Fusarium headblight (FHB). There was an early up-regulation of leucine-rich repeat receptor-like kinases (LRR-RKs), which differed between susceptible and resistant genotypes, and this was followed by up-regulation of diverse sets of genes linked with defensive responses. The resistant genotypes' expression patterns differ, indicating genotype specific defensive mechanisms. The transcriptomics of HC374 and Nyubai are more similar in our study, which is consistent with their sharing of two FHB resistance QTLs on 3BS and 5AS, compared to Wuhan 1, which shares one QTL on 2DL with HC374. Similarly, previously, Xiao et al. (2013) used high-throughput RNA-Seq based on NGS technology to gain an overview of Wangshuibai transcriptome features during infection with *F. graminearum*. As determined by a Blastx search, 165,499 unigenes were generated and assigned to known protein databases, and KEGG gene expression in resistant Wangshuibai and susceptible NAUH117 was profiled and compared at two infection phases by inoculation of *F. graminearum* at 24 and 48 hours using another NGS-based platform, a digital gene expression (DGE) system, with the goal of discovering genes implicated in FHB resistance.

Ren et al. (2020) investigated the transcriptome profile of spikes of the sensitive wheat cultivar Dongxuan 3 to see how *Tilletia controversa* affects wheat, which produces wheat dwarf bunt. The findings of a differential expression study of *T. controversa*-infected plants vs mock-infected plants revealed that 10,867 and 10,487 genes were up- and downregulated, respectively, with these genes concentrated in 205 distinct pathways. KEGG enrichment analysis revealed that 20 genes were expressed differently during the *T. controversa* infection of wheat. After infection, wheat plants' transcriptomes showed significant alterations. The findings of this study might aid in elucidating the mechanism regulating the pathogen's interactions with wheat plants, as well as the creation of new techniques to boost wheat resistance to *T. controversa*, such as overexpression of defence-related genes.

9.4.3 Transcriptomics in Maize

Microarrays have been widely utilized to identify changes in gene expression in maize after inoculation with various diseases since their inception. Because *Fusarium verticillioides* is one of the most economically important diseases of maize, microarray experiments have been extensively used to investigate the Fusarium-maize pathosystem (Lanubile et al., 2017; Campos-Bermudez et al., 2013).

Adhikari et al. (2021) discovered functionally similar co-expressed modules in maize and sorghum in response to maize-specific and sorghum-specific *Exerohilum turcicum* isolates producing turcicum leaf blight in maize and sorghum by using RNA-sequencing approach. He discovered a number of defence-related, shared differentially expressed orthologs (DEOs) with similar expression patterns in these crops, implying a parallel evolution of those genes in both crops.

Besides other diseases, Liu et al. (2015) created a genome-wide transcription profile for systemic symptom development in *Bipolaris zeicola* (Causal organism for Northern corn leaf spot)-infected maize leaves using mRNA sequencing and assessing differential gene expression throughout four disease induction phases. Furthermore, the molecular activities of several genes and their related pathways shed light on the molecular mechanisms underlying the development of maize leaf symptoms in response to *B. zeicola* infection. Furthermore, the pattern of TF accumulation in response to *B. zeicola* infection throughout leaf development highlighted that these TFs are likely involved in plant hormone signal transduction. Further, to identify DEGs and suggest potential genes mapping into Gibberella ear rot (GER) resistance quantitative trait loci (QTL), RNA-Seq transcriptome profiles of fungal- and mock-inoculated, developing kernel tissues of two maize-inbred lines were used. This work assessed the plant response to disease in maize and found fungal-responsive genes mapping into chromosomal areas linked with GER resistance. This information aided in identifying potential genes that may be involved in the defensive response, which will aid our understanding of GER resistance mechanisms. Similarly, Lambarey et al. (2020) reported the changes in gene expression after *F. verticillioides* infection using whole-transcriptome RNA-Seq.

Recently, based on multi-omics analysis, including RNA-Seq, Ge et al. (2021) identified DEGs and DAMs implicated in *Rp1-D21*-mediated HR using transcriptome and metabolomic studies, which provide a valuable resource for studying maize disease resistance genes. Grey leaf spot (GLS) is a severe disease in maize and leads to an extreme reduction in grain yield. He et al. (2021) performed the transcriptomic profile during stages of disease development and analysed it after infection with *Cercospora zeina*. Combining QTL mapping and WGCNA, they studied candidate genes and the homologs of those that are related to immune response and disease resistance. Other than the functional genes, TFs, such as MYBs, and hormone-related genes like JA and ethylene-related genes were found to be involved in GLS response and resistance.

9.5 PROTEOMICS APPROACHES FOR CEREAL DISEASE MANAGEMENT

The applications of proteomics in cereal disease management include the characterization of resistance proteins and their cognate (avirulence) Avr counterparts, the identification of proteins forming the defensome complex, and a network of plant proteins involved in imparting pathogen responses (Devanna et al., 2014; Ray et al., 2016). SDS and native-PAGEs for protein profiling and structural analysis, respectively; Western blot hybridization for antibody-based identification of candidate protein; and isobaric tags for relative and absolute quantification (iTRAQ)-labelling-based mass spectrometry (MS) and MS/MS analysis for proteome characterization (Ji et al., 2019; Yang et al., 2021b).

9.5.1 Rice-Pathogen Interaction

The Rice-*Magnaporthe oryzae* (Mo) interaction provides several basic pieces of information for understanding the host-pathogen interaction. Using 2-DE, N-terminal protein sequencing followed by Western blotting (WB), Konishi et al. (2001) identified the expression of several proteins, such as PR proteins, during the Rice-Mo interaction on rice leaves. Similarly, Kim et al. (2003, 2004, 2009) have identified 12 different proteins during Rice-Mo interaction using PEG fractionation, 2-DE, N-terminal sequencing, TOF-MS, MALDI-TOF PMF, mLC-ESI-MS/MS, and WB. Using total protein extraction, 2DE, and nano-LC-MS/MS, Liao et al. (2009) identified 21 proteins: 11 upregulated, three downregulated, and 7 newly induced proteins during the Rice-Mo interaction.

In the case of the rice-*Xanthomonas oryzae* pv. *oryzae* (Xoo) interaction, several proteomics approaches were used to study the Rice-Xoo interaction. Using cytosolic and membrane protein fractions, 2DE, N-terminal aa sequencing, and MALDI-TOF PMF, several proteins related to energy, metabolism, and defence (PR5, PBZ1, SOD, and peroxiredoxin) were identified (Mahmood et al., 2006). Later, tandem mass spectrometry (MS/MS) coupled with 2-DE was used to identify PM-associated H1-ATPase, phosphatase, hypersensitive-induced response protein (OsHIR1), prohibitin (OsPHB2), zinc finger and C2 domain protein, universal stress protein (USP), and heat shock protein during the Rice-Xoo interaction. Similarly, MALDI-TOF-MS analyses and tandem mass tag (TMT) coupled with LC-MS/MS (Zhang et al., 2020) have been used to identify different proteins involved in Rice-Xoo interaction.

Sheath blight of rice caused by *R. solani* (RS) is also an important limiting factor for rice production. Several proteins were identified during the Rice-RS interaction using several proteomic approaches, such as 2DE and MALDI-TOF-MS/MS (Prathi et al., 2018). Rice false smut disease caused by *Ustilaginoidea virens* (Uv) is an emerging biotic stress of rice, and significant progress has been made in understanding the Rice-Uv interaction through genomics and proteomic approaches (Han et al., 2015; Sharanabasav et al., 2021). A TMT-based quantitative proteomic analysis was used to identify the several differentially expressed proteins during Uv infection (Chen et al., 2021). Uv produces several mycotoxins, such as ustilaginodins, in the infected rice grains and is characterized using high-performance liquid chromatography (HPLC), LC-MS, and lateral flow immunoassays (LFIA) (Jiehua et al., 2019). Several viruses infect rice and cause significant yield losses. Ventelon-Debout et al. (2004) identified several abiotic stress response proteins during the Rice-Rice yellow mottle virus interaction.

9.5.2 Wheat-Pathogen Interaction

Stripe rust of wheat caused by *P. striiformis* f.sp. *tritici* (Pst) is an important wheat disease. The protein expression in the wheat system was studied under the stress of Pst and found the production of peroxisomal membrane proteins PEX2, PEX14-like, and dehydroascorbate reductase along with subtilisin-chymotrypsin inhibitor-2A, p23 (a co-chaperone of Hsp90), CBL-interacting protein kinase 17, SUN domain-containing protein 3-like, and serine/threonine-protein phosphatase PP1 isozyme 9 as resistance response-related proteins (Zhang et al., 2019).

Another important foliar disease of wheat is leaf rust caused by *P. triticina*, a serious threat to durum and bread wheat production. In the pathogen, the spore germination and fungal growth are not synchronous, and hence it is difficult to demonstrate the structure involved in the secretion of proteins. However, the haustoria are found to be the main source for the secreted proteins, as reported by Voegele and Mengden (2003) and Rampitsch et al. (2015) by measuring the transcript levels and proteome of purified rust haustoria, respectively. The production of proteins increased at five days post-inoculation with increased numbers of haustoria in the incompatible interaction between Thatcher wheat and *P. triticina* race-1 (Rampitsch et al., 2015), as the compatible interaction leads to the death of cells prior to the haustoria formation (Cloutier et al., 2007).

The temporal changes in the apoplastic fluid proteome were studied using a label-free LC-MS-based approach involving susceptible and resistant wheat leaves infected with *P. triticina* race-1. There was a negligible difference in the apoplastic proteomes until the formation of haustoria, and once the haustoria were formed, there was a spike in the proteome production from 46 host proteins (up to 72 hours after inoculation) to 150 (post 5 days of inoculation), with the majority being chitinases, peroxidases, β-1–3-endoglucanases, and other PR proteins (Rampitsch et al., 2019).

9.5.3 Pearl Millet-Pathogen Interaction

Pearl millet is affected by downy mildew caused by an obligate biotrophic pathogen, *Sclerospora graminicola*, which causes more than 80% yield loss (Yadav et al., 2002). The proteomic study reported more than 70% of the disease reduction under field conditions due to the resistance conferred by elicitors (Anup et al., 2015). In order to find the molecular basis of host defence during the early phase of infection in resistant pearl millet cultivars, a 2DE and nano-LC-MS/MS-based proteomic approach was performed (Anup and Kini, 2016). The up-regulation of defence proteins such as catalase, ascorbate peroxidase, NBS-LRR resistance protein, and calcium-permeable stress-gated cation channel was expressed at 4.5, 20, 2.8, and 13.5 folds, respectively. The chitinase (PR protein) was upregulated by threefolds; however, glutathione *S*-transferase and xylanase inhibitor protein associated with plant defence mechanisms, were downregulated by 0.35 and 0.47 folds, respectively (Anup and Kini, 2016).

9.5.4 Maize-Pathogen Interaction

Maize (*Zea mays* L.) is cultivated across the world and is greatly affected by ear and stalk rot disease caused by *Fusarium moniliforme* (=*F. verticillioides*) [teleomorph *Gibberella moniliformis*], resulting in extensive yield losses (Aoki et al., 2014). The proteome of the maize apoplastic fluid was analysed to understand the maize-*Fusarium* interaction using LC-MS/MS coupled with label-free quantification. A total of 742 proteins were identified in response to *F. moniliforme*, among which 158 were differentially accumulated proteins (DAPs). In these 158 proteins, 35 were upregulated and 18 were downregulated, whereas 40 and 65 unique proteins were identified from un-inoculated and inoculated plants, respectively. Based on the enrichment analyses, the proteins were categorized into four pathways, of which amino acid biosynthesis was the most prominent, followed by the metabolic pathways. Among these, 86 DAPs were predicted as secretory proteins and were related to defence response pathways, including cell wall modification, signal transduction, and cell redox homeostasis (Haseeb et al., 2022).

9.5.5 Protein-Based Fungicide Designing

Plant disease control based on classical fungicides produces environmental pollution and multidrug resistance in the target fungi. Due to the advancements in genomics and proteomics, it is possible to make changes in the design of chemical fungicides. Techniques for biosynthetic fungicide design have been established as a new strategy in developing fungicides (Collado et al., 2007). Based on clear knowledge about the pathogen biology and life cycle, the use of modified protein-based inhibitors provides a potential species-specific controlling method for plant pathogens by specifically inhibiting certain proteins involved in the disease cycle (Pinedo et al., 2008). These modified protein inhibitors reduce environmental pollution as they are biodegradable and poorly integrated with the food chain. More advancement in the area of pathogen proteomics has helped to develop several eco-friendly target-specific fungicides. Some protein-based fungicides include peptide aptamers, peptide probes, etc. (Fernandez Acero et al., 2011).

From the work carried out till today in proteomics, we can conclude that the combination of proteomics with other techniques makes a powerful tool in understanding host-pathogen interaction, pathogenicity, and virulence, thereby using this in pathogen detection and disease management.

9.6 METABOLOMICS APPROACH FOR CEREAL DISEASE MANAGEMENT

In barley, Bollina et al. (2010), using liquid chromatography coupled to mass spectrometry (LC-MS), reported 496 metabolites in resistant cultivars during the *Fusarium* head blight disease infection. Doehlemann et al. (2008) analysed a tissue-specific infection of a maize crop with a fungal attack, *Ustilago maydis*, and, in response to pathogen colonization, revealed a prominent induction of the shikimate and flavonoid pathways. The hemibiotrophic fungus *Colletotrichum graminicola* infecting maize during infection reported an organ-specific metabolome change, uncovering higher levels of defence-associated metabolites, including flavonoids, in roots compared to leaves (Balmer et al., 2013).

Studies have shown involvement of phosphorylatic and phosphatidyl glycerol (PG) in imparting resistance in rice against *M. oryzae* (Parker et al., 2009). Rice blast fungus infection on ABR1 (susceptible) and ABR5 (resistant) rice cultivars identified phospholipids (PLs) as important substances, and fatty acids were found to be the most important metabolites of the antagonistic species. Hamzehzarghani et al. (2005) and Swarbrick et al. (2006) studied the non-target metabolomics profile of resistant and susceptible cultivar of different cereals and identified several metabolite biomarkers linked to disease resistance. William Allwood et al. (2006) elucidated the key changes in the metabolite profile during infection of *M. grisea* in susceptible and resistant accessions of *Brachypodium distachyon*; it's the alternate host for this pathogen. Parker et al. (2009) analysed changes in metabolomic profiles of rice, *Brachypodium*, barley, and grass species before and after the inoculation of rice *M. grisea*. It reported the metabolic variation of *Rhizoctonia solani* in different stages of its growth using UPLC-QTOF-MS metabolomics analysis. The results also showed that some metabolites may act as biomarkers for the developmental stages of *R. solani* AG-1-IA. It also revealed the mechanisms of sclerotium formation and mycelium differentiation at the metabolic level.

9.7 PHENOMICS APPROACH FOR DISEASE MANAGEMENT

Visual progressions of plant-pathogen interactions, including hypersensitive response (HR) or disease progression, are common. However, at different levels of infection, this approach may be imprecise with varying levels of accuracy and a non-constant error rate (Bock et al., 2010). Evaluations using phenotyping tools have the potential to reduce or eliminate the subjectivity associated with sensory assessments. Furthermore, sensor-based phenotyping may be amenable to automation, allowing for more rapid evaluations, lower labour costs, and more

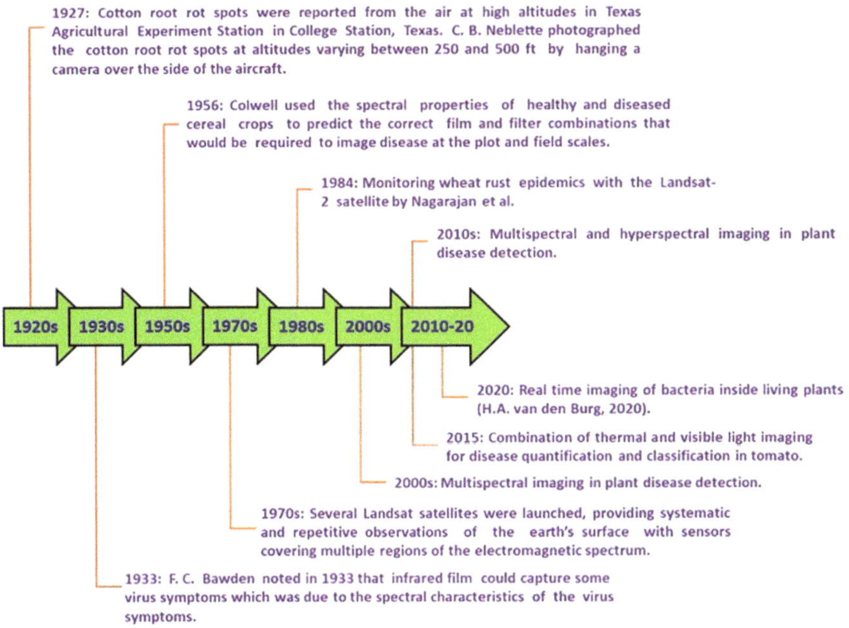

FIGURE 9.5 Timeline of research accomplishments in the last eight decades by employing different non-invasive sensors for plant disease management.

frequent measurements of the same plants over time. If the right methods and tools are used, phenomics may be able to detect disease-related changes in plants before they are visible (presymptomatic detection) or identify changes that are not visible. Such timely and accurate assessments of plant reactions are critical when studying plant-pathogen interactions, developing new cultivars in breeding programmes, and ultimately in crop management, where these techniques could be applied on a large scale in an agronomic context. A summary of different research accomplishments in the last eight decades that led to understanding of plant disease detection and management by using different non-invasive sensors is presented in Figure 9.5.

Innovative phenotyping methods like the use of optical and non-invasive sensors with the potential to overcome the phenotyping bottleneck and improve the breeding process have been well characterized. Sensors such as RGB (Fiorani and Schurr, 2013), fluorescence, HSI (Ustin and Gamon, 2010), and thermography are able to characterize different plant parameters using various optical imaging methods, which could potentially be implemented in high-throughput automated phenotyping pipelines. Optical sensors have the potential to be useful tools for non-invasive disease detection and diagnosis. There is an increasing availability of imaging and non-invasive sensors that can aid in diagnosis and plant disease detection (Figure 9.6).

FIGURE 9.6 Different non-invasive sensors employed for plant disease management.

The advancement of sensor and information technologies, as well as the expansion of geographic information systems, creates new opportunities for precision agriculture and plant phenotyping. Table 9.3 summarizes the recent developments in the applications of highly sensitive sensors and multiple data analysis pipelines in the plant-pathosystem of cereal crops.

9.8 CONCLUSIONS AND FUTURE PERSPECTIVES

Multi-omics tools in association with affective plant breeding techniques are being explored for the development of various stress-tolerant plants, including vitally important cereal grain crops such as wheat, rice, maize, barley, oat sorghum, and millets. The omics tools have enabled the researchers to better understanding the host-pathogen interactions in cereals. These approaches are helping in deciphering the disease resistance mechanism in cereal crops and developing novel approaches by enhancing our understanding of the underlying networks at the cell, individual, and ecology scale, altering diseases in plants, and also by providing us with technological skills to quickly react to changing, emerging, and re-emerging plant pathogens in a dynamic environment. The CRISPR-Cas-mediated genome-editing tool with its multiplexing ability also has the potential to be converted into another important 'omics' tool with a scope beyond the current horizon of plant research.

TABLE 9.3
Applications of Imaging Technique in Different Crops to Understand Plant-Pathogen Interactions

Sl. No.	Disease	Causal Organism	Crop	Stage of the Crop	Imaging Technique	References
1	Fusarium head blight	*Fusarium graminearum* species complex (FGSC)	Durum Wheat	Early flowering	Combination of thermal and RGB	Francesconi et al. (2021)
2	Fusarium head blight	*Fusarium graminearum* species complex (FGSC)	Durum Wheat	Late infection stage	Thermographic imaging	Oerke and Steiner (2010)
3	Blast	Magnoporthe oryzae	Rice	Leaf health for detection at early stage	Thermographic imaging	Yang et al. (2012)
4	Blast	Magnoporthe oryzae	Rice	Early infection	Hyperspectral imaging	Munns et al. (2010)
5	Yellow rust	*Puccinia striformis* va. triticii	Wheat	Real-time detection of yellow rust	Hyperspectral imaging	Moshou et al. (2005)
6	All the diseases		Wheat	Early detection of disease incidence	Chlcrophyll fluorescence	Moshou et al. (2005)
7	Fusarium head blight	*Fusarium graminearum* species complex (FGSC)	Wheat	Resistance screening	Dual Mask-RCNN Deep Learning Frameworks in Computer Vision	Su et al. (2021)
8	Fusarium blight	Fusarium graminearum	Barley	determination of deoxynivalenol and resistance screening	Hyperspectral imaging	Su et al. (2020)
9	Fusarium head blight	*Fusarium graminearum* species complex (FGSC)	Wheat	Disease resistance screening	Imaging	Stack and McMullen (1998)
10	Yellow rust and fusarium head blight	Puccinia striformis and Fusarium graminearum	Cereals	Disease assessment	Hyperspectral	Whetton et al. (2018)

(Continued)

TABLE 9.3 (*Continued*)
Applications of Imaging Technique in Different Crops to Understand Plant-Pathogen Interactions

Sl. No.	Disease	Causal Organism	Crop	Stage of the Crop	Imaging Technique	References
11	Fusarium head blight	*Fusarium graminearum* species complex (FGSC)	Wheat	Disease resistance screening	Hyperspectral imaging	Alisaac et al. (2018)
12	Fusarium head blight	*Fusarium graminearum* species complex (FGSC)	Wheat -	Disease assessment	Combination of thermal, fluorescence, and hyperspectral imaging	Mahlein et al. (2019)
13	Fusarium head blight	*Fusarium graminearum* species complex (FGSC)	Wheat	Disease detection	Hyperspectral imaging	Zhang et al. (2010)
14	Fusarium head blight	*Fusarium graminearum* species complex (FGSC)	Wheat	Disease detection	Multispectral image analysis	Dammer et al. (2011)
15	Fusarium head blight	*Fusarium graminearum* species complex (FGSC)	Wheat	Disease detection	Deep neural network and colour imaging	Qiu et al. (2019)
16	Root knot nematode	Meloidogyne incognita	Potato	Detection of tuber Infection	Hyperspectral remote sensing and real-time PCR	Žibrat et al. (2021)
17	Yellow rust	*Puccinia striformis* va. triticii	Wheat	Disease detection	Hyperspectral imaging	Guo et al. (2020)

ACKNOWLEDGEMENTS

TRS is thankful to the SERB-DSR dor for financial assistance under C Bose National Fellowship.

REFERENCES

Adhikari P, Mideros SX, Jamann TM. Differential regulation of maize and sorghum orthologs in response to the fungal pathogen Exserohilum turcicum. *Front. Plant Sci.* 2021 May 25;12:675208. doi:10.3389/fpls.2021.675208.

Aizat WM, Hassan M. "Proteomics in systems biology," In *Omics Applications for Systems Biology. Advances in Experimental Medicine and Biology*, eds. W. Aizat, H. H. Goh, and S. Baharum (Cham: Springer), 2018, 31–49. doi:10.1007/978-3-319-98758-3_3

Alberts B, Johnson A, Lewis J, Raff M, Roberts K, Walter P. "Analyzing protein structure and function," In *Molecular Biology of the Cell* (Garland Science), 4th edition, 2002.

Alisaac E, Behmann J, Kuska MT, Dehne HW, Mahlein AK. Hyperspectral quantification of wheat resistance to Fusarium head blight: Comparison of two Fusarium species. *Eur. J. Plant Pathol.* 2018 Dec;152(4):869–84.

Anup CP, Kini KR. Analysis of dynamics of proteome in resistant cultivar of pearl millet seedlings during Sclerospora graminicola infection. *J. App. Biol. Biotechnol.* 2016 Mar;4:67–71.

Anup CP, Melvin P, Shilpa N, Gandhi MN, Jadhav M, Ali H, Kini KR. Proteomic analysis of elicitation of downy mildew disease resistance in pearl millet by seed priming with β-aminobutyric acid and Pseudomonas fluorescens. *J. Proteom.* 2015;120:58–74.

Aoki T, O'Donnell K, Geiser DM. Systematics of key phytopathogenic Fusarium species: Current status and future challenges. *J. Gen. Plant Pathol.* 2014;80(3):189–201.

Arora K, Rai AK, Devanna BN, Dubey H, Narula A, Sharma TR. Deciphering the role of microRNAs during Pi54 gene mediated Magnaporthe oryzae resistance response in rice. *Physiol. Mol. Biol. Plants.* 2021;27(3):633–47.

Aslam B, Basit M, Atif NM, Khurshid M, Rasool MH. Proteomics: Technologies and their applications. *J. Chromatogr. Sci.* 2017;55:182–96.

Baldrich P, San Segundo B. MicroRNAs in rice innate immunity. *Rice.* 2016 Dec;9(1):1–9.

Balint-Kurti PJ, Johal GS. "Maize disease resistance," In *Handbook of Maize: Its Biology* (New York: Springer), 2009, 229–50.

Balmer D, de Papajewski DV, Planchamp C, Glauser G, Mauch-Mani B. Induced resistance in maize is based on organ-specific defence responses. *Plant J.* 2013;74(2):213–25.

Bao J, Lee S, Chen C, Zhang X, Zhang Y, Liu S, Clark T, Wang J, Cao M, Yang H, Wang SM. Serial analysis of gene expression study of a hybrid rice strain (LYP9) and its parental cultivars. *Plant Physiol.* 2005;138(3):1216–31.

Baroncelli R, Sanz-Martín JM, Rech GE, Sukno SA, Thon MR. Draft genome sequence of Colletotrichum sublineola, a destructive pathogen of cultivated sorghum. *Genome Announc.* 2014 Jun 26;2(3):e00540–14.

Bartels B, Svatoš A. Spatially resolved in vivo plant metabolomics by laser ablation-based mass spectrometry imaging (MSI) techniques: LDI-MSI and LAESI. *Front. Plant Sci.* 2015 Jul 10;6:471.

Blumenberg M. "Introductory chapter: Transcriptome analysis," In *Transcriptome Analysis*, 2019 Apr 8;370:1–5.

Bock CH, Poole GH, Parker PE, Gottwald TR. Plant disease severity estimated visually, by digital photography and image analysis, and by hyperspectral imaging. *Crit. Rev. Plant Sci.* 2010;29(2):59–107.

Bollina V, Kumaraswamy GK, Kushalappa AC, Choo TM, Dion Y, Rioux S, Faubert D, Hamzehzarghani H. Mass spectrometry-based metabolomics application to identify quantitative resistance-related metabolites in barley against Fusarium head blight. *Mol. Plant Pathol.* 2010;11(6):769–82.

Broberg M, Dubey M, Sun MH, Ihrmark K, Schroers HJ, Li SD, Jensen DF, Brandström Durling M, Karlsson M. Out in the cold: Identification of genomic regions associated with cold tolerance in the biocontrol fungus Clonostachys rosea through genome-wide association mapping. *Front. Microbiol.* 2018 Nov 22;9:2844.

Campo S, Peris-Peris C, Siré C, Moreno AB, Donaire L, Zytnicki M, Notredame C, Llave C, San Segundo B. Identification of a novel micro RNA (miRNA) from rice that targets an alternatively spliced transcript of the N ramp6 (N atural resistance-associated macrophage protein 6) gene involved in pathogen resistance. *New Phytol.* 2013;199(1):212–27.

Campos-Bermudez VA, Fauguel CM, Tronconi MA, Casati P, Presello DA, Andreo CS. Transcriptional and metabolic changes associated to the infection by Fusarium verticillioides in maize inbreds with contrasting ear rot resistance. *PLoS One.* 2013 Apr 18;8(4):e61580.

Cantu D, Govindarajulu M, Kozik A, Wang M, Chen X, Kojima KK, Jurka J, Michelmore RW, Dubcovsky J. Next generation sequencing provides rapid access to the genome of Puccinia striiformis f. sp. tritici, the causal agent of wheat stripe rust. *PLoS One.* 2011 Aug 31;6(8):e24230.

Chang QL, Xu HJ, Peng YL, Fan J. Subtractive hybridization-assisted screening and characterization of genes involved in the rice-Magnaporthe oryzae interaction. *Phytopathol. Res.* 2019 Dec;1(1):1–3.

Chen F, Ma R, Chen XL. Advances of metabolomics in fungal pathogen-plant interactions. *Metabolites.* 2019 Aug;9(8):169.

Chen L, Zhu QH, Kaufmann K. Long non-coding RNAs in plants: Emerging modulators of gene activity in development and stress responses. *Planta.* 2020;252(5):1–4.

Chen S, Zhang Y, Zhan Q. TMT-Based proteomics analysis of LPS-induced acute lung injury. *Exp. Lung Res.* 2021;47(8):402–15.

Cheng AP, Chen SY, Lai MH, Wu DH, Lin SS, Chen CY, Chung CL. Transcriptome analysis of early defenses in rice against *Fusarium fujikuroi*. *Rice.* 2020;13(1):1–5.

Chien CH, Chow CN, Wu NY, Chiang-Hsieh YF, Hou PF, Chang WC. EXPath: A database of comparative expression analysis inferring metabolic pathways for plants. *BMC Genomics.* 2015 Dec;16(2):1–10.

Chivasa S, Simon WJ, Yu XL, Yalpani N, Slabas AR. Pathogen elicitor induced changes in the maize extracellular matrix proteome. *Proteomics.* 2005;5:4894–904.

CIMMYT. 2019: https://www.cimmyt.org/news/the-cereals-imperative-of-future-food-systems/

Cloutier S, McCallum BD, Loutre C, Banks TW, Wicker T, Feuillet C, Keller B, Jordan MC. Leaf rust resistance gene Lr1, isolated from bread wheat (Triticum aestivum L.) is a member of the large psr567 gene family. *Plant Mol. Biol.* 2007 Sep;65(1):93–106.

Cobo-Díaz JF, Baroncelli R, Le Floch G, Picot A. Combined metabarcoding and co-occurrence network analysis to profile the bacterial, fungal and Fusarium communities and their interactions in maize stalks. *Front. Microbiol.* 2019 Feb 18;10:261.

Collado IG, Sánchez AJ, Hanson JR. Fungal terpene metabolites: Biosynthetic relationships and the control of the phytopathogenic fungus Botrytis cinerea. *Nat. Prod. Rep.* 2007;24(4):674–86.

Corredor-Moreno P, Minter F, Davey PE, Wegel E, Kular B, Brett P, Lewis CM, Morgan YM, Macías Pérez LA, Korolev AV, Hill L. The branched-chain amino acid aminotransferase TaBCAT1 modulates amino acid metabolism and positively regulates wheat rust susceptibility. *Plant Cell.* 2021 May;33(5):1728–47.

Dammer KH, Möller B, Rodemann B, Heppner D. Detection of head blight (Fusarium ssp.) in winter wheat by color and multispectral image analyses. *Crop Prot.* 2011;30(4):420–8.

Das A, Soubam D, Singh PK, Thakur S, Singh NK, Sharma TR. A novel blast resistance gene, Pi54rh cloned from wild species of rice, Oryza rhizomatis confers broad spectrum resistance to Magnaporthe oryzae. *Funct. Integr. Genomics.* 2012;12(2):215–28.

Dash S, Van Hemert J, Hong L, Wise RP, Dickerson JA. PLEXdb: Gene expression resources for plants and plant pathogens. *Nucleic Acids Res.* 2012;40(D1):D1194–201.

Daviss B. Growing pains for metabolomics: The newest'omic science is producing results--and more data than researchers know what to do with. *The Scientist.* 2005;19(8):25–9.

Dean RA, Talbot NJ, Ebbole DJ, Farman ML, Mitchell TK, Orbach MJ, Thon M, Kulkarni R, Xu JR, Pan H, Read ND. The genome sequence of the rice blast fungus Magnaporthe grisea. *Nature.* 2005;434(7036):980–6.

Devanna BN, Hosahatti R, Raghu S, Singh PK, Jain P, Parameswaran C, Singh J, Arora K, Samantaray S, Katara JL, Verma RL. "Advances in genetics and genomics for management of blast disease in cereal crops," In *Blast Disease of Cereal Crops* (Cham: Springer), 2021, 173–181.

Devanna NB, Vijayan J, Sharma TR. The blast resistance gene Pi54of cloned from *Oryza officinalis* interacts with Avr-Pi54 through its novel non-LRR domains. *PLoS One.* 2014;9(8):e104840.

Doehlemann G, Wahl R, Horst RJ, Voll LM, Usadel B, Poree F, Stitt M, Pons-Kühnemann J, Sonnewald U, Kahmann R, Kämper J. Reprogramming a maize plant: Transcriptional and metabolic changes induced by the fungal biotroph Ustilago maydis. *Plant J.* 2008;56(2):181–95.

Doucouré H, Pérez-Quintero AL, Reshetnyak G, Tekete C, Auguy F, Thomas E, Koebnik R, Szurek B, Koita O, Verdier V, Cunnac S. Functional and genome sequence-driven characterization of TAL effector gene repertoires reveals novel variants with altered specificities in closely related Malian Xanthomonas oryzae pv. oryzae strains. *Front. Microbiol.* 2018 Aug 6;9:1657.

FAOSTAT. 2021: https://www.fao.org/worldfoodsituation/csdb/en/

Fernandez Acero FJ, Carbú M, El-Akhal MR, Garrido C, González-Rodríguez VE, Cantoral JM. Development of proteomics-based fungicides: New strategies for environmentally friendly control of fungal plant diseases. *Int. J. Mol. Sci.* 2011;12(1):795–816.

Fernie A, Usadel B. The plant transcriptome-from integrating observations to models. *Front. Plant Sci.* 2013 Mar 11;4:48.

Figueroa M, Hammond-Kosack KE, Solomon PS. A review of wheat diseases-A field. *Perspective. Mol. Plant Pathol.* 2018;19(6):1523–36.

Fiorani F, Schurr U. Future scenarios for plant phenotyping. *Annu. Rev. Plant Biol.* 2013;64:267–91.

Francesconi S, Harfouche A, Maesano M, Balestra GM. UAV-based thermal, RGB imaging and gene expression analysis allowed detection of Fusarium head blight and gave new insights into the physiological responses to the disease in durum wheat. *Front. Plant Sci.* 2021;2:551.

Gao S, Li Y, Gao J, Suo Y, Fu K, Li Y, Chen J. Genome sequence and virulence variation-related transcriptome profiles of Curvularia lunata, an important maize pathogenic fungus. *BMC Genomics.* 2014 Dec;15(1):1–8.

Ge C, Wang YG, Lu S, Zhao XY, Hou BK, Balint-Kurti PJ, Wang GF. Multi-omics analyses reveal the regulatory network and the function of ZmUGTs in maize defense response. *Front. Plant Sci.* 2021;12:738261.

Gerlai, R. Phenomics: Fiction or the future? *Trends Neurosci.* 2002;25(10):506–9.

Groen SC, Ćalić I, Joly-Lopez Z, Platts AE, Choi JY, Natividad M, Dorph K, Mauck WM, Bracken B, Cabral CL, Kumar A. The strength and pattern of natural selection on gene expression in rice. *Nature.* 2020;578(7796):572–6.

Großkinsky DK, Svensgaard J, Christensen S, Roitsch T. Plant phenomics and the need for physiological phenotyping across scales to narrow the genotype-to-phenotype knowledge gap. *J. Exp. Bot.* 2015;66(18):5429–40.

Guo A, Huang W, Ye H, Dong Y, Ma H, Ren Y, Ruan C. Identification of wheat yellow rust using spectral and texture features of hyperspectral images. *Remote Sensing.* 2020 Jan;12(9):1419.

Gygi SP, Rist B, Gerber SA, Turecek F, Gelb MH, Aebersold R. Quantitative analysis of complex protein mixtures using isotope-coded affinity tags. *Nat. Biotechnol.* 1999;17(10):994–9.

Hamzehzarghani H, Kushalappa AC, Dion Y, Rioux S, Comeau A, Yaylayan V, Marshall WD, Mather DE. Metabolic profiling and factor analysis to discriminate quantitative resistance in wheat cultivars against fusarium head blight. *Physiol. Mol. Plant Pathol.* 2005;66(4):119–33.

Han Y, Zhang K, Yang J, Zhang N, Fang A, Zhang Y, Liu Y, Chen Z, Hsiang T, Sun W. Differential expression profiling of the early response to Ustilaginoidea virens between false smut resistant and susceptible rice varieties. *BMC Genomics.* 2015 Dec;16(1):1–5.

Hao Y, Wang T, Wang K, Wang X, Fu Y, Huang L, Kang Z. Transcriptome analysis provides insights into the mechanisms underlying wheat plant resistance to stripe rust at the adult plant stage. *PLoS One.* 2016 Mar 18;11(3):e0150717.

Haseeb HA, Zhang J, Guo YS, Gao MX, Wei GU. Proteomic analysis of pathogen-responsive proteins from maize stem apoplast triggered by *Fusarium verticillioides. J. Integr. Agric.* 2022;21(2):446–59.

He W, Zhu Y, Leng Y, Yang L, Zhang B, Yang J, Zhang X, Lan H, Tang H, Chen J, Gao S. Transcriptomic analysis reveals candidate genes responding maize gray leaf spot caused by Cercospora zeina. *Plants.* 2021 Nov;10(11):2257.

Hirano H, Islam N, Kawasaki H. Technical aspects of functional proteomics in plants. *Phytochemistry.* 2004;65(11):1487–98.

Hittalmani S, Mahesh HB, Shirke MD, Biradar H, Uday G, Aruna YR, Lohithaswa HC, Mohanrao A. Genome and transcriptome sequence of finger millet (Eleusine coracana (L.) Gaertn.) provides insights into drought tolerance and nutraceutical properties. *BMC Genomics.* 2017;18(1):1–16.

Houle D, Govindaraju DR, Omholt S. Phenomics: The next challenge. *Nat. Rev. Genet.* 2010;11(12):855–66.

International Wheat Genome Sequencing Consortium (IWGSC), Mayer KF, Rogers J, Doležel J, Pozniak C, Eversole K, Feuillet C, et al. A chromosome-based draft sequence of the hexaploid bread wheat (Triticum aestivum) genome. *Science.* 2014; 345(6194):1251788.

Irie M, Fujimura Y, Yamato M, Miura D, Wariishi H. Integrated MALDI-MS imaging and LC-MS techniques for visualizing spatiotemporal metabolomic dynamics in a rat stroke model. *Metabolomics*, 2014;10(3):473–83.

Issaq HJ, Chan KC, Janini GM, Conrads TP, Veenstra TD. Multidimensional separation of peptides for effective proteomic analysis. *J. Chromatogr. A B.* 2005;817(1):35–47.

Jaenicke S, Bunk B, Wibberg D, Spröer C, Hersemann L, Blom J, Winkler A, Schatschneider S, Albaum SP, Kölliker R, Goesmann A. Complete genome sequence of the barley pathogen Xanthomonas translucens pv. translucens DSM 18974T (ATCC 19319T). *Genome Announc.* 2016 Dec 1;4(6):e01334–16.

Jain P, Singh PK, Kapoor R, Khanna A, Solanke AU, Krishnan SG, Singh AK, Sharma V, Sharma TR. Understanding host-pathogen interactions with expression profiling of NILs carrying rice-blast resistance Pi9 gene. *Front. Plant Sci.* 2017 Feb 23;8:93.

Ji Z, Zeng Y, Liang Y, Qian Q, Yang C. Proteomic dissection of the rice-Fusarium fujikuroi interaction and the correlation between the proteome and transcriptome under disease stress. *BMC Genomics.* 2019;20(1):1–1.

Jiehua Q, Shuai M, Yizhen D, Shiwen H, Yanjun K. Ustilaginoidea virens: A fungus infects rice flower and threats world rice production. *Rice Sci.* 2019;26(4):199–206.

Johnson C, Kasprzewska A, Tennessen K, Fernandes J, Nan GL, Walbot V, Sundaresan V, Vance V, Bowman LH. Clusters and superclusters of phased small RNAs in the developing inflorescence of rice. *Genome Res.* 2009;19(8):1429–40.

Kamper J, Kahmann R, Bölker M, Ma LJ, Brefort T, Saville BJ, Banuett F, Kronstad JW, Gold SE, Müller O, Perlin MH. Insights from the genome of the biotrophic fungal plant pathogen Ustilago maydis. *Nature.* 2006;444(7115):97–101.

Kim ST, Cho KS, Yu S, Kim SG, Hong JC, Han CD, Bae DW, Nam MH, Kang KY. Proteomic analysis of differentially expressed proteins induced by rice blast fungus and elicitor in suspension-cultured rice cells. *Proteomics.* 2003;3(12):2368–78.

Kim ST, Kang YH, Wang Y, Wu J, Park ZY, Rakwal R, Agrawal GK, Lee SY, Kang KY. Secretome analysis of differentially induced proteins in rice suspension-cultured cells triggered by rice blast fungus and elicitor. *Proteomics.* 2009;9(5):1302–13.

Kim ST, Yu S, Kim SG, Kim HJ, Kang SY, Hwang DH, Jang YS, Kang KY. Proteome analysis of rice blast fungus (Magnaporthe grisea) proteome during appressorium formation. *Proteomics.* 2004;4(11):3579–87.

Kiran K, Rawal HC, Dubey H, Jaswal R, Bhardwaj SC, Prasad P, Pal D, Devanna BN, Sharma TR. Dissection of genomic features and variations of three pathotypes of Puccinia striiformis through whole genome sequencing. *Sci. Rep.* 2017;7(1):1–6.

Kiran K, Rawal HC, Dubey H, Jaswal R, Devanna BN, Gupta DK, Bhardwaj SC, Prasad P, Pal D, Chhuneja P, Balasubramanian P. Draft genome of the wheat rust pathogen (Puccinia triticina) unravels genome-wide structural variations during evolution. *Genome Biol. Evol.* 2016;8(9):2702–21.

Klosterman SJ, Rollins JR, Sudarshana MR, Vinatzer BA. Disease management in the genomics era-summaries of focus issue papers. *Phytopathology.* 2016;106(10):1068–70.

Konishi H, Ishiguro K, Komatsu S. A proteomics approach towards understanding blast fungus infection of rice grown under different levels of nitrogen fertilization. *Proteomics.* 2001;1(9):1162–71.

Kumagai T, Ishii T, Terai G, Umemura M, Machida M, Asai K. Genome sequence of Ustilaginoidea virens IPU010, a rice pathogenic fungus causing false smut. *Genome Announc.* 2016;4(3):e00306–16.

Kumar G, Dasgupta I. Comprehensive molecular insights into the stress response dynamics of rice (Oryza sativa L.) during rice tungro disease by RNA-seq-based comparative whole transcriptome analysis. *J. Biosci.* 2020;45(1):1–23.

Kumar M, Mahato A, Kumar S, Mishra VK. "Phenomics-assisted breeding: An emerging way for stress management," In *New Frontiers in Stress Management for Durable Agriculture* (Singapore: Springer), 2020, 295–310.

Kumari A, Das A, Devanna BN, Thakur S, Singh PK, Singh NK, Sharma TR. Mining of rice blast resistance gene Pi54 shows effect of single nucleotide polymorphisms on phenotypic expression of the alleles. *Eur. J. Plant Pathol.* 2013;137(1):55–65.

Kuruvilla FG, Shamji AF, Sternson SM, Hergenrother PJ, Schreiber SL. Dissecting glucose signalling with diversity-oriented synthesis and small-molecule microarrays. *Nature.* 2002;416:653–7.

Laha GS, Singh R, Ladhalakshmi D, Sunder S, Prasad MS, Dagar CS. "Babu VR Importance and management of rice diseases: A global perspective," In *Rice Production Worldwide* (Cham: Springer), 2017, 303–360.

Lambarey H, Moola N, Veenstra A, Murray S, Suhail Rafudeen M. Transcriptomic analysis of a susceptible African maize line to Fusarium verticillioides infection. *Plants.* 2020 Sep;9(9):1112.

Lanubile A, Maschietto V, Borrelli VM, Stagnati L, Logrieco AF, Marocco A. Molecular basis of resistance to Fusarium ear rot in maize. *Front. Plant Sci.* 2017 Oct 12;8:1774.

Li Y, Zhao SL, Li JL, Hu XH, Wang H, Cao XL, Xu YJ, Zhao ZX, Xiao ZY, Yang N, Fan J. Osa-miR169 negatively regulates rice immunity against the blast fungus Magnaporthe oryzae. *Front. Plant Sci.* 2017 Jan 17;8:2.

Liao M, Li Y, Wang Z. Identification of elicitor-responsive proteins in rice leaves by a proteomic approach. *Proteomics.* 2009;9(10):2809–19.

Liu M, Gao J, Yin F, Gong G, Qin C, Ye K, Zhang M, Sun X, Zhou Y, Zhang Y. Transcriptome analysis of maize leaf systemic symptom infected by Bipolaris zeicola. *Plos One.* 2015 Mar 17;10(3):e0119858.

Lu Q, Wang C, Niu X, Zhang M, Xu Q, Feng Y, Yang Y, Wang S, Yuan X, Yu H, Wang Y. Detecting novel loci underlying rice blast resistance by integrating a genome-wide association study and RNA sequencing. *Mol. Breed.* 2019;39(6):1–10.

Lund TC, Anderson LB, McCullar V, Higgins L, Yun GH, Grzywacz B, Verneris MR, Miller JS. iTRAQ is a useful method to screen for membrane-bound proteins differentially expressed in human natural killer cell types. *J. Proteome Res.* 2007;6(2):644–53.

Ma L, Bajic VB, Zhang Z. On the classification of long non-coding RNAs. *RNA Biol.* 2013;10(6):924–33.

Mahlein AK, Alisaac E, Al Masri A, Behmann J, Dehne HW, Oerke EC. Comparison and combination of thermal, fluorescence, and hyperspectral imaging for monitoring fusarium head blight of wheat on spikelet scale. *Sensors.* 2019 Jan;19(10):2281.

Mahmood T, Jan A, Kakishima M, Komatsu S. Proteomic analysis of bacterial-blight defense-responsive proteins in rice leaf blades. *Proteomics.* 2006;6(22):6053–65.

Manikandan R, Harish S, Karthikeyan G, Raguchander T. Comparative proteomic analysis of different isolates of Fusarium oxysporum f. sp. lycopersici to exploit the differentially expressed proteins responsible for virulence on tomato plants. *Front. Microbiol.* 2018;9:420.

Marouga R, David S, Hawkins E. The development of the DIGE system: 2D fluorescence difference gel analysis technology. *Anal. Bioanal. Chem.* 2005;382(3):669–78.

Martin FM. *The Ecological Genomics of Fungi.* Wiley Blackwell, 2014.

Maughan PJ, Lee R, Walstead R, Vickerstaff RJ, Fogarty MC, Brouwer CR, et al. Genomic insights from the first chromosome-scale assemblies of oat (Avena spp.) diploid species. *BMC Biology*, 2019;17(1):1–19.

Mayer KF, Waugh R, Langridge P, Close TJ, Wise RP, Graner A, et al. A physical, genetic and functional sequence assembly of the barley genome. *Nature.* 2012;491:711–6.

Mehta A, Silva LP. MALDI-TOF MS profiling approach: How much can we get from it? *Front. Plant Sci.* 2015 Mar 23;6:184.

Miller NA, Gong Q, Bryan R, Ruvolo M, Turner LA, LaBrie ST. Cross-hybridization of closely related genes on high-density macroarrays. *Biotechniques.* 2002;32(3):620–5.

Mosa KA, Ismail A, Helmy M. "Omics and system biology approaches in plant stress research," In *Plant Stress Tolerance: An Integrated Omics Approach*, eds K. A. Mosa, A. Ismail, and M. Helmy (Cham: Springer), 2017, 21–34. doi:10.1007/978-3-319-59379-1_2

Moshou D, Bravo C, Oberti R, West J, Bodria L, McCartney A, Ramon H. Plant disease detection based on data fusion of hyper-spectral and multi-spectral fluorescence imaging using Kohonen maps. *Real-Time Imaging.* 2005;11(2):75–83.

Moustafa K, Cross JM. Genetic approaches to study plant responses to environmental stresses: An overview. *Biology.* 2016 Jun;5(2):20.

Munns R, James RA, Sirault XR, Furbank RT, Jones HG. New phenotyping methods for screening wheat and barley for beneficial responses to water deficit. *J. Exp. Bot.* 2010;61(13):3499–507.

Muthamilarasan M, Singh NK, Prasad M. Multi-omics approaches for strategic improvement of stress tolerance in underutilized crop species: A climate change perspective. *Adv. Genet.* 2019;103:1–38.

Niall HD. "Automated edman degradation: The protein sequenator," In *Methods in Enzymology* (Academic Press), 1973 Jan 1, Vol. 27, 942–1010.

Oerke EC, Steiner U. "Potential of digital thermography for disease control," In *Precision Crop Protection-The Challenge and Use of Heterogeneity* (Dordrecht: Springer), 2010, 167–182.

O'Farrell PH. High resolution two-dimensional electrophoresis of proteins. *J. Biol. Chem.* 1975;250(10):4007–21.

Pan Y, Liu Z, Rocheleau H, Fauteux F, Wang Y, McCartney C, Ouellet T. Transcriptome dynamics associated with resistance and susceptibility against fusarium head blight in four wheat genotypes. *BMC Genomics.* 2018 Dec;19(1):1–26.

Parker D, Beckmann M, Zubair H, Enot DP, Caracuel-Rios Z, Overy DP, Snowdon S, Talbot NJ, Draper J. Metabolomic analysis reveals a common pattern of metabolic re-programming during invasion of three host plant species by Magnaporthe grisea. *Plant J.* 2009;59(5):723–37.

Parkinson J, Blaxter M. Expressed sequence tags: An overview. *Methods Mol. Biol.* 2009;533:1–2.

Paterson AH, Bowers JE, Bruggmann R, Dubchak I, Grimwood J, Gundlach H, et al. The Sorghum bicolor genome and the diversification of grasses. *Nature.* 2009;457(7229):551–6.

Pinedo C, Wang CM, Pradier JM, Dalmais B, Choquer M, Le Pêcheur P, Morgant G, Collado IG, Cane DE, Viaud M. Sesquiterpene synthase from the botrydial biosynthetic gene cluster of the phytopathogen Botrytis cinerea. *ACS Chem. Biol.* 2008;3(12):791–801.

Poretti M, Sotiropoulos AG, Graf J, Jung E, Bourras S, Krattinger SG, Wicker T. Comparative transcriptome analysis of wheat lines in the field reveals multiple essential biochemical pathways suppressed by obligate pathogens. *Front. Plant Sci.* 2021;12:720462.

Prathi NB, Palit P, Madhu P, Ramesh M, Laha GS, Balachandran SM, Madhav MS, Sundaram RM, Mangrauthia SK. Proteomic and transcriptomic approaches to identify resistance and susceptibility related proteins in contrasting rice genotypes infected with fungal pathogen Rhizoctonia solani. *Plant Physiol. Biochem.* 2018;130:258–66.

Qiu R, Yang C, Moghimi A, Zhang M, Steffenson BJ, Hirsch CD. Detection of fusarium head blight in wheat using a deep neural network and color imaging. *Remote Sens.* 2019;11(22):2658.

Rabanus-Wallace MT, Hackauf B, Mascher M, Lux T, Wicker T, Gundlach H, et al. Chromosome-scale genome assembly provides insights into rye biology, evolution and agronomic potential. *Nat. Genet.* 2021;53(4):564–73.

Raghu S, Devanna BN, Baite MS, Yadav MK, Rathinam PS, Umapathy K, Pati P, Rath PC. "Integrative advances in rice research," In *Emerging Minor Diseases of Rice in India: Losses and Management Strategies* (IntechOpen), 2021. doi:10.5772/intechopen.99898

Rai AK, Kumar SP, Gupta SK, Gautam N, Singh NK, Sharma TR. Functional complementation of rice blast resistance gene Pi-k h (Pi54) conferring resistance to diverse strains of Magnaporthe oryzae. *J. Plant Biochem. Biotechnol.* 2011;20(1):55–65.

Rampitsch C, Bykova NV. Proteomics and plant disease: Advances in combating a major threat to the global food supply. *Proteomics.* 2012;12(4–5):673–90.

Rampitsch C, Günel A, Beimcik E, Mauthe W. Proteome of monoclonal antibody-purified haustoria from Puccinia triticina Race-1. *Proteomics,* 2015;15(7):1307–15.

Rampitsch C, Huang M, Djuric-Cignaovic S, Wang X, Fernando U. Temporal quantitative changes in the resistant and susceptible wheat leaf apoplastic proteome during infection by wheat leaf rust (Puccinia triticina). *Front. Plant Sci.* 2019;10:1291.

Ray S, Singh PK, Gupta DK, Mahato AK, Sarkar C, Rathour R, Singh NK, Sharma TR. Analysis of Magnaporthe oryzae genome reveals a fungal effector, which is able to induce resistance response in transgenic rice line containing resistance gene, Pi54. *Front. Plant Sci.* 2016 Aug 8;7:1140.

Ren Z, Liu J, Din GM, Zhang H, Du Z, Chen W, Liu T, Zhang J, Zhao S, Gao L. Transcriptome analysis of wheat spikes in response to Tilletia controversa Kühn which cause wheat dwarf bunt. *Sci. Rep.* 2020 Dec 9;10(1):1–3.

Rose JK, Bashir S, Giovannoni JJ, Jahn MM, Saravanan RS. Tackling the plant proteome: Practical approaches, hurdles and experimental tools. *Plant J.* 2004;39(5):715–33.

Różewicz M, Wyzińska M, Grabiński J. The most important fungal diseases of cereals-problems and possible solutions. *Agronomy.* 2021;11(4):714.

Sahebi M, Hanafi MM, Azizi P, Hakim A, Ashkani S, Abiri R. Suppression subtractive hybridization versus next-generation sequencing in plant genetic engineering: Challenges and perspectives. *Mol. Biotechnol.* 2015;57(10):880–903.

Sali A, Glaeser R, Earnest T, Baumeister W. From words to literature in structural proteomics. *Nature.* 2003;422:216–25.

Sari E, Cabral AL, Polley B, Tan Y, Hsueh E, Konkin DJ, Knox RE, Ruan Y, Fobert PR. Weighted gene co-expression network analysis unveils gene networks associated with the Fusarium head blight resistance in tetraploid wheat. *BMC Genomics.* 2019;20(1):1–24.

Sasaki T. The map-based sequence of the rice genome. *Nature.* 2005;436(7052):793–800.

Schnable PS, Ware D, Fulton RS, Stein JC, Wei F, Pasternak S, Liang C, Zhang J, Fulton L, Graves TA, Minx P. The B73 maize genome: Complexity, diversity, and dynamics. *Science.* 2009 Nov 20;326(5956):1112–5.

Schork NJ. Genetics of complex disease: Approaches, problems, and solutions. *Am. J. Respir. Crit. Care Med.* 1997;156(4):S103–9.

Sharanabasav H, Pramesh D, Prasannakumar MK, Chidanandappa E, Yadav MK, Ngangkham U, Parivallal B, Raghavendra BT, Manjunatha C, Sharma SK, Karthik N. Morpho-molecular and mating-type locus diversity of Ustilaginoidea virens: An incitant of false smut of rice from Southern parts of India. *J. Appl. Microbiol.* 2021;131(5):2372–86.

Sharma TR, Das A, Thakur S, Devanna BN, Singh PK, Jain P, Vijayan J, Kumar S. Oscillating transcriptome during rice-Magnaporthe interaction. *Curr. Issues Mol. Biol.* 2016;19(1):99–120.

Sharma TR, Devanna BN, Kiran K, Singh PK, Arora K, Jain P, Tiwari IM, Dubey H, Saklani BK, Kumari M, Singh J. Status and prospects of next-generation sequencing technologies in crop plants. *Curr. Issues Mol. Biol.* 2018;27:1–36.

Shoemaker DD, Linsley PS. Recent developments in DNA microarrays. *Curr. Opin. Microbiol.* 2002;5:334–7.

Singh J, Gupta SK, Devanna BN, Singh S, Upadhyay A, Sharma TR. Blast resistance gene Pi54 over-expressed in rice to understand its cellular and sub-cellular localization and response to different pathogens. *Sci. Rep.* 2020;10(1):1–13.

Singh PK, Mahato AK, Jain P, Rathour R, Sharma V, Sharma TR. Comparative genomics reveals the high copy number variation of a retro transposon in different Magnaporthe isolates. *Front. Microbiol.* 2019 May 6;10:966.

Soanes DM, Chakrabarti A, Paszkiewicz KH, Dawe AL, Talbot NJ. Genome-wide transcriptional profiling of appressorium development by the rice blast fungus Magnaporthe oryzae. *PLoS Pathog.* 2012;8(2):e1002514.

Stack RW, McMullen MP. A visual scale to estimate severity of Fusarium head blight in wheat. 1998.

Su WH, Yang C, Dong Y, Johnson R, Page R, Szinyei T, Hirsch CD, Steffenson BJ. Hyperspectral imaging and improved feature variable selection for automated determination of deoxynivalenol in various genetic lines of barley kernels for resistance screening. *Food Chem.* 2021 May 1;343:128507.

Su WH, Zhang J, Yang C, Page R, Szinyei T, Hirsch CD, Steffenson BJ. Automatic evaluation of wheat resistance to fusarium head blight using dual mask-RCNN deep learning frameworks in computer vision. *Remote Sens.* 2020 Dec 23;13(1):26.

Sumner LW, Yang DS, Bench BJ, Watson BS, Li C, Jones AD. Spatially resolved plant metabolomics. *Ann Plant Rev*. 2011;43:343–66.

Swarbrick PJ, Schulze-Lefert PA, Scholes JD. Metabolic consequences of susceptibility and resistance (race-specific and broad-spectrum) in barley leaves challenged with powdery mildew. Plant *Cell Environ*. 2006;29(6):1061–76.

Tong AH, Evangelista M, Parsons AB, Xu H, Bader GD, Pagé N, et al. 2001. Systematic genetic analysis with ordered arrays of yeast deletion mutants. *Science*. 294:2364–8.

Twyman RM. *Principles of Proteomics* (Abingdon: Garland Science Press), 2013.

Tyers M, Mann M. From genomics to proteomics. *Nature*. 2003;422(6928):193–7.

Ünlü M, Morgan ME, Minden JS. Difference gel electrophoresis. A single gel method for detecting changes in protein extracts. *Electrophoresis*. 1997;18(11):2071–7.

Ustin SL, Gamon JA. Remote sensing of plant functional types. *New Phytolog*. 2010; 186(4):795–816.

Varshney RK, Shi C, Thudi M, Mariac C, Wallace J, Qi P, et al. Pearl millet genome sequence provides a resource to improve agronomic traits in arid environments. *Nat. Biotechnol*. 2017;35(10):969–76.

Ventelon-Debout M, Delalande F, Brizard JP, Diemer H, Van Dorsselaer A, Brugidou C. Proteome analysis of cultivar-specific deregulations of Oryza sativa indica and O. sativa japonica cellular suspensions undergoing rice yellow mottle virus infection. *Proteomics*. 2004;4(1):216–25.

Vijayan J, Jain S, Jain N, Devanna BN, Rathour R, Variar M, Prashanthi SK, Singh AK, Singh UD, Singh NK, Sharma TR. Identification of differentially expressed genes in rice during its early phases of interaction with Magnaporthe oryzae. *Indian J. Genet*. 2013;73(3):233–43.

Voegele RT, Mendgen K. Rust haustoria: Nutrient uptake and beyond. *New Phytolog*. 2003;159(1):93–100.

Wang C, Tariq R, Ji Z, Wei Z, Zheng K, Mishra R, Zhao K. Transcriptome analysis of a rice cultivar reveals the differentially expressed genes in response to wild and mutant strains of Xanthomonas oryzae pv. oryzae. *Sci. Rep*. 2019 Mar 6;9(1):1–3.

Wang M, Lu X, Xu G, Yin X, Cui Y, Huang L, Rocha PS, Xia X. OsSGL: A novel pleiotropic stress-related gene enhances grain length and yield in rice. *Sci. Rep*. 2016 Dec 5;6(1):1–2.

Wang Z, Gerstein M, Snyder M. RNA-Seq: A revolutionary tool for transcriptomics. *Nat. Rev. Genet*. 2009;10(1):57–63.

Whetton RL, Waine TW, Mouazen AM. Hyperspectral measurements of yellow rust and fusarium head blight in cereal crops: Part 2: On-line field measurement. *Biosyst. Eng*. 2018;167:144–58.

Wilkins MR, Pasquali C, Appel RD, Ou K, Golaz O, Sanchez JC, Yan JX, Gooley A, Hughes G, Humphery-Smith I, Williams KL. From proteins to proteomes: Large scale protein identification by two-dimensional electrophoresis and amino acid analysis. *Biotechnol*. 1996;14:61–5.

William Allwood J, Ellis DI, Heald JK, Goodacre R, Mur LA. Metabolomic approaches reveal that phosphatidic and phosphatidyl glycerol phospholipids are major discriminatory non-polar metabolites in responses by Brachypodium distachyon to challenge by *Magnaporthe grisea*. *Plant J*. 2006;46(3):351–68.

Woolfson M. The development of structural X-ray crystallography. *Phys. Scr*. 2018;93:1–32.

Xiao J, Jin X, Jia X, Wang H, Cao A, Zhao W, Pei H, Xue Z, He L, Chen Q, Wang X. Transcriptome-based discovery of pathways and genes related to resistance against Fusariumhead blight in wheat landrace Wangshuibai. *BMC Genomics*. 2013 Dec;14(1):1–9.

Xu D, Mou G, Wang K, Zhou G. MicroRNAs responding to southern rice black-streaked dwarf virus infection and their target genes associated with symptom development in rice. *Virus Res*. 2014 Sep 22;190:60–8.

Yadav RS, Hash CT, Bidinger FR, Cavan GP, Howarth CJ. Quantitative trait loci associated with traits determining grain and stover yield in pearl millet under terminal drought-stress conditions. *Theor. Appl. Genet.* 2002;104(1):67–83.

Yang D, Li S, Xiao Y, Lu L, Zheng Z, Tang D, Cui H. Transcriptome analysis of rice response to blast fungus identified core genes involved in immunity. Plant *Cell Environ.* 2021a;44(9):3103–21.

Yang M, Wang X, Dong J, Zhao W, Alam T, Thomashow LS, Weller DM, Gao X, Rustgi S, Wen S. Proteomics reveals the changes that contribute to Fusarium head blight resistance in wheat. *Phytopathol.* 2021b;111(2):386–97.

Yang Y, Chai R, He Y. Early detection of rice blast (Pyricularia) at seedling stage in Nipponbare rice variety using near-infrared hyper-spectral image. *Afr. J. Biotechnol.* 2012;11(26):6809–17.

Yu J, Hu S, Wang J, Wong GKS, Li S, Liu B, et al. A draft sequence of the rice genome (Oryza sativa L. ssp. indica). *Science.* 2002;296(5565):79–92.

De Yuan P, Xu XF, Hong WJ, Wang ST, Jia XT, Liu Y, Li S, Li ZM, Sun Q, Mei Q, Li S. Transcriptome analysis of rice leaves in response to Rhizoctonia solani infection and reveals a novel regulatory mechanism. Plant Biotech. Rep. 2020 Oct;14(5):559–73.

Zhang DY, Chen G, Yin X, Hu RJ, Gu CY, Pan ZG, Zhou XG, Chen Y. Integrating spectral and image data to detect Fusarium head blight of wheat. *Comput. Electron. Agric.* 2020 Aug 1;175:105588.

Zhang H, Fu Y, Guo H, Zhang L, Wang C, Song W, Yan Z, Wang Y, Ji W. Transcriptome and proteome-based network analysis reveals a model of gene activation in wheat resistance to stripe rust. *Int. J. Mol. Sci.* 2019 Jan;20(5):1106.

Zhang J, Chen L, Fu C, Wang L, Liu H, Cheng Y, Li S, Deng Q, Wang S, Zhu J, Liang Y. Comparative transcriptome analyses of gene expression changes triggered by Rhizoctonia solani AG1 IA infection in resistant and susceptible rice varieties. *Front. Plant Sci.* 2017 Aug 17;8:1422.

Zhang J, Lu H, Li X, Li Y, Cui H, Wen CK, Tang X, Su Z, Zhou JM. Effector-triggered and pathogen-associated molecular pattern-triggered immunity differentially contribute to basal resistance to Pseudomonas syringae. Mol. *Plant-Microbe Interact.* 2010;23(7):940–8.

Zhao C, Zhang Y, Du J, Guo X, Wen W, Gu S, Wang J, Fan J. Crop phenomics: Current status and perspectives. *Front. Plant Sci.* 2019 Jun 3;10:714.

Zheng A, Lin R, Zhang D, Qin P, Xu L, Ai P, Ding L, Wang Y, Chen Y, Liu Y, Sun Z. The evolution and pathogenic mechanisms of the rice sheath blight pathogen. *Nat. Commun.* 2013a Jan 29;4(1):1–10.

Zheng W, Huang L, Huang J, Wang X, Chen X, Zhao J, Guo J, Zhuang H, Qiu C, Liu J, Liu H. High genome heterozygosity and endemic genetic recombination in the wheat stripe rust fungus. *Nat. Commun.* 2013b Oct 23;4(1):1.

Zhou Y, Tang J, Walker MG, Zhang X, Wang J, Hu S, Xu H, Deng Y, Dong J, Ye L, Lin L. Gene identification and expression analysis of 86,136 Expressed Sequence Tags (EST) from the rice genome. *GPB.* 2003 Feb 1;1(1):26–42.

Zhu QH, Stephen S, Kazan K, Jin G, Fan L, Taylor J, Dennis ES, Helliwell CA, Wang MB. Characterization of the defense transcriptome responsive to Fusarium oxysporum-infection in Arabidopsis using RNA-seq. *Gene.* 2013;512(2):259–66.

Žibrat U, Gerič Stare B, Knapič M, Susič N, Lapajne J, Širca S. Detection of root-knot nematode Meloidogyne luci infestation of potato tubers using hyperspectral remote sensing and real-time PCR molecular methods. *Remote Sens.* 2021 Jan;13(10):1996.

10 Invasive Species – Impediment to Crop Biosecurity

Policy Perspectives and Management Strategies

Pranjib K. Chakrabarty
ASRB; ICAR-National Research Centre for Grapes

Sujoy Saha and Ratna U. Thosar
ICAR-National Research Centre for Grapes

S. Selvakumaran
ASRB; ICAR-National Research Centre for Grapes

10.1 INTRODUCTION

An invasive species can be defined as any kind of harmful living organism (like an amphibian, e.g. the cane toad, insect, fish, plant, bacteria, fungus, or even an organism's seeds or eggs) that is not native to an ecosystem. They can harm the environment, the economy, or even human health. In other words, species that grow and reproduce quickly, spread aggressively, and have the potential to cause harm are identified as "invasive." An invasive alien species is one that is established outside its natural past or present distribution, introduction, and/or spread and that threatens biological diversity. The majority of invasive species are alien, but some native species may also become invasive under altered environmental conditions such as changes in nutrient regimes, colonization by an invasive species, or other alterations. India is one of the 17 "megadiverse" countries and is composed of a diversity of ecological habitats such as forests, grasslands, wetlands, coastal and marine ecosystems, and desert ecosystems (convention on biological diversity, CBD). About 40% of the Indian flora is alien, of which 25% are IAS (Raghubanshi et al., 2005). Invasive alien species (IAS) make endemic species defenceless, uncompetitive, and may result in the world's ecosystem being dominated by a few ultra-competitive "super species" (Mandal and Nandi, 2009). Besides, invasive species are one of the most important food security issues, as they are a major cause of crop loss and can adversely affect

DOI: 10.1201/9781003403142-10

food security (Cook et al., 2011). Despite the existence of various management practices (Khetrapal and Gupta, 2007; Mandal, 2011), their effective management is a major concern for crop biosecurity. They may also serve as potential bioweapons by non-state actors in agroterrorism (Monke, 2007).

The Government of India, in collaboration with the Norwegian Government, has established a "Centre for Biodiversity Policy and Law (CEBPOL)" in the National Biodiversity Authority (NBA), Chennai, to deal with emerging and current biodiversity governance and policy-related issues (http://nbaindia.org/cebpol). NBA's invasive species expert committee has suggested adopting the important invasive attributes, viz., invasiveness, impacts, range of extension, and others, on a graded scale to designate the alien species as invasive in India. Besides, the committee also took into account the personal experiences of the researchers and their views on declaring a species invasive even in case where the literature on the species is unavailable or scanty (Sandilyan, 2019) (Table 10.1).

10.2 MODES OF THE TRANSPORTATION/ PATHWAYS OF INVASIVE SPECIES

A pathway is broadly defined as the means (e.g. aircraft, vessel, or person), purpose or activity (e.g. farming, shipping, or pet trade), or commodity (e.g. fisheries) by which an IAS may be transported to a new location, either intentionally or unintentionally. Species can also expand their range through natural means. Some species or their propagules can be moved to new locations by wind, currents, or animals. This is referred to as natural dispersal and is not an introduction. Natural dispersal may play a significant role in the subsequent spreading of an alien species once introduced to a new region or country. International spread or agroterrorism could serve as a non-conventional bioweapon harmful to the biosafety or economy of the country.

TABLE 10.1
Invasive Attributes Used to Confirm the Invasive Status of the Species Reported in Indian Ecosystems (Sandilyan, 2019)

S. No.	Invasive Attributes
Invasiveness	
1.	**IE** – Invasive Elsewhere
2.	**RMS** – Rapid Multiplication and Spread in different ecosystems
3.	**MMR** – Multiple Modes of Reproduction
4.	**MMD** – Multiple Modes of Dispersion
Impacts	
1.	**B1** – Affecting ecosystem functions and services
2.	**B2** – Biodiversity loss
3.	**B3** – Economic loss and health hazard
Invasion areas (continues spread)	
	RE – Range Extension

10.3 CAUSES OF INVASIVE ALIEN SPECIES

While all species compete to survive, invasive species appear to have specific traits or specific combinations of traits that allow them to outcompete native species. In some cases, the competition is about rates of growth and reproduction. In other cases, species interact with each other more directly.

Common invasive species traits include the following:

- Fast growth
- Rapid reproduction
- High dispersal ability
- Phenotype plasticity (the ability to alter growth form to suit current conditions)
- Tolerance of a wide range of environmental conditions (ecological competence)
- Ability to live off of a wide range of plants
- Association with humans
- Prior successful invasions

10.4 ECONOMIC IMPACT OF IAS

Invasive species are responsible for major economic losses in agriculture, forestry, and several other segments of the world economy. Over the past few decades (1970–2017), total invasion costs reached a minimum of US$1.288 trillion with an annual mean cost of US$26.8 billion (Diagne et al. 2021). They also negatively impact ecological integrity (Pimentel et al., 2000). Human population growth and increased human activities have escalated the rate and risk associated with alien species introductions as they aid in alterations of the environment (Pimentel et al., 2000; Charles et al., 2007; Pyšek and Richardson, 2010). In India, crop losses due to pests, diseases, and weeds are approximately assessed to range between 10% and 30% of crop productions (Sanjeev, 2014). Plant pathogens reduce crop production by approximately 16%, worth US$48 billion per year (Singh et al., 1996). Approximately 30,000 species of plant pathogens attack Indian crops, including 23,000 species of fungi (Pandotra, 1997) and 650 species of plant viruses (Patel and Patel, 1985). Approximately 74% of the major plant pathogens in India are considered alien species, based on the predominant infections in vegetable crops (Singh, 1985). Some of the IAS that got introduced and/or domesticated in India are documented by Sandilyan (2019) in Table 10.2. The estimated cost of alien plant pathogens causing a loss to Indian crops amounts to about US$35.5 billion per year (Pimentel et al., 2001).

The first epidemic of wheat blast in South Asia dramatically reduced the yield of wheat in the affected field in Bangladesh up to 100% (Islam et al., 2020). The climate analogue model suggested that out of the 40.85 million ha (Anonymous, 2017a) wheat area in the Indian subcontinent, 6.99 million ha are vulnerable to wheat blast, and in case the disease is domesticated, there may be a loss worth US$132 million considering the 5% blast-inflicted loss of wheat (Mottaleb et al., 2018).

TABLE 10.2
Invasive Alien Species of Microbes/Strains in Agriculture Ecosystems in India (Sandilyan, 2019)

Sr. No.	Name	Common Name	References
Fungus			
1.	*Hemileia vastatrix*	Coffee rust	Kushalappa and Eskes (2003)
2.	*Phytophthora infestans*	Late blight of potato	Butler (1918)
3.	*Urocystis tritici*	Flag smut of wheat	Sydow et al. (1906)
4.	*Puccinia carthami*	Rust of chrysanthemum	Sydow et al. (1906)
5.	*Venturia inequalis*	Apple Scab	Rajak et al. (1974)
6.	*Plasmopara viticola*	Downy mildew of grapes	CMI (1988)
7.	*Sclerospora phillipinensis*	Downy mildew of maize	Payak and Renfro (1967)
8.	*Pyricularia grisea*	Blast of paddy	Padmanabhan (1965)
9.	*Fusarium moniliforme*	Foot rot of rice	Padmanabhan (1959)
10.	*Phyllachora sorghi*	Leaf spot of sorghum	Ramakrishnan and Sundaram (1953)
11.	*Oidium heavea*	Powdery mildew of rubber	Mitra and Mehta (1938); Ramakrishnan and Radhakrishna Pillay (1963)
12.	*Phytophthora nicotianae var. nicotianae*	Tobacco black shank	Govindarao and Koteswararao (1956)
13.	*Sphaeropsis spp.*	Canker of apple	Mundkur and Kheshwala (1943)
14.	*Synchytrium endobioticum*	Potato wart	Ganguly and Paul (1953)
15.	*Plasmopara halstedii*	Downy mildew of sunflower	Mayee and Patil (1986)
16.	*Magnaporthe oryzae sub. sp. triticum (MoT)*	Blast of Wheat	
Bacteria			
1.	*Xanthomonas campestris pv. campestris*	Black rot of crucifers	Patel et al. (1949)
2.	*Agrobacterium tumefaciens*	Crown gall of apple/pear	Sharma et al. (2002)
3.	*Erwinia amylovora*	Fire blight of pear	Papdiwal and Deshpande (1978)
4.	*Xanthomonas oryzae pv. oryzae*	Bacterial leaf blight of Paddy	Srivastava and Rao (1964)
Virus			
1.	*Banana bunchy top virus (Babu virus)*	Banana bunchy top	Varghese (1945)
2.	*Sunflower necrosis illar virus*	Sunflower necrosis	Prasada Rao et al. (2000)
Nematode			
1.	*Globodera rostochiensis*	Potato golden nematode	Jones (1961)

10.5 IAS INCURSIONS INTO INDIA

The plants and seeds imported into India are subjected to pre-entry, entry, and post-entry quarantine regulations. More than 1,200 pests are regulated in the Plant Quarantine (Regulation of Import into India) Order, 2003 (hereinafter referred to as the PQ Order, 2003), which includes insect pests/nematodes/mites (563), pathogens (639), parasitic plants (2), and noxious weeds (57). Incursions of pests/pathogens on seeds/planting material imported for research by NBPGR are rather easier to intercept and manage. Recently, a number of new incursions of pathogens and pests were intercepted on the imported germplasm (Table 10.3). These imported germplasms were either destroyed or salvaged, as per the needs (Bhalla et al., 2018; Chalam et al., 2017a; Chalam and Maurya, 2018). However, incursions through the large volume introduction of plant/seed material for commercial purposes may pose a serious threat to agriculture. Some latent incursions were suspected while importing large volumes of apple saplings in HP from Italy in 2015–2016 and 2016–2017 (Anonymous, 2021c; and Personal communication). There is an urgent need for the Indian NPPO to revise the list of quarantine pests (in PQ Order, 2003) based on the present status of invasive species, to make biosecurity strategies more robust and relevant.

TABLE 10.3
Quarantine Pests Intercepted in Germplasm Imported into India During 2016–2020

S. No.	Pest	Crop	Source
Fungi			
1.	*Diaporthe eres*	*Robinia pseudoacacia*	Hungary
2.	*Diaporthe longicolla*	*Glycine max*	USA
3.	*Peronospora manshurica*	*Glycine max*	Canada
Viruses			
4.	*Barley stripe mosaic virus*	*Zea mays*	Philippines
5.	*Bean pod mottle virus*	*Glycine max*	USA
6.	*Broad bean stain virus*	*Glycine max*	Japan, USA
7.	*Cherry leaf roll virus*	*Glycine max*	AVRDC (Taiwan), USA
8.	*High plains virus*	*Zea mays*	Chile
9.	*Maize chlorotic mottle virus*	*Zea mays*	Chile
10.	*Peanut stunt virus*	*Glycine max*	USA
11.	*Pepper mild mottle virus*	*Capsicum annuum*	Netherlands, Taiwan
12.	*Pepino mosaic virus*	*Capsicum annuum*	Netherlands, Taiwan
		Solanum lycopersicum	Netherlands
13.	*Raspberry ringspot virus*	*Glycine max*	USA
		Vigna radiata	Afghanistan
		Vigna unguiculata	Mali

(*Continued*)

TABLE 10.3 (*Continued*)
Quarantine Pests Intercepted in Germplasm Imported into India During 2016–2020

S. No.	Pest	Crop	Source
14.	*Tomato bushy stunt virus*	*Capsicum annuum*	Netherlands, Taiwan
		Solanum lycopersicum	Netherlands
15.	*Tomato ringspot virus*	*Capsicum annuum*	Netherlands, Taiwan
		Glycine max	USA
		Solanum lycopersicum	Netherlands
Insects			
16.	*Bruchus dentipes*	*Vicia faba*	Lebanon
17.	*Bruchus ervi*	*Lens culinaris*	Lebanon
18.	*Bruchus tristis*	*Lathyrus sativus*	Lebanon
Weeds			
19.	*Atriplex patula*	*Carthamus tinctorius*	USA
20.	*Avena barbata*	*Hordeum vulgare*	Morocco
21.	*Bromus secalinus*	*Hordeum vulgare*	Morocco, USA
22.	*Centaurea melitensis*	*Hordeum vulgare*	Morocco
23.	*Centaurea iberica*	*Peganum harmala*	Afghanistan
24.	*Convolvulus erubescens*	*Cicer arietinum*	Lebanon
25.	*Echinochloa crus-pavonis*	*Oryza sativa*	China, The Philippines, Switzerland, Vietnam, USA
	Fallopia convolvulus	*Triticum aestivum*	Poland
27.	*Galium tricornutum*	*Hordeum vulgare, Lens culinaris Triticum aestivum*	Lebanon, Morocco
26.	*Galium aparine*	*Hordeum vulgare, Triticum aestivum*	Lebanon, Morocco
29.	*Lolium rigidum*	*Hordeum vulgare*	Morocco
30.	*Phalaris paradoxa*	*Hordeum vulgare, Triticum aestivum*	Morocco
31.	*Polygonum cuspidatum*	*Hordeum vulgare, Solanum lycopersicum, Triticum aestivum*	Morocco, Poland, USA
32.	*Sinapsis arvensis*	*Hordeum vulgare, Triticum aestivum*	Lebanon

10.6 RECENT IAS INCURSIONS AND THEIR ECONOMIC IMPLICATIONS

10.6.1 Wheat Blast (*Magnaporthe oryzae* Pathovar *Triticum*)

First discovered in Northern Paraná State, Brazil, in 1985 (Igarashi et al., 1986), wheat blast disease caused by *Magnaporthe oryzae* pathovar *Triticum* (MoT) was restricted to South America, where sporadic outbreaks in the tropical, warmer, growing regions of Brazil and Bolivia were reported (Duveiller et al., 2011). After the devastating blast outbreaks in March of 2016, the pathogen was found for the first time in Bangladesh, where it devastated fields in 10 districts (Bishnoi et al., 2021). During the crop season 2016–2017, some stray fields in the international border regions in Murshidabad and Nadia districts of West Bengal, India, also experienced blast disease. The onslaught was proactively addressed as a result of the coordinated biosecurity vigil of ICAR-IIWBR, State Agriculture Universities (SAUs), and State Department of Agriculture. Crops in infected fields were burned in an attempt to prevent inoculum survival, but blasts reoccurred in Bangladesh, in 2017. The Wheat Research Center (WRC) of Bangladesh warned that wheat blast disease had the chance to spread to China, India, and Pakistan, which rank second, third, and seventh, respectively, in global wheat production. The magnitude of the outbreak, due to the lack of management, can be imagined by the fact that the wheat blast-vulnerable area of India is equal to 21% of the total area under wheat cultivation in India. The situation becomes more alarming in the absence of effective control measures, including a lack of information about resistant varieties. However, as a result of a timely and proactive coordinated effort by SAUs, State Department of Agriculture, ICAR, IIWBR, and DAC&FW, Government of India, the disease was successfully managed and prevented from spreading in the major wheat areas of West Bengal and adjoining states. This is one of the notable success stories regarding the management of invasive disease in India (Singh et al., 2019).

10.7 BIOSECURITY MEASURES ADOPTED

To prevent the disease from domesticating in Indian territories, especially in the Bangladesh bordering districts of West Bengal, a wheat crop holiday, i.e. banning wheat cultivation in Murshidabad and Nadia districts of West Bengal, was declared by Govt of India. Farmers were provided support and incentives for growing alternate crops (non-poaceaous crops such as gram, *urad*, and oilseed crops such as rapeseed, mustard, and potatoes). A "no wheat zone" was recommended in a 5-km area along the international border (Anonymous, 2017b). This is one of the most proactive roles, ICAR-Indian Institute of Wheat and Barley Research (IIWBR) and *National Plant Protection Officers (NPPO)* under the aegis of the Indian Council of Agricultural Research (ICAR) has played to thwart any chance of threat to wheat biosecurity in the county. In addition to this, the Government of India has initiated research and development on wheat blast since 2016 with the characterization of Indian wheat lines for blast resistance in collaboration with CIMMYT, Mexico.

The border regions in West Bengal were planted with "wheat trap nurseries" to curb the pathogen movement in every season (Bishnoi et al., 2021). Also, in 2017, the Department of Agricultural Research and Education (DARE) under the Government of India approved Bolivia as the hotspot for wheat blast, which helped ICAR-IIWBR to constitute a set of 40 Indian popular varieties and advanced lines. The lines were sent to the National Institute for Agricultural and Forestry Innovation (INIAF), Bolivia, for screening under endemic conditions through CIMMYT, Mexico, USA, Jessore, and Bangladesh for screening under containment facilities. Since then, sets of Indian wheat varieties have been routinely sent for screening at the precision phenotyping platforms in Bolivia and Bangladesh, resulting in identification of the same blast-resistant varieties. These included HD 3249, DBW 187, and HD 2967 for irrigated and timely sown conditions, and HD 3293, DBW 252, and HD 3171 for restricted irrigation and timely sown conditions. Among these, HD 2967 is a popular variety in the North West Plain Zone (NWPZ), and DBW 187 (Karan Vandana) is the latest and highest-yielding variety recently released and recommended for the North East Plain Zone (NEPZ). Apart from this, the available potential donors (BH 1146, Milan, SHA7, *Aegilops tauschi* (derivatives), varieties carrying *Lr34*, and genotypes possessing 2NS translocation) are being exploited for introgression breeding for resistance in Indian cultivars at ICAR-IIWBR as part of the Wheat Blast anticipatory breeding research initiative (Bishnoi et al., 2021).

10.8 WHEAT STEM RUST – UG99 (RACE TTKSK): A LOOMING THREAT TO INDIAN WHEAT INDUSTRY

Stem rust (*Puccinia graminis* f.sp. *tritici*) is historically the most damaging disease of wheat. In 1999, a new virulent race of stem rust, popularly known as Ug99 (named after the year and country of discovery), was identified in wheat fields in Uganda. Ug99 (Race TTKSK) is a cause for concern as it exhibits unique virulence patterns. No other race of stem rust has been observed to overcome so many wheat resistance genes, including the broad-spectrum gene *Sr31*. By 2007, strains of TTKSK in Ug99 had spread via wind movements out of East Africa (Wanyera et al., 2006) into Yemen and as far as Iran (Nazari et al., 2009). The subsequent confirmation of Ug99 (race TTKSK) in Iran was declared by FAO (Food and Agriculture Organization) in March 2008 (FAO, 2008; Pretorius et al., 2010). Rust pathogens change rapidly, often by mutation. Six additional variants are now recognized in the Ug99 lineage. Additional important resistance genes, e.g. *Sr24* and *Sr36* have now been defeated by variants of Ug99. Due to favourable environmental conditions for spread and outbreak, the wheat stem rust race Ug99 (TTKSK) is predicted to pose a serious threat to the wheat-producing countries (Singh et al., 2011).

10.9 BIOSECURITY MEASURES ADOPTED

The Indian wheat programme has a collaboration with CIMMYT to identify various resistant cultivars so as to strategize their deployment in wheat zones in case of any possible outbreak of Ug99. A set of 19 Indian wheat varieties and 3 genetic stocks were screened under the natural outbreak of Ug99 at Njoro (Nakuru), Kenya, in the summer nursery in 2005 (Joshi et al., 2008). Until 2011, a total of 22 wheat varieties resistant to the Ug99 race were developed by India, which included DBW 17, PBW 550, and Lok 1. Other resistant varieties are GW273, GW322, HI1500, HD2781, MP4010, HUW510, MACS 2846 (durum), HI8498 (durum), UP2338, DL153-2, HW1085, Raj4120, and PBW343 (Roy et al., 2011). A variety HW 1085 developed by IARI regional station, Wellington, and three genetic stocks (FLW2, FLW6, and FLW8 at DWR Regional Station, Shimla, India) also possessed resistance against race Ug99. The existing data suggests that varieties such as HI 1531, HI 8627, HD 4672, MACS 2846, NI 5439, 9498, WH 147, GW 322, HI 8663, UAS 321, DL 788-2, MPO 1215, NIDW 295, Lok 1, HI, and UAS 431, which are under cultivation in

Central and Peninsular India, possess substantial resistance to Ug99 variants (Sudha et al., 2020). To date, Ug99 pathotypes have not been confirmed/reported from India (Rautela and Dwivedi, 2018). Hence, the Ug99 race, despite the existing threat, can be regarded as another example of successful biosecurity efforts by scientists and the Government of India.

10.10 PANAMA DISEASE OF BANANA (*FUSARIUM OXYSPORUM* F.SP. *CUBENSE* TROPICAL RACE 4)

The pathogen was first detected approximately 50 years ago in Southeast Asia and spread to countries within the greater Mekong subregion and to Australia. In the Middle East, TR4 was first reported from Jordan and Lebanon and later from Israel in 2016 (Maymon et al., 2020). Recently, the pathogen was reported in India, Pakistan, Oman, and Mozambique (Africa) and was identified on the South American continent in Colombia in 2019. In India, the disease was first reported from Bihar (Thangavelu et al., 2019). Subsequently, the symptoms of *Fusarium* wilt of banana caused by the TR4 strain were observed in a block of Grand Naine in the Faizabad district of Uttar Pradesh (Damodaran et al., 2019a). The affected plants exhibited distinct yellowing symptoms on mature leaves, progressing towards the younger leaves. The laminae of the emerging leaves were markedly reduced and shrivelled. The infected leaves collapse gradually and often break at the petiole.

The TR4 strain of *F.o. cubense* is more severe in Cavendish banana plantations. The variety was responsible for reviving the banana industry that was earlier threatened by race 1 of Fusarium wilt, way back in 1874. Exact global estimates of losses are not available, but in the Philippines, the disease is reported to have affected around 15,700 ha of banana plantations (out of a total acreage of 4,40,000 ha). About 70% of plantations in China's Guangdong and Hainan provinces have also been reported to be affected by the TR4 strain of Panama wilt. Annual economic losses caused by TR4 have been estimated at US$121 million in Indonesia, US$253 million in Taiwan, and US$14 million in Malaysia. In Mozambique, TR4 caused such severe damage to a 1,500-ha plantation within four years of the first detection of the disease that the farm was forced to cease operations (Altendorf, 2019).

India being the largest producer of bananas in the world, a mere 10% loss would amount to approximately US$321 million.

10.11 MANAGEMENT OF BANANA WILT

Successful management was reported in India in biological management through the use of ICAR-FUSICONT. The Bio consortium FUSICONT, composed of the CSR-T-3 strain of *Trichoderma reesei* along with a strain of *Pseudomonas fluourescens*, was effective in managing the disease by 85%. Bio-immunization of banana plantlets using biomolecules characterized from the biocontrol agents in the consortium imparted enhanced host tolerance, resulting in the control of *F.o. cubense* TR4 strain to the extent of 95% when coupled with a two-time application of ICAR-FUSICONT (Damodaran et al., 2019b, 2021).

10.12 THE NEGATIVE IMPACT OF IAS

- **Biodiversity loss:** IAS are a major cause of biodiversity loss. According to analysis of the International Union for Conservation of Nature (IUCN) Red List, IAS are the second most common threat associated with species that have gone completely extinct, e.g. *Parthenium hysterophorus* weed, which came to India accidentally with a consignment of wheat from the United States and has spread over large forest areas at the expense of native species.

  (IUCN Red List is a catalogue of taxa that are facing the risk of extinction with the purpose of developing awareness about the threat of loss of biodiversity, identification and documentation of endangered species, providing a global index of the decline of biodiversity and defining conservation).
- **Change in the ecosystem:** IAS can also lead to changes in the structure and composition of ecosystems, leading to significant detrimental impacts on ecosystem services and affecting economies and human well-being. For example, coffee rust caused the shifting of coffee plantations with tea in Ceylon.
- **Loss of habitat:** Invasive species can have a number of negative impacts on the areas they invade. Perhaps the most significant of these is the widespread loss of habitat, e.g. water hyacinth, native to tropical America, is an aggressive colonizer and grows abundantly in still or slow-flowing waters. The species is a nuisance for an aquatic ecosystem.
- **May kill a large number of endemic species or may lead to extinction:** The introduction of non-native species and range expansions of native species with changing land use and climate may have profound consequences for the ecosystems they occupy. Invasive species drive ecological dynamics at multiple spatial scales and levels of organization through local and regional extinctions of native species, e.g. chestnut blight (Mack et al., 2000; Crowl et al., 2008). The disease has resulted in wiping out the chestnut from eastern USA.
- **Lead to resource crunch:** Invasive species can also threaten native species by outcompeting them for resources. For example, Asian carp introduced into the United States outcompete native fish for both food and space, leading to large declines in native fish populations.
- **Impact on human health:** Invasive species can also impact human health when they accumulate toxins in their tissues. When other organisms prey on these species, the toxins are passed up the food chain and can also enter animals consumed by humans. Infectious disease agents may themselves be invasive species or may be introduced by invasive vectors, e.g. introduction of species of mosquitoes carrying malaria parasites.
- **Economic cost:** In addition to these impacts, invasive species can also have enormous economic costs. For example, water hyacinth poses a threat to fisheries.

10.13 STEPS TO CONTROL IAS

IAS are a global issue that requires international cooperation and actions. Preventing the international movement of IAS and rapid detection at borders are less costly than control and eradication. DRDO-DIHAR (Defence Institute of High-Altitude Research) has recorded some species of fungi, bacteria, and viruses in the Ladakh regions that have not been documented so far (Chakrabarty, 2021). Preventing the entry of IAS is carried out through inspections of international shipments, customs checks, and proper quarantine regulations.

- Article 8(h) of CBD and Aichi Target 9 aim to control or eradicate alien species that threaten ecosystems, habitats, and species.
- The Global Invasive Species Programme is supporting the implementation of Article 8(h) of CBD with IUCN as a partner organization and is also working to address the global threat to IAS.
- IUCN's Invasive Species Specialist Group has also been working to promote and facilitate the exchange of IAS information and knowledge across the globe and ensure linkages between policymaking and flow of knowledge.
- IUCN has also developed a number of global databases that provide critical information on IAS, such as the Global Invasive Species Database and the Global Register of Introduced and Invasive Species.
- Intervention (legally binding punitive action).

10.14 POLICIES FOR THE MANAGEMENT OF INVASIVE ALIEN SPECIES

The versatile struggle for the management of IAS has been at the apex of policy programme all over the world. Methods to control, eradicate, and prevent IAS are one of the main challenges that need to be addressed in policy formation. In developing countries, the policy challenge posed by invasive species is a serious issue as they arise due to prevailing complex social, economic, and political problems (Chenje and Mohamed-Katerere, 2006; Rajasekaran et al., 2020).

10.14.1 International Policies Related to IAS

Several international policies have been developed to deal with IAS, both directly and indirectly.

10.14.1.1 Convention on Biological Diversity

CBD was formed in 1993 at the UN Conference held in Rio de Janeiro in 1992 on Environment and Development and has been ratified by nearly 180 countries. It calls on 178 governments as of 2000 to "prevent the introduction, control, or eradicate those alien species that threaten ecosystems, habitats, or species" (Article 8h).

Major goals of CBD:

1. To take appropriate measures to conserve biological diversity,
2. To ensure the sustainable use of biological resources, and
3. To promote the fair and equitable sharing of benefits arising from the utilization of genetic resources.

Under the CBD, governments agree to prepare national biodiversity strategies and action plans; identify genomes, species, and ecosystems crucial for conservation and sustainable use; monitor biodiversity and factors that are affecting biological systems; establish effectively managed systems of protected areas (PAs); rehabilitate degraded ecosystems; exchange information; conduct public information programmes; and undertake various other activities for implementing the objectives of the CBD. The convention's governing body is the Conference of the Parties (COP), consisting of all governments (and regional economic integration organizations) that have ratified the treaty. This ultimate authority reviews progress under the Convention, identifies new priorities, and sets work plans for members. The COP can also make amendments to the Convention, create expert advisory bodies, review progress reports by member nations, and collaborate with other international organizations and agreements. The other main bodies of CBD are the Secretariat, based in Montreal, Canada, Subsidiary Body for Scientific, Technical and Technological Advice (SBSTTA) and Subsidiary Body on Implementation (SBI). The Subsidiary Body on Scientific Technical and Technological Advice (SBSTTA) of the CBD meets periodically and addresses key issues, including invasive aliens. The CBD has also established an interim financial mechanism, the Global Environment Facility (GEF), which provides approximately US$100 million per year to projects for implementing the Convention in developing countries (http://www.biodiv.org; McNeely et al., 2001).

10.14.1.2 The Conference of the Parties

The CBD has recognized that there is an urgent need to address the impact of IAS and established IAS as a cross-cutting issue at its fourth meeting. The decision of COP 6 included the adoption of Guiding Principles for the Prevention, Introduction, and Mitigation of Impacts of Alien Species that Threaten Ecosystems, Habitats, or Species (decisionVI/23); https://www.cbd.int/decision/cop/?id=7197.

The guiding principles are as follows:

1. ***Precautionary approach***: If an alien species has a suspected risk of causing harm to the ecosystem, habitat, or native species, the proof of burden that "the introduction is NOT harmful" falls on those who introduce the alien species. In order to protect the environment, the precautionary approach shall be widely applied by states according to their capabilities. Where there are threats of serious or irreversible damage, lack of full scientific certainty shall not be used as a reason for postponing cost-effective measures to prevent environmental degradation.
2. ***Three-stage hierarchical approach***: Response to IAS based on prevention is the first line of defence. Early detection and rapid action when prevention

fails, eradication if feasible being the second line, and, finally, management and containment of established invasions is the third and final approach.

3. ***Ecosystem approach***: Management of invasive species is not solely based on the removal or eradication of invading species but also considers the invaded ecosystem, involves communities and stakeholders, and adopts a multi-sectoral approach. The integrated management of land, water, and living resources to render them unfavourable for the invasive pest will effectively support the implementation of the CBD and its Strategic Plan for Biodiversity 2011–2020.
4. ***State responsibility***: Provision of measures that fall under each responsible authority on environment, agriculture, trade, transport, industry, science, and health can reduce the risks and impacts of IAS, including pests and diseases.
5. ***Research and monitoring***: The concept is to develop an adequate knowledge base to address the problem, and appropriate research on and monitoring of alien invasive species should be undertaken. This should document the history of invasions (origin, pathways, and time period), characteristics of the alien invasive species, the ecology of the invasion, and the associated ecological and economic impacts and how they change over time. Monitoring is the key to the early detection of new alien species. It requires targeted and general surveys, which can benefit from the involvement of local communities.
6. ***Education and public awareness***: Citizens and stakeholders in biodiversity can take effective steps with sound information on IAS, pests, and diseases collectively.
7. ***Border control and quarantine measures***: International trade, transport, and tourism are pathways of IAS. Border controls and quarantine measures for pests and diseases can include IAS in the work of border controls to stop entries of invasive and suspected alien species.
8. ***Exchange of information***: Surveillance (monitoring and reporting on invasions) is a fundamental mechanism to manage IAS, including pests and diseases. Information on invasive species with the correct taxonomic name and geographic reference is critically important for prevention.
9. ***Cooperation, including capacity-building***: Co-operative efforts between two or more countries, such as:
   a. Where a state of origin is aware that a species being exported has the potential to be invasive in the receiving state, the exporting state should provide information, as available, on the potential invasiveness of the species to the importing state. Particular attention should be paid where exporting parties have similar environments.
   b. Agreements between countries on a bilateral or multilateral basis should be developed and used to regulate trade in certain alien species, with a focus on particularly damaging invasive species.
   c. States should support capacity-building programmes for states that lack the expertise and resources, including financial, to assess the risks of introducing alien species. Such capacity-building may involve technology transfer and the development of training programmes.

10. ***Intentional introduction***: The intentional introduction of a known invasive species in a recipient/importing country could be intercepted by the national authority. Suspected alien species could be subjected to an appropriate risk analysis process prior to importation/introduction (burden of proof is with the proposer of the introduction or be assigned as appropriate by the recipient state).
11. ***Unintentional introduction***: Common pathways of unintentional introduction include: escape from a confined condition; transports to a way (e.g. ballast water, biofauling, and hitchhikers); transport-contamination (e.g. wood packaging, contaminated timber, soil, or media). Appropriate pathway management measures need to be in place.
12. ***Mitigation of impacts***: Once the establishment of an IAS has been detected, states, individually and cooperatively, should take appropriate steps, such as eradication, containment, and control, to mitigate adverse effects.
13. ***Eradication***: The best opportunity for eradicating IAS is in the early stages of invasion, when populations are small and localized; hence, early detection systems focused on high-risk entry points can be critically useful, while post-eradication monitoring may be necessary.
14. ***Containment***: When eradication is not feasible, limiting the spread (containment) of IAS is an appropriate strategy in cases where the range of the organisms or of the population is small enough. Regular monitoring is essential, and it needs to be linked with rapid response to eradicate new area of infestation.
15. ***Control***: When complete eradication or containment is possible, reducing the damage caused as well as the number of IAS would be an option.

10.14.1.3 The International Plant Protection Convention

The International Plant Protection Convention (IPPC) is an intergovernmental, multilateral treaty that has been in force since 1952 and has been signed by over 180 countries. The Convention aims to protect the world's plant resources from the spread and introduction of pests and promote safe trade. The IPPC Secretariat, housed at FAO in Rome, facilitates the development and use of International Standards for Phytosanitary Measures (ISPMs) as its instrument to achieve the above objectives. IPPC is thus the sole global standard setting organization for plant health and to promote appropriate measures for their control. The implementation of the Convention was applied mainly to crops, but it also extended to the protection of natural flora. Thus, the scope of the IPPC covers any IAS that may be considered to be a plant pest.

The IPPC is one of the three entities of the Sanitary and Phytosanitary Measures (SPS) Agreement recognized by the World Trade Organization (WTO) (http://www.ippc.int; McNeely et al., 2001). The other two are the Codex Alimentarius Commission for food safety standards and the World Organization for Animal Health (OIE) for animal health standards.

The Commission on Phytosanitary Measures (CPM) is the main governing body of the IPPC and holds its sessions annually. The CPM promotes implementation of the Convention's objectives, including reviewing the state of plant protection in the world and actions to control the international spread of pests and their introduction into endangered areas; establishing and keeping under review the necessary institutional arrangements and procedures for the development and adoption of international standards; and adopting international standards.

Besides the CPM, the IPPC is aided by a number of committees and planning groups, viz. the Standards Committee (SC), the Strategic Planning Group (SPG), and the Financial Committee (FC).

Despite the existence of such international conventions, the growing impact of IAS equally on global economies and the environment suggests that these international instruments need robust implementation and strengthening to prevent and combat IAS effectively. Furthermore, expanding international trade is moving ever more organisms more rapidly around the world, thereby increasing the threat of these species to native ecosystems and potentially overwhelming government efforts to prevent unwanted invasions. In response to these concerns, the global scientific community established the Global Invasive Species Programme (GISP) in 1997.

10.14.1.4 The Global Invasive Species Programme

The GISP is an independent, not-for-profit association. Its mission is to conserve biodiversity and sustain human livelihoods by minimizing the spread and impact of IAS, which is located in the Centre for Agriculture and Biosciences International (CABI) in Nairobi, Kenya (Kelly, 2009).

It was initially developed in January 1996 and established in 1997 to address the global threats caused by IAS and to provide support for the implementation of Article 8(h) of the Convention on Biological Diversity. GISP is operated by a consortium of the Scientific Committee on Problems of the Environment (SCOPE), CAB International (CABI), and the World Conservation Union (IUCN), in partnership with the United Nations Environment Programme (UNEP). GISP is a component of DIVERSITAS (an international programme on biodiversity science). GISP seeks to improve the scientific basis for decision-making on invasive species; develop capacities to employ early warning and rapid assessment and response systems; enhance the ability to manage them; reduce the economic impacts of invasive species and control methods; develop better risk assessment methods; and strengthen international agreements. GISP also strives to develop public education about invasive species, improve understanding of the ecology, examine legal and institutional frameworks to control them, develop new codes of conduct for the movement of the species, and design new tools to quantify their impact. GISP involves the voluntary contribution of a substantial group of scientists, lawyers, and managers from all parts of the world (http://www.stanford.jasper.edu/gisp/).

GISP has contributed widely to the knowledge and awareness of IAS and has prepared a guide, "Invasive Alien Species: A Toolkit of Best Prevention and Management" to address the problem. This guide is designed to aid in the elaboration and adoption of an effective national strategy by pointing to experiences in various nations (Wittenberg and Cock, 2001). Many challenges, however, still need to be further addressed, and a continuation of the programme is recommended. They have proposed activities to implement five global initiatives:

1. Global access to information on IAS threats and their prevention and management;
2. Directed action at key pathways of IAS introduction through public and private sector cooperation;
3. Acceleration of critical research and its dissemination;
4. Awareness-raising and support for policy development; and
5. Building cooperation between institutions towards a global biosecurity platform to mitigate the threat of IAS.

This global programme aims to help governments and development agencies identify and initiate national and regional projects in order to mitigate threats posed by IAS. It will also provide support to the existing projects and initiatives, develop national and international capacity, and facilitate international networking through the utilization and adaptation of the GISP Toolkit of Prevention and Best Management Practices and other elements developed in Phase I. It will also use outputs generated by international GISP activities (McNeely et al., 2001).

10.14.1.5 The International Union for Conservation of Nature (IUCN) and IAS

IUCN is an international organization located in Gland, Switzerland, working in the field of nature conservation and sustainable use of natural resources (Anonymous, 2021a) with a mission to "influence, encourage and assist societies throughout the world to conserve nature and to ensure that any use of natural resources is equitable and ecologically sustainable." The IUCN is widely known for compiling and publishing the IUCN Red List of Threatened Species, which assesses the conservation status of species worldwide (Anonymous, 2021a).

The efforts of IUCN regarding IAS are mainly focused on achieving Aichi Target 9, which states that "by 2020, IAS and pathways are identified and prioritized, priority species are controlled or eradicated, and measures are in place to manage pathways to prevent their introduction and establishment" (https://www.cbd.int/sp/targets/). In order to achieve this, IUCN has been working in three major areas: providing scientific knowledge, engaging in and supporting national and regional policy development, and taking action on the ground. A large amount of the work IUCN undertakes on IAS is through the Invasive Species Specialist Group (ISSG), which is a global network of scientific and policy experts on invasive species. Organized by the IUCN Species Survival Commission, the ISSG aims to reduce threats to natural ecosystems and the native species they contain by increasing awareness of IAS and ways to prevent, control, or eradicate them (https://www. iucn.org/theme/species/our-work/invasive-species).

There are a number of knowledge databases developed by IUCN that offer critical support to decision-making processes related to IAS, like "The Global Invasive Species Database" and "The Global Register of Introduced and Invasive Species." Besides, some others, like the Island Biodiversity and Invasive Species Database the

IAS Pathways Management Resource, and classifying the environmental impact of different IAS, are under development.

IUCN is the leading provider of conservation data, assessments, and analysis and an implementing agency for conservation projects around the world. IUCN champions nature's role in achieving the Sustainable Development Goals (SDGs). It also serves as an official agency monitoring progress towards biodiversity-related targets. The SDGs objective was to produce a set of universal goals that meet the urgent environmental, political, and economic challenges facing the world.

10.14.2 National Conventions and Legislations to Support Management of IAS in India

India is a signatory to different international conventions, treaties, and agreements dealing with invasive species, such as

1. Convention on Biological Diversity (Nairobi, 1992);
2. Cartagena Protocol on Biosafety to the Convention on Biological Diversity (Montreal, 2000);
3. United Nations Convention on the Law of the Sea (Montego Bay, 1982);
4. The Convention on Wetlands of International Importance, especially as Waterfowl Habitat (Ramsar, 1971);
5. IUCN Guidelines for the Prevention of Biodiversity Loss caused by Alien Invasive;
6. Agenda 21-United Nations Conference on Environment and Development (Rio, 1992); and
7. Asia Pacific Forest Invasive Species Network and FAO Regional Office, Bangkok (http://apfisn.net/sites/default/files/India.pdf).

India is also engaged with several organizations that deal with invasion research, such as the World Conservation Union (WCU), GISP, Invasive Species Advisory Council (ISAC), Invasive Species Specialist Group (ISSG), and National Invasive Species Council, 2001. Furthermore, Article 8(h) of the CBD calls on member governments to "as far as possible and appropriate, prevent the introduction of, control or eradicate those alien species which threaten ecosystems, habitats or species."

India has a number of different legislations relating to microbially invasive species, such as

- The Destructive Insects and Pests Act, 1914, and amendments
- Indian Forest Act, 1927
- Forest (Conservation) Act, 1980
- Environment Protection Act, 1986
- The Plants, Fruits and Seeds (Regulation of Import into India) Order 1989 (PFS Order 1989)
- National Policy and Macro level Action Strategy on Biodiversity, 1999
- The Biological Diversity Act, 2002

- The Plant Quarantine (Regulation of Import into India) Order, 2003
- National Environment Policy, 2004

Also, India has a number of agencies that are engaged to prevent the introduction of invasive species and their management and control. Some of these include:

1. DPPQS (Directorate of Plant Protection, Quarantine, and Storage), Ministry of Agriculture and Farmer's Welfare
2. ICAR – NBPGR (National Bureau of Plant Genetic Resources)
3. ICAR – NBAIR (National Bureau of Agricultural Insect Resources)
4. ICAR – NBAIM (National Bureau of Agriculturally Important Microorganisms)
5. ICAR – NBAGR (National Bureau of Animal Genetic Resources)
6. ICAR – NBFGR (National Bureau of Fish Genetic Resources)
7. The Ministry of Environment Forests and Climate Change

However, concerted efforts and effective coordination among concerts interministorial depth are lacking in effective management of the IAS. India lacks a robust core policy and legislative framework to manage the IAS. Moreover, one of the major bottlenecks in policy development and implementation has been a lack of awareness of the causes and consequences of biological invasions taking place in different regions.

10.15 ROLE AND PREPAREDNESS OF INDIA TO THWART ATTEMPTS OF AGROTERRORISM

Deliberate use of biological agents to destroy agriculture is broadly termed as "agroterrorism" (Monke, 2007). Agroterrorism can be used to seriously destabilize socioeconomic matters in an agriculture-based state by targeting its food crops, cash crops, agriculture-based industries, etc. Different state and non-state (terrorist) organizations have purposefully used biological agents for various nefarious purposes (Sibnarayan et al. 2020). Agriculture or agriculture-based sectors are relatively easy targets, attacks are difficult to detect, and they have far-reaching effects as compared to biological warfare against humans. At present, agroterrorism is primarily aimed at disrupting food security or crippling the economy (Anonymous, 2008; Chakrabarty, 2021).

The large population, dependence on the agriculture sector, ill-developed surveillance system for crop and animal diseases, and time gap between disease diagnosis and adoption of remedial measures make Indian agriculture highly vulnerable to attack. The development of infrastructure required for an efficient surveillance system with a rapid communication system, quick diagnosis of crop and animal diseases, development of a pathogen database for major crops, development of resistance varieties, and critical assessment of genetically modified crops are required to be developed for sustainable economic growth and quality food production. In addition, a sound democratic political system to ensure the distribution of life-sustaining food at times of scarcity, a capable legal system to check and punish the offender, and international cooperation for information exchange and extradition of bioterrorists are also required to tackle the threat of agroterrorism (Banerjee, 2007).

10.16 BIOLOGICAL AND TOXIN WEAPONS CONVENTION

During the 1970s, sensing the impending dangers of rapidly proliferating chemical and biological weapons, the international community constituted the Biological and Toxin Weapons Convention (BTWC) in 1972 to "prohibit the development, production, acquisition, transfer, retention, stockpiling and use of biological and toxin weapons" as invasive microbial species can also serve as biological weapons (Falkenrath et al., 1998). India is a signatory to the BWC and is a member among 183 countries. It includes the following articles:

- **Article I**: Never under any circumstances to develop, produce, stockpile, acquire, or retain biological weapons.
- **Article II**: To destroy or divert to peaceful purposes biological weapons and associated resources prior to joining.
- **Article III**: Not to transfer, or in any way assist, encourage, or induce anyone else to acquire or retain biological weapons.
- **Article IV**: To take any national measures necessary to implement the provisions of the BWC domestically.
- **Article V**: Undertaking to consult bilaterally and multilaterally and cooperate in solving any problems that may arise in relation to the objective, or application, of the BWC.
- **Article VI**: Right to request the United Nations Security Council to investigate alleged breaches of the BWC, and undertaking to cooperate in carrying out any investigation initiated by the Security Council.
- **Article VII**: To assist states that have been exposed to danger as a result of a violation of the BWC.
- **Article X**: Undertaking to facilitate, and having the right to participate in, the fullest possible exchange of equipment, materials, and information for peaceful purposes.

In order to strengthen the convention, State Parties have focused on articles VI, VII, and V. Article VII of the BWC provides for assistance; however, the Convention "contains no operational capacity to respond to biological attacks." Rendering assistance is a legal obligation under Article VII of the Convention. It is clear that the absence of clearly laid out procedures for seeking assistance and the associated techno-political-legal questions related to Article VII have led to a deficit in the ability of the global community to respond effectively and provide assistance to States Parties exposed to dangers as a result of violations of the BWC. In this context, India and France have proposed the establishment of a database for assistance in the framework of Article VII.

To clear the deficit in implementation of Article III of the BWC and by recognizing the importance of building broad-based support for strengthening implementation of Article III, India, along with the United States, has submitted a Working Paper titled "Strengthening Implementation of Article III of the BTWC." Effective implementation of Article III would also ensure that the cooperation envisaged under Article X is taken forward in mutual confidence. Equitable benefits for States

Parties, in particular developing countries, resulting from effective implementation of Article X will also help in promoting the universalization of the Convention. India has maintained that issues related to the alleged use of BWTC (Article VI) should be addressed by the States Parties in accordance with the provisions of the Convention (Anonymous, 2021b).

10.17 RECOMMENDATIONS FOR MANAGEMENT OF IAS

India needs to devise a clear and stringent policy on IAS, assess landscape dynamics in an intensive manner, and conduct detailed research for practical solutions by leveraging the resources and coordination among different research organizations/ institutions such as DPPQS, DAC, ICAR-NBPGR, CSIR, DRDO, DBT, and Ministry of commerce, along with public participation.

- To better monitor and control the most possible routes by which species arrive, as well as enable a risk assessment of potential invasive species prior to introduction. India not only needs a coordinated national effort to develop an inventory and document invasive species, but also the pathways by which they are introduced. This comprehensive documentation would also enable pre-emptive assessment of alien species yet to become invasive (Hiremath and Krishnan, 2016).
- India should seriously consider to depute biosecurity personnel in the international airports/other ports of entry to screen incoming passengers for any plants/food materials as sources of IAS and disallow any potential sources to prevent their introduction.
- For the assessment of invasive species, mapping the extent of species invasion can be achieved using the Geographical Information System, remote sensing, and time series analysis. Adhikari et al. (2015) illustrated the usefulness of open source data and software in generating a knowledge base with practical implications for IAS management by understanding and analysing the alien species invasion problem. Such an effort will provide guidance during the formulation of an effective policy and management strategy for controlling the IAS in the country.
- India should aim to identify PAs and strengthen guidelines to enforce domestic quarantine to safeguard PAs from the serious scourge of invasive species. There is an urgent need to relook at the policies governing PA management in the country. Comprehensive international and national action is required: Numerous global and regional policies are already addressing the problem of IAS. Coordination of implementation and practical cooperation among those responsible for these instruments, however, are highly insufficient and need to be strengthened. A practical prevention, eradication, and control measures are also inadequate. A consolidated action plan is earnestly recommended and is the need of the hour to preserve the environment and agri-biosecurity. The Convention on Biodiversity (CBD) and the IPPC could take the lead, but trade, transport, travel, and other economic

sectors must be closely involved and engaged. Other institutions, including the United Nations Environment Programme (UNEP), the WTO, the FAO, and the International Maritime Organization (IMO), are key components at the international level. These institutions are supported by international non-governmental organizations (NGOs) such as the World Conservation Union (IUCN), World Wildlife Fund (WWF), Wetlands International, Conservation International, and the Nature Conservancy (TNC). Likewise, at the national level, consolidated and coordinated action is required. This could be part of a National Biodiversity Strategy and Action Plan, with close involvement of the economic sectors and identifying people responsible for operative actions involving potential IAS as a key prerequisite. Clear responsibilities for each relevant sector should be identified/fixed.

- Insurance mechanisms and liability of regulations for the spread of IAS are almost non-existent, presenting a major deficiency in controlling the problem. Governments should cooperate with the insurance sector to find solutions, beginning with feasibility studies.
- The capacity and expertise to deal with IAS are highly insufficient in many countries including, India. Capacity-building and further research on the biology and control of IAS and biosecurity issues should therefore be given attention and priority. This also relates to financial institutions and other organizations responsible for environmental and development cooperation at national and international levels.
- A national information system regarding the biology and control of IAS is urgently needed.

10.18 CONCLUSION

Invasive species are new species with the potential to modify and cause harm to the existing structure and function of ecosystems and the ecosystem services upon which people rely. The introduction of IAS, including pests and pathogens, through various pathways alters the productivity of crops, thus causing environmental as well as economic losses. The introduction of IAS can happen intentionally or unintentionally. Monitoring is required to determine the location/early detection of problematic species, including weeds, pathogens, and insect pests. Monitoring can be done through qualitative approaches such as species inventory and quantitative approaches using phyto-sociological methods and mapping using ground-based methods. Prevention of introductions is the only cost-effective option. The measures involving early detection and contingency plans for the eradication/containment of invasive species can be efficiently implemented and would have a higher benefit cost outcome. In India, the economic costs of the management of IAS are yet to be analysed in detail. Specific legislation is needed in all countries to slow or prevent non-native species introductions. All introductions of non-native species, irrespective of purpose, should be strictly regulated. In addition, the government should make every effort to inform the public concerning the serious environmental and economic threats that are associated with alien species introductions.

REFERENCES

Adhikari, U., Nejadhashemi, A. P., Woznicki, S. A. 2015. Climate change and eastern Africa: A review of impact on major crops. *Food and Energy Security*, 4(2), 110–132.

Altendorf, S. 2019. Banana Fusarium Wilt Tropical Race 4: A Mounting Threat to Global Banana Markets? In *2019 Food Outlook – Biannual Report on Global Food Markets*, FAO, Rome, pp. 13–20.

Anonymous, 2008. Guidelines for Management of Agro-terrorism. In *National Disaster Management Guidelines-Management of Biological Disasters*, National Disaster Management Authority, Government of India, New Delhi, India, pp. 93–107..

Anonymous, 2017a. *Crop Production Statistics for Selected States, Crops, and Range of Years. Crop Production Statistics Information System. Special Data Dissemination Standard Division*, Directorate of Economics & Statistics, Ministry of Agriculture and Farmers Welfare, New Delhi. https://aps.dac.gov.in/APY/Public_Report1.aspx (accessed on June 27, 2017).

Anonymous, 2017b. Tackling wheat blast: Bengal government bans wheat cultivation within 5 km of B'desh border. Hindustan Times, July 19, 2017.

Anonymous, 2021b. Statement delivered by Ambassador (Dr.) Pankaj Sharma, Permanent Representative of India to the Conference on Disarmament during the Meeting of Experts (MX5) of the Biological Weapons Convention on Institutional Strengthening of the Convention on August 8, 2019 in Geneva. Biological and Toxin Weapons Convention. https://meaindia.nic.in/cdgeneva/?8332?000 (accessed on Nov 22–26, 2021).

Anonymous, 2021c. Apple plants imported from Italy hit by disease in Himachal. The Tribune.

Banerjee, K. 2007. *India and Agricultural Bioterrorism. Science and Technology to Counter Terrorism: Proceedings of an Indo-U.S. Workshop*. The National Academies Press, pp. 111–120.

Bhalla, S., Chalam, V. C., Singh, B., Gupta, K., Dubey, S. C. 2018. *Biosecuring Plant Genetic Resources in India: Role of Plant Quarantine*, ICAR-National Bureau of Plant Genetic Resources, New Delhi.

Bishnoi, S. K., Kumar, S., Singh, G. P. 2021. Wheat blast readiness of the Indian wheat sector. *Current Science*, 120(2), 262–263.

Butler, E. J. 1918. *Fungi and Diseases in Plants*, Thacker Spink and Co., Culcutta.

Chakrabarty, P. K. 2021. Agroterrorism – A threat to food safety and national biosecurity. *Indian Phytopathological Society Newsletter*, 4(1), 1–3.

Chalam, V. C., Maurya, A. K. 2018. Role of quarantine in ensuring biosecurity against trans-boundary plant viruses. *Agricultural Research Journal*, 55(4), 612–626.

Chalam, V. C., Parakh, D. B., Khetarpal, R. K. 2017b. Biosecurity from Plant Viruses in India. In *A Century of Plant Virology in India*, Springer, Singapore, pp. 757–782.

Charles Lis, H., Dukes, J. 2007. Impacts of Invasive Species on Ecosystem Services. In *Biological Invasions, Ecological Studies*, W. Nentwig (Ed.), Springer-Verlag, Berlin, Heidelberg, Vol. 193, pp. 217–237. doi:10.1007/978-3-540-36920-2_13.

Chenje, M., Mohamed-Katerere, J. 2006. Invasive Alien Species. In *Africa Environment Outlook 2-Our Environment, Our Wealth.*

CMI. 1988. *Distribution Maps of Plant Diseases*, Fourth Edition. Common Wealth Mycological Institute Publication, London, Map 221.

Cook, D. C., Fraser, R. W., Paini, D. R., Warden, A. C., Mark Lonsdale, W., De Barro, P. J. 2011. Biosecurity and yield improvement technologies are strategic complements in the fight against food insecurity. *PLoS One*, 6(10), e26084.

Crowl, T. A., Thomas, O. C., Robert, R. P., Belovsky, G., Lugo, A. E. 2008. The spread of invasive species and infectious disease as drivers of ecosystem change. *Frontiers in Ecology and Environment*, 6(5), 227–288.

Damodaran, T., Mishra, V. K., Jha, S. K., Gopal, R., Rajan, S., Ahmed, I. 2019a. *First Report of Fusarium Wilt in Banana Caused by Fusarium oxysporum f. sp. Cubense Tropical Race 4 in India*. Disease Notes, APS Publications. https://apsjournals.apsnet.org.

Damodaran, T., Rajan, S., Gopal, R., Yadav, A., Yadav, K., Shukla, P. K., Muthukumar, M., Kumari, N., Ahmad, I., Jha, S. K., Nayak, D., Mishra, V. K. 2019b. Successful community-based management of banana wilt caused by *Fusarium oxysporum* f. sp. *cubense* tropical race-4 through ICAR-FUSICONT. *Journal of Applied Horticulture*, 21(1), 37–41.

Damodaran, T., Rajan, S., Misra, M., Muthukumar, M., Sharma, P. C., Gopal, R., Yadav, K. 2021. In-vitro engineering of bio-immune formulation for inducting Fusarium wilt tolerance in banana plant. (Application No: 202111003761, Patent filed).

Diagne, C., Leroy, B., Vaissière, A. C., Gozlan, R. E., Roiz, D., Jarić, I., Salles, J.-M., Bradshaw, C. J. A., Courchamp, F. 2021. High and rising economic costs of biological invasions worldwide. *Nature*, 592, 571–576. doi:10.1038/s41586-021-03405-6.

Duveiller, E., Hodson, D., Sonder, K., von Tiederman, A. 2011. Wheat blast-A potential threat to global wheat production. *Phytopathology*, 101, S220.

Falkenrath, R. A., Newman, R. D., Thayer, B. A. 1998. *America's Achilles' Heel: Nuclear, Biological, and Chemical Terrorism and Covert Attack*, MIT Press, Cambridge, USA, 380 p. https://www.jstor.org/stable/j.ctt1fq9wk8 (accessed on June 20, 2019).

FAO. 2008. Wheat killer detected in Iran. https://www.fao.org/newsroom/en/news/2008/1000805/index.html (accessed on December 14, 2009).

Ganguly, A., Paul, D. K. 1953. Wart disease of potatoes in India. *Science and Culture*, 18, 605–606.

Govindarao, P., Koteswararao, D. 1956. Seedling blight of tobacco caused by *Phytophthora parasitica var. nicotianae* (Van breda de Haan), Tucker. *Indian Phytopathology*, 9, 145.

Hiremath, J. A., Krishnan, S. 2016. India knows its invasive species problem but this is why nobody can deal with it properly. The Wire (11 December 2016).

Igarashi, S., Utiamada, C. M., Igarashi, L. C., Kazuma, A. H., Lopes, R. S. 1986. Pyricularia em trigo. 1. Ocorrência de Pyricularia sp. no estado do Paraná. *Fitopatologia Brasileira*, 11, 351–352.

Islam, M. T., Gupta, D. R., Hossain, A., Roy, K. K., He, X., Muhammad, R., Kabir, M. R., Singh, P. K., Khan, M. A. R., Rahman, M., Wang, G. 2020. Wheat blast: A new threat to food security. *Phytopathology Research*, 2, 28.

Jones, F. G. W. 1961. Potato root eelworm *Heterodera rostochiensis* Woll. in India. *Current Science: Fortnightly Journal of Research, Bangalore*, 30, 187.

Joshi, A. K., Mishra, B., Prashar, M., Tomar, S. M. S., Singh, R. P. 2008. Ug99 race of stem rust pathogen: Challenges and current status of research to sustain wheat production in India. *Indian Journal of Genetics and Plant Breeding*, 68, 231–241.

Kelly, L. 2009. The Global Invasive Species Program. In *Global Program Review*, Vol. 3, The World Bank, ISBN: 978-1-60244-122-4.

Khetrapal, R. K., Gupta, K. 2007. Plant biosecurity in India: Status and strategy. *Asian Biotechnology and Development Review*, 20, 83–107.

Kushalappa, A., Eskes, A. 2003. Advances in coffee rust research. *Annual Review of Phytopathology*, 27, 503–531. doi:10.1146/annurev.py.27.090189.002443.

Mack, R., Simberloff, D., Lonsdale, W. M., Evans, H., Clout, M., Bazzaz, F. 2000. Biotic invasions: Causes, epidemiology, global consequences and control. *Ecological Applications*, 10, 2–20.

Mandal, F. B. 2011. The management of alien species in India. *International Journal of Biodiversity and Conservation*, 3(9), 467–473.

Mandal, F. B., Nandi, N. C. 2009. *Biodiversity: Concept, Conservation and Biofuture*, Asian Books Private Limited, New Delhi.

Mayee, C. D., Patil, M. A. 1986. Downy Mildew of sunflower. *Indian Phytopathology*, 39, 314.

Maymon, M., Sela, N., Shpatz, U., Galpaz, N., Freeman, S. 2020. The origin and current situation *of Fusarium oxysporum f. sp. cubense* tropical race 4 in Israel and the Middle East. *Scientific Reports*, 10, 1590. doi:10.1038/s41598-020-58378-9.

McNeely, J. A., Mooney, H. A., Neville, L. E., Schei, P. J., Waage, J. K. 2001. Global Strategy on Invasive Alien Species. In *Collaboration with the Global Invasive Species Programme*, IUCN, Cambridge, UK.

Mitra, M., Mehta, P. R. 1938. Some leaf diseases of *Hevea brasiliensis* new to India. *The Indian Journal of Agricultural Sciences*, 8, 185–188.

Monke, J. 2007. Agroterrorism: Threats and preparedness. In CRS Report for Congress, RL32521. The Library of Congress, Congressional Research Service, Homeland Security Digital Library. https://www.hsdl.org/?abstract&did=472158 (accessed on August 12, 2004).

Mottaleb, K. A., Singh, P. K., Sonder, K., Kruseman, G., Tiwari, T. P., Barma, N. C. D., Malaker, P. K., Braun, H.-J., Erenstein, O. 2018. Threat of wheat blast to South Asia's food security: An ex-ante analysis. *PLoS One*, 13, e0197555.

Mundkur, B. B., Kheshwala, K. F. 1943. A canker of apple trees in Mysore. *Indian Journal of Agricultural Science*, 13(4), 397–398, pl. 16.

Nazari, K., Mafi, M., Yahyaoui, A., Singh, R. P., Park, R. F. 2009. Detection of wheat stem rust (*Puccinia graminis f. sp. tritici*) race TTKSK (Ug99) in Iran. *Plant Disease*, 93(3), 317. doi:10.1094/PDIS-93-3-0317B.

Padmanabhan, S. Y. 1959. The present position and control of rice diseases in India. *Proceedings of the Indian Academy of Sciences – Section B*, 49, 349–362. doi:10.1007/BF03052057.

Padmanabhan, S. Y. 1965. Estimating Losses from Rice Blast in India. In *The Rice Blast Disease*, John Hopkins Press, Baltimore, MD, pp. 203–221.

Pandotra, V. R. 1997. *Illustrated Fungi of North India with Special Reference to J & K State*, International Book Distributors, Dehra Dun, India.

Papdiwal, P. B., Deshpande, K. B. 1978. A survey of plant bacterial diseases of Aurangabad. *Natural Science Journal, Marathwada University*, 17(10), 77–82.

Patel, B. N., Patel, G. J. 1985. Virus Diseases of Tobacco and Their Management. In *Integrated Pest and Diseases Management*. Proceedings of the National Seminar, S. Jayaraj (Ed.), Tamil Nadu Agricultural University, Coimbatore, India, pp. 261–266.

Patel, M. K., Abhyankar, S. G., Kulkarni, Y. S. 1949. Black rot of cabbage. *Indian Phytopathology*, 2, 58–61.

Payak, M. M., Renfro, B. L. 1967. A new downy mildew disease of maize. *Phytopathology*, 57, 394–397.

Pimentel, D., Lach, L., Zuniga, R., Morrison, D., 2000. Environmental and economic costs associated with non-indigenous species in the United States. *BioScience*, 50(1), 53–65.

Pimentel, D., McNair, S., Janecka, J., Wightman, J., Simmonds, C., O'Connell, C., Wong, E., Russel, L., Zern, J., Aquino, T., Tsomondo, T. 2001. Economic and environmental threats of alien plant, animal, and microbe invasions. *Agriculture, Ecosystems and Environment*, 84(1), 1–20.

Prasada Rao, R. D. V. J., Reddy, A. S., Chander Rao, S., Varaprasad, K. S., Thirumala-Devi, K., Nagaraju, Muniyappa, V., Reddy, D. V. R. 2000. Tobacco streak ilarvirus as causal agent of sunflower necrosis disease in India. *Journal of Oilseeds Research*, 17, 400–401.

Pretorius, Z. A., Bender, C. M., Visser, B., Terefe, T. 2010. First report of a *Puccinia graminis f. sp. tritici* race virulent to the Sr24 and Sr31 wheat stem rust resistance genes in South Africa. *Plant Disease*, 94(6), 784. doi:10.1094/PDIS-94-6-0784C.

Pyšek, P., Richardson, D. M. 2010. Invasive species, environmental change and management, and health. *Annual Review of Environment and Resources*, 35(1), 25–55. doi:10.1146/annurev-environ-033009-095548.

Raghubanshi, A. S., Rai, L. C., Gaur, J. P., Singh, J. S. 2005. Invasive alien species and biodiversity in India. *Current Science*, 88(4), 539–540.

Rajasekaran, A., Singh, M., Shantaram, P. 2020. Policies, instruments and institutions pertaining to invasive alien species management: A synthesis. In *Invasive Species-I. C. F. R. E. A. Handbook*, pp. 141–152.

Ramakrishnan, T. S., Radhakrishna Pillay, P. N. 1963. *Jatropha curcas* L.: A collateral host for *Oidium heveae* Stein. *Current Science*, 32, 428.

Ramakrishnan, T. S., Sundaram, N. V. 1953. Notes on some fungi from South India-I. *Indian Phytopathology*, 5(2), 110–115.

Rautela, A., Dwivedi, M. 2018. Wheat stem rust race Ug99: A shifting enemy. *International Journal of Current Microbiology and Applied Sciences*, 7(1), 1262–1266.

Roy, P., Majumder, S., Datta, S. 2011. Ug99-Threat to agriculture. *Science Reporter*, 48, 23–24.

Sanjeev, K. 2014. Plant disease management in India: Advances and challenges. *African Journal of Agricultural Research*, 9(15), 1207–1217.

Sharma, R.C., Kamal, B. and Gupta, A.K. 2004. Occurance of crown gall of stone fruits in warmer areas of India and its management. *Acta Hortic.* 662, 411–417. DOI: 10.17660/ActaHortic.2004.662.62.

Sibnarayan, D., Ilmuaka, V., Sanjai Kumar, D. 2020. Agroterrorism in Indian context. *Defence Life Science Journal*, 5(2), 125–132, doi: 10.14429/dlsj.5.15568.

Singh, C. M., Angras, N. M., Kumar, S. 1996. *Weed Management*, MD Publications, New Delhi.

Singh, G. P., He, X., Phuke, R. M., Joshi, A. K., Bishnoi, S. K., Singh, P. K. 2019. Wheat blast readiness of the Indian wheat sector. In *Current Trends in Wheat and Barley Re-search Development*, S. V. Sai Prasad, A. N. Mishra, G. P. Singh (Eds.), ICAR-Indian Agricultural Research Institute, Regional Station, Indore, India, p. 214.

Singh, R. P., Hodson, D. P., Huerta-Espino, J., Jin, Y., Bhavani S., Njau, P., Herrera-Foessel, S., Singh, P. K., Singh, S., Govindan, V. 2011. The emergence of Ug99 races of the stem rust fungus is a threat to world wheat production. *Annual Review of Phytopathology*, 49(1), 465–481.

Singh, R. S. 1985. *Diseases of Vegetable Crops*, Oxford & IBH Publishers, New Delhi, India.

Srivastava, D. N., Rao, Y. P. 1964. A simple technique for detecting *Xanthomonas oryzae* in rice seeds. *Seed Science and Technology*, 5, 123–127.

Sudha, N., Devanshu, D., Singh, K. P. 2020. Ug99: Wheat Stem Rust Race: Exploring the World. In *Diseases of Field Crops Diagnosis and Management: Volume 1: Cereals, Small Millets, and Fiber Crops*, J. N. Srivastava, Singh, A. K. (Eds.), Apple Academic Press, CRC Press, Taylor and Francis Group, pp. 81–88.

Sydow, H., Sydow, P., Butler, E. J. 1906. Fungi Indiae orientalis. *Journal of Mycology*, 4, 427.

Thangavelu, R., Mostert, D., Gopi, M., Ganga Devi, P., Padmanaban, B., Molina, A. B., Viljoen, A. 2019. First detection of *Fusarium oxysporum* f. sp. cubense tropical race 4 (TR4) on Cavendish banana in India. *European Journal of Plant Pathology*, 154, 777–786.

Varghese, M. K. 1945. Bunchy top disease of plantain. *Indian Farming.*, 5, 239–240.

Wanyera, R., Kinyua, M. G., Jin, Y., Singh, R. P. 2006. The spread of stem rust caused by *Puccinia graminis* f. sp. *tritici*, with virulence on Sr31 in wheat in Eastern Africa. *Plant Diseases*, 90(1), 113. doi:10.1094/PD-90-0113A.

Wittenberg, R., Cock, M. J. W. 2001. *Invasive Alien Species: A Toolkit of Best Prevention and Management Practices*, CAB International, Wallingford, Oxon, UK, pp. xvii–228. https://www.researchgate.net (Accessed on January 2001).

11 Plant Biosecurity in Asia-Pacific
Status and Challenges

Ravinder K. Khetarpal
Asia-Pacific Association of Agricultural Research Institutions (APAARI)

V. Celia Chalam
ICAR-National Bureau of Plant Genetic Resources

Pranjib K. Chakrabarty
Agricultural Scientists Recruitment Board

Kavya Dashora
Indian Institute of Technology Delhi
Priya YadavICAR-National Bureau of Plant Genetic Resources

11.1 INTRODUCTION

Biosecurity is a strategic and integrated approach that encompasses a policy and regulatory framework to analyse and manage risks associated with food safety, plant and animal health and the environment. It covers the introduction of plant and animal pests and diseases, and zoonosis; the introduction and management of invasive alien species and genotypes; and the introduction and release of genetically modified organisms (GMOs) and their products (http://www.fao.org). Thus, biosecurity is a holistic approach to the sustainability of agriculture, food safety, and the protection of the environment, including biodiversity. Plant biosecurity encompasses the policy and regulatory framework to analyse and manage risks in the sectors of plant life and health and related environmental concerns (Khetarpal and Gupta 2007; Waage and Mumford 2008; Khetarpal et al. 2009).

The Asia-Pacific region comprises all of Asia and Australasia as well as small, medium and large Pacific Island nations. It has countries experiencing vast differences in topography, climates, livelihood systems and economic development. Agriculture forms the backbone of rural livelihoods and the national economy in most of the Asia-Pacific countries. The dynamic growth in agriculture trade in the Asia-Pacific region since the last few years has shifted the world economy's centre of gravity from the Atlantic to the Pacific basin. The major crops of the region, such as

 DOI: 10.1201/9781003403142-11

rice, wheat, maize, cassava, coffee, coconut, banana, cotton, tea, papaya, mango and others, are attacked by a number of destructive pests threatening local or regional food security. The global threat to plant biosecurity in the present day is further exacerbated by over 1 billion international travellers worldwide per year moving through airports and 18 million containers making over 200 million trips through seaports as part of the trade across the world every year (Sharma and Wightman 2015).

The use of the term pest in the article is as per International Standards of Phytosanitary Measures (ISPM) No. 5, which defines pest as any species, strain or biotype of plant, including animal an agent injurious to plants or plant products (https://www.ippc.int/largefiles/adopted.../en/ISPM_05_2007_En_2007-07-26.pdf)/(https://www.ippc.int/largefiles/adopted_ISPMs_previousversions/en/ISPM_05_2007_En_2007-07-26.pdf), thus including all pathogens, insects, mites and even weeds.

The present review deals strictly with the status of plant biosecurity in Asia in reference to pest outbreaks and spread in agricultural and commodity crops, the role of national and international organizations in dealing with such pests, and the challenges encountered in ensuring plant biosecurity in light of emerging pests. The article does not deal with (1) invasive alien weeds, which are also classified as pests in ISPM 5 and have ecological, economic and social impacts; (2) plant pests used as bio-weapons; and (3) biosafety issues that relate to genetically modified crops developed through recombinant DNA technology.

11.2 STATUS OF PLANT PESTS AND BIOSECURITY IN ASIA-PACIFIC

The important crops of Asia-Pacific are host to a large number of devastating pests, inflicting severe losses to the tune of billions of dollars annually, though the exact figure is not available. A global estimate of 30%–40% loss in developing countries annually from 'field-to-fork' is common (The Royal Society 2009; Flood 2010) and may be even higher when post harvest loss is taken fully into account. Historically, a number of economically important plant pests have also transgressed across the region, either as migratory or quarantine pests. The fastest-flying transboundary plant pest, the desert locust (*Schistocera gregaria*), is the world's largest migratory insect, travelling long distances. Among the pests of high economic importance Brown plant hoppers (BPH) of rice, not considered a serious threat in tropical Asia until the 1970s, suddenly emerged as a significant pest of rice, with grave outbreaks in tropical Asia-Pacific countries. Rice cultivation has grown up to 158 million ha, out of which 90% is being carried out in developing countries (Poolprasert and Jongjitvimol 2014). The sudden emergence of BPH as a major pest of rice has been linked to the advance of the Green Revolution in countries such as Bangladesh, Brunei, China, Fiji, Korea, Malaysia, Papua New Guinea, the Philippines, Solomon Islands, Sri Lanka, Thailand and Vietnam (Heinrichs 1977; Dyck and Thomas 1979). Serious outbreaks of this pest have been reported in different Asia-Pacific countries (Lim et al. 1978; Mochida 1979; Soejitno 1999; Asghar et al. 2013). In Vietnam alone, the area affected increased from about 0.35 m ha in 2005 to 1.1 m ha in 2010. In China during 2008, when 7.5 million ha in several provinces were infested, it resulted in a loss of 2.8 million t of paddy, and the losses are now being compounded with 10.4 million ha reported to be infested in 2010 (Cheng 2009).

Wheat is an important food security crop in many countries in the region and is extensively grown in Australia, China, India, Pakistan and Indonesia. The wheat rusts, stem rust (Ug99) and yellow rust are perpetual threats to the countries that rely on the wheat crop as the source of their subsistence and livelihood. Since its detection, new variants of Ug99 have evolved, with the result that over 80% of all commercially grown wheat varieties are now highly susceptible to stem rust. It has subsequently spread from Uganda to Kenya (1999), to Ethiopia (2003), to Yemen and Sudan (2003) and to Iran (2008). Because of the susceptibility of 90% of the wheat varieties grown worldwide, the Ug99 group of races was recognized as a major threat to wheat production and food security. Its spread, either wind-mediated or human-aided, to other countries in Africa, Asia, and beyond is evident (Singh et al. 2011). Depending on wind movement and the presence of susceptible cultivars, countries in Central and South Asia are now exposed to a high risk of the disease. Besides, the emergence and establishment of yellow rust races with virulence for Yr9 (a common resistance gene of high-yield wheat varieties around the world) to stripe or yellow rust resulted in large-scale epidemics that completely eliminated the wheat crop. A region-wide breakdown of this gene was reported in South Asia during 2002–2004. In addition, since 2000, two new aggressive strains of yellow or stripe rust capable of adapting to warmer temperatures have been identified (Hodson 2011). Another pest of high economic importance, Jack Beardsley mealybug, *Pseudococcus jackbeardsleyi*, established in Hawaii in 1959 and other Pacific Islands in the 1970s, has now spread to many countries in Southeast Asia (Williams 2004; Muniappan et al. 2011) and India (Shylesha and Joshi 2011; Mani et al. 2013).

11.2.1 Country-Wise List of Outbreaks of Major Pests

A detailed list of major pest outbreaks in different countries, including their phytosanitary legislation, is given in Table 11.1. Also listed in Table 11.1 are pests for which surveillance and control measures are being taken.

The majority of the countries in Asia-Pacific are members of the WTO and are parties to the Agreement on Application of Sanitary and Phytosanitary Measures (known as the SPS Agreement). This prompted the majority to amend their legislative measures or to append additional decrees to include key provisions of pest reporting, survey and surveillance, pest risk analysis prior to imports, and identification of areas with low pest prevalence or pest-free areas as per international or regional standards. However, the implementation of legislative provisions, leaving aside some developed countries, appears to be far from satisfactory in the region due to economic and other reasons.

Table 11.1 reveals that during the last few years, about 90 serious plant pest outbreaks have occurred in 36 countries in the region. Of these 25 outbreaks in multiple countries, the outbreak of BPH on rice alone occurred in 8 different countries. This has enhanced pest surveillance activities in 32 countries for important pests (Table 11.1). Some countries have adopted the IPPC standards for declaring a specified zone as a pest-free area for a total of 16 pests. The pest-free status of a country facilitates the export of a given produce from the pest-free areas, even if the pest is known to occur in the country.

TABLE 11.1
Legislative Provisions, List of Pest Outbreaks, Pests Under Surveillance and Pests with Control Measures in Asia-Pacific Countries

Country Legislative Provisions	Outbreaks	Surveillance Programme/ Pest-Free Status	Economically Important Pests Meriting Control Options
Afghanistan Plant Quarantine Act	Desert locust (*Schistocerca gregaria*); Moroccan locust (*Dociostaurus maroccanus*)	Desert locust (*S. gregaria*)	Desert locust (*S. gregaria*); wheat sunn pest (*Eurygaster integriceps*; *Aelia rostrata*)
Australia Australian Biosecurity Act 2015	Chestnut blight (*Cryphonectria parasitica*); giant pine scale (*Marchalina hellenica*); red imported fire ants (*Solenopsis invicta*); Russian wheat aphid (*Diuraphis noxia*)	Australian plague locust (*Chortoicetes terminifera*); migratory locust (*Locusta migratoria*); spur-throated locust (*Austracris guttulosa*)	Chestnut blight (*C. parasitica*); citrus canker (*Xanthomonascitri*); giant pine scale (*M. hellenica*); red imported fire ant (*S. invicta*)
Bangladesh Plant Quarantine Act 2011	Papaya mealybug (*Paracoccus marginatus*); rice blast (*Magnaporthe grisea*); rice bacterial leaf blight (*X. oryzae*); rice brown plant hopper (*Nilaparvata lugens*); Stem borer common cutworm (*Spodoptera litura*); white fly (*Bemisia* sp.)	Rice pest; fruit fly (*Bactrocera* sp.)	Jute anthracnose (*Glomeralla cingulata*); Gummosis of Jack fruit (*Phytophthora* sp.); jute stem rot (*Macrophomína phaseolina*); white mould (*Sclerotinia sclerotiorum*)
Bhutan The Plant Quarantine Act of Bhutan 1993	Army worm (*Mythimna separata*)	Chilli pod borer (*Helicoverpa armigera*); red palm weevil (*Rhyncophorus ferrugineus*), rice blast (*M. grisea*)	Army worm (*M. separata*); chilli pod borer (*H. armigera*); Chinese fruit fly (*B. minax*); citrus trunk borer (*Anoplophora versteegi*); rice case worm (*Nymphula depunctalis*)
Brunei Agricultural Pests and Noxious Plants Act Section 24(1)(F) Revised edition 1984	—	—	—

(*Continued*)

TABLE 11.1 (*Continued*)

Legislative Provisions, List of Pest Outbreaks, Pests Under Surveillance and Pests with Control Measures in Asia-Pacific Countries

Country Legislative Provisions	Outbreaks	Surveillance Programme/ Pest-Free Status	Economically Important Pests Meriting Control Options
Cambodia Cambodian Phytosanitary Regulation 2005	Army worm (*M. separata*); coconut rhinoceros beetle (*Oryctes rhinoceros*); golden apple snail (*Pomacea canaliculata*); pink mealybug (*Maconellicoccus hirsutus*); rice brown plant hopper (*N. lugens*)	Citrus; grain insect; potato; rice; sugarcane; tomato	Army worm (*M. separata*); rice brown plant hopper (*N. lugens*); golden apple snail (*P. canaliculata*)
China* Law of the People's Republic of China on the Entry and Exit Animals and Plants Quarantine (Order of President No.53) 2016	Army worm (*M. separata*); potato late blight (*Phytophthora infestans*); rice blast (*M. grisea*); rice brown plant hopper (*N. lugens*); rice leaf folder (*Cnaphalocrocis medinalis*); wheat scab (*Gibberrellazeae*)	Coconut black headed caterpillar (*Opisina arenosella*); migratory locust (*L. migratoria*); potato beetle (*Leptinotarsa decemlineata*); rice brown plant hopper (*N. lugens*) *pest-free area declared for Codling moth (*Cydia pomonella*)	Army worm (*M. separata*); coconut black headed caterpillar (*O. arenosella*); fall webworm (*Hyphantria cunea*); Huanglongbing (*Diaphorina citri*); meadow moth (*Loxostege sticticalis*); migratory locust (*L. migratoria*); pine wilt disease (*Bursaphelenchus xylophilus*); rice brown plant hopper (*N. lugens*); rice leaf roller (*Cnaphalocrocis medinalis*); rice sheath blight (*Corticium sasakii*); rice blast (*M. grisea*); sirex woodwasp (*Sirex noctilio*); wheat scab (*Gibberrellazeae*); wheat stripe rust (*Puccinia striiformis*)

(*Continued*)

TABLE 11.1 (*Continued*)
Legislative Provisions, List of Pest Outbreaks, Pests Under Surveillance and Pests with Control Measures in Asia-Pacific Countries

Country Legislative Provisions	Outbreaks	Surveillance Programme/ Pest-Free Status	Economically Important Pests Meriting Control Options
Cook Islands Biosecurity Act 2008	Sweet potato whitefly (*B. tabaci*)	Queensland fruit fly (*B. tryoni*)	Sweet potato scarabee (*Euscepes postfasciatus*); sweet potato weevil (*Cylas formicarius*)
Fiji Fiji BAF Legislation 2015	Chilli anthracnose (*Colletotrichum* sp.); coconut rhinoceros beetle (*Oryctes rhinoceros*); coconut stick insect (*Graeffea crouanii*); Taro beetle (*Eucopidocaulus*)	Fruit flies (*Bactocera* sp.)	Chilli anthracnose (*Colletotrichum* sp.); coconut rhinoceros beetle (*O. rhinoceros*); coconut stick insects (*G. crouanii*); fruit flies (*B. passiflorae, B. xanthodes*)
India* Plant Quarantine (Regulation of Import into India) Order; 2003 Draft Agricultural Biosecurity Bill 2013	Rice armyworm (*S. litura*); desert locust (*S. gregaria*); papaya mealybug (*P. marginatus*); whitefly (*B. tabaci*)	Desert locust (*S. gregaria*); fruit fly (*B. dorsalis*); potato golden nematode (*Globodera rostochiensis*); tomato leaf miner (*Tuta absoluta*) *Pest-free area for Brown rot (*Ralstonia solanacearum*) and ring rot (*Clavibacter michiganensis*) of potato	Banana bunchy top virus; coconut eriophyid mite (*Aceria gurrreronis*); coffee berry borer (*Hypothenemus hampei*); cotton mealy bug (*Phenococcus solenopsis*); desert locust (*S. gregaria*); papaya mealybug (*P. marginatus*); spiralling white fly (*Aleurodicus disperses*); sunflower downy mildew (*Plasmopara halstedii*)
Indonesia Animal, Fish and Plant Quarantine Law 1992	Rice bacterial leaf blight (*X. oryzae*); rice brown plant hopper (*N. lugens*); rice leaf blast (*M. grisea*); rice stem borer (*Scirpophaga incertulas*); *Rice tungro virus*	Avocado; bell pepper; dragon fruit; durian; ginger; lansium; mango; mangosteen; orchids; papaya; pineapple; raphis; salacca	Chilli anthracnose (*Colletotrichum* sp.); soybean downy mildew (*Peronospora manshurica*); khapra beetle (*Trogoderma granarium*); raspberry ringspot virus; stenocarpella ear rot (*Stenocarpella macrospora*)

(*Continued*)

TABLE 11.1 (*Continued*)
Legislative Provisions, List of Pest Outbreaks, Pests Under Surveillance and Pests with Control Measures in Asia-Pacific Countries

Country Legislative Provisions	Outbreaks	Surveillance Programme/ Pest-Free Status	Economically Important Pests Meriting Control Options
Japan Plant Protection Law 2010	Rice brown plant hopper (*N. lugens*); rice white-backed planthpper (*Sogatella frucifera*)	Japanese cedar longhorn beetle (*Callidiellumrufipenne*)	Fruit bugs (*Glaucias subpunctatus*, *Halyomorpha halys*, *Plautia crossota*)
Laos PDR Plant Protection and Plant Quarantine Law No. 06/NA dated 9 December 2008	Yellow-spined bamboo locust (*Ceracris kiangsu*)	Banana; cassava; locusts; maize; rice; rubber; tobacco; water melon	Citrus canker (*X. citri*); coconut hispine beetle (*Brontispa longissima*); cassava pink mealybug (*Phenacoccus manihoti*); coffee berry borer (*Hypothenemus hampei*); sugarcane white leaf (phytoplasma); yellow-spined bamboo locust (*C. kiangsu*)
Malaysia Plant Quarantine Act 1976 (2016); MAQIS Act 2016	Green paddy leafhopper (*Nephotettix virescens*); malayan black bug (*Scotinophora coarctata*); rice armyworm (*S. mauritia*); rice brown plant hopper (*N. lugens*); rice caseworm (*Nymphula depunctalis*); rice ear bug (*Leptocorisa oratorius*); rice leaf folder (*Cnaphalocrocis medinalis*); rice white-backed planthopper (*S. furcifera*)	Mango seed weevil (*Sternochetus mangiferae*); South American Leaf Blight (*Microcyclus ulei*)	Green paddy leafhopper (*N. virescens*); rice ear bug (*L. oratorius*); rice brown plant hopper (*N. lugens*); rice leaf folder (*C. medinalis*)

(*Continued*)

TABLE 11.1 (*Continued*)
Legislative Provisions, List of Pest Outbreaks, Pests Under Surveillance and Pests with Control Measures in Asia-Pacific Countries

Country Legislative Provisions	Outbreaks	Surveillance Programme/ Pest-Free Status	Economically Important Pests Meriting Control Options
Myanmar Plant Pest Quarantine Law 2015	Rice brown plant hopper (*N. lugens*); diamondback moth (*Plutella xylostella*); golden apple snail (*P. canaliculata*); rice stem borer (*Scirpophaga incertulas*)	Army worm (*H. armigera*)	Rice nematodes (*Ditylenchus angustus*, *Hirschmanniella oryzea*); white fly (*B. tabaci*)
Marshall Islands Quarantine Restrictions Act	Coconut rhinoceros beetle (*O. rhinoceros*)	Fruit fly (*Bactrocera* sp.)	Corn southern leaf blight (*Cochliobolus heterostrophus*); leaf spot of seedling (*Bipolaris incurvata*); sweet potato scab (*Elsinoe batatas*)
Micronesia Chapter 4 of Title 22 of the Code of Federated States of Micronesia-Quarantines	Aphids; cucumber beetle (*Diabrotica undecimpunctata*)	Fruit fly (*Bactrocera* sp.)	Coconut hispid (*Brontispa* sp.); coconut rhinoceros beetle (*O. rhinoceros*); coconut scale (*Aspidiotus destructor*); new Guinea sugarcane weevil (*Rhabdoscelus obscurus*); red coconut scale (*Furcaspis oceanica*)
Mongolia Environmental Protection Law of Mongolia	Grasshopper (Loxostegesticticalis); Mongolian locust (Oedaleusasiaticus)	—	Mongolian locust (*O. asiaticus*)
Nauru Agriculture Quarantine Bill (Plant & Animal Quarantine Regulation 2004)	Melon fly (*B. cucurbitae*)	Fruit fly (*Bactrocera* sp.)	Mango fly (*B. frauenfeldi*); oriental fruit fly (*B. dorsalis*); pacific fruit fly (*B. xanthodes*)

(*Continued*)

TABLE 11.1 (*Continued*)
Legislative Provisions, List of Pest Outbreaks, Pests Under Surveillance and Pests with Control Measures in Asia-Pacific Countries

Country Legislative Provisions	Outbreaks	Surveillance Programme/ Pest-Free Status	Economically Important Pests Meriting Control Options
Nepal* Plant Protection Act 2007	Army worm (*Mythimna* sp.); rice brown plant hopper (*N. lugens*); sugarcane plant hopper (*Pyrilla perpusilla*)	Citrus pests; fruit flies; *Pest-Free area for Mandrin orange and sweet orange fruit fly	Citrus fruit fly (*Bactrocera* sp.); cotton bollworm (*H. armigera*); epilachna beetle (*Epilachna vigintioctopunctata*); black blister beetle (*Epicauta vittata*); green peach aphid (*Myzus persicae*); potato cut worm (*Agrotis segetum*); tomato semi-looper (*Chrysodeixis chalcites*); potato tuber moth (*Phthorimaea operculella*); red ant (*Pogonomyrmex barbatus*); white grubs (*Cyclocephala* sp.)
New Zealand New Zealand's Biosecurity Act and Regulations 2007	Brown marmorated stink bug (*Halyomorpha halys*); fruit fly (*Bactrocera* sp.)	Brown marmorated stink bug (*H. halys*); fruit fly (*Bactrocera* sp.); gypsy moth (*Lymantria dispar*)	Fruit flies (*Bactrocera* sp.); potato psyllids (*Bactericera cockerelli*)
Niue Draft Biosafety (Genetically Modified Organisms) Regulations 2006; Draft Environment (Amendment) Bill 2006	Aphid (*Aphis craccivora*); pineapple mealybug (*Dysmicoccus brevipes*); southern green stink bug (*Nezara viridula*)	Fruit fly (*Bactrocera* sp.)	Citrus leaf miner (*Phyllocnistis citrella*)

(*Continued*)

TABLE 11.1 (*Continued*)
Legislative Provisions, List of Pest Outbreaks, Pests Under Surveillance and Pests with Control Measures in Asia-Pacific Countries

Country Legislative Provisions	Outbreaks	Surveillance Programme/ Pest-Free Status	Economically Important Pests Meriting Control Options
North Korea Plant Protection Act 2012	Diamondback moth (*P. xylostella*); small white butterfly (*Pieris rapae*)	Fruit fly (*Bactrocera* sp.)	Armyworm (*H. armigera*); maize smut (*Ustilago maydis*); maize stem borer (*Atherigona orientalis*); diamondback moth (*P. xylostella*); small white butterfly (*Pierisrapae, P. ramorum*)
Pakistan Phytosanitary Requirements According to Law of the Islamic Republic of Pakistan in the Field of Plant Quarantine for Importing Countries 2016	Desert locust (*S. gregaria*); dubas bug (*Ommatissus lybicus*)	Fruit fly (*Bactrocera* sp.)	American boll worm (*H. armigera*); army worm (*S. litura*); cotton leaf folder (*Sylepta derogata*); dubas bug (*O. lybicus*); pink boll worm (*Pectinophora gossypiella*); spotted boll worm (*Earias insulana*); thrip (*Thrips tabaci*); white fly (*B. tabaci*)
Papua New Guinea National Agriculture Quarantine and Inspection Authority Act 1997	Mango fly (*B. frauenfeldi*); melon fly (*B. cucurbitae*); potato late blight (*Phytophthora infestans*)	Fruit fly (*Bactrocera* sp.); khapra beetle (*Trogoderma granarium*)	Coconut hispine beetle (*B. longissimi*); crazy ant (*Anoplolepis gracilipes*); fruit fly (*Bactrocera* sp.); ratoon stunting disease (*Leifsonia xyli*)

(*Continued*)

TABLE 11.1 (*Continued*)
Legislative Provisions, List of Pest Outbreaks, Pests Under Surveillance and Pests with Control Measures in Asia-Pacific Countries

Country Legislative Provisions	Outbreaks	Surveillance Programme/ Pest-Free Status	Economically Important Pests Meriting Control Options
Philippines* Philippine Plant Quarantine Law 1978	Cassava phytoplasma disease; coconut scale insect (*A. destructor*)	Army worm (*M. separata*); coconut scale insect (*A. destructor*); rice grain bug (*Leptocorisa acuta*); rice bacterial leaf blight (*X. oryzae*); *Rice tungro virus* *Pest-free area for Banana Xanthomonas wilt (*X. campestris* pv. *musacearum*) and *Banana bract mosaic virus*	Armyworm (*M. separata*); rice bacterial leaf blight (*X. oryzae*); rice black bug (*Scotinophara coarctata*); rice grain bug (*L. acuta*)
Samoa Phytosanitary Requirements/Restrictions/ Prohibitions 2015	Taro leaf blight (*Phytophthora colocasiae*)	Coconut rhinoceros beetle (*O. rhinoceros*); fruit flies (*B. kirki*, *B. xanthodes*)	Fruit flies (*B. kirki*, *B. xanthodes*)
Singapore Control of Plants Act 2000	Golden apple snail (*P. canaliculata*); rice brown plant hopper (*N. lugens*)	Coconut rhinoceros beetle (*O. rhinoceros*)	Cotton bollworm (*H. armigera*); golden apple snail (*P. canaliculata*); rice brown plant hopper (*N. lugens*)
Solomon Islands Biosecurity Act 2013	Taro beetles (*Papuanawoodlarkiana*, *P. biroi*, *P. huebneri*, *P. trinodosa*)	Little fire ant (*Wasmannia auropunctata*); surinam cockroach (*Pycnoscelus surinamensis*)	Dasheen mosaic virus; Taro leaf blight (*P. colocasiae*)
South Korea Korea's Plant Protection Act	Fire blight (*Erwinia amylovora*)	Cabbage white butterfly larvae (*P. brassicae*)	Cabbage white butterfly larvae (*P. brassicae*)

(*Continued*)

TABLE 11.1 (*Continued*)
Legislative Provisions, List of Pest Outbreaks, Pests Under Surveillance and Pests with Control Measures in Asia-Pacific Countries

Country Legislative Provisions	Outbreaks	Surveillance Programme/ Pest-Free Status	Economically Important Pests Meriting Control Options
Sri Lanka Plant Protection Act of Sri Lanka 2016	Rice brown plant hopper (*N. lugens*); rice thrips (*Stenchaetothrips biformis*)	Mealy bug; rice pest	Black cut worm (*Agrotis* sp.); cabbage caterpillar (*Plutella xylostella*); mealybug (*Planococcus lilacinus, P. marginatus*, *P. cryptus*); rice sheath blight (*Rhizoctonia solani*); root knot (*Meloidogyne* spp.)
Thailand* Plant Quarantine Act B.E.2507 and amended 2011	Coconut blackheaded caterpillar (*O. arenosella*)	Corn downy mildew (*Peronosclerospora philippinensis*); maize leaf blight (*Bipolaris maydis*); maize vascular wilt (*Pseudomonas avenae*; *P. syringae*) *Pest-free area for Corn downy mildew (*P. philippinensis*); maize leaf blight (*B. maydis*); maize vascular wilt (*P. avenae*; *P. syringae*)	Cassava pink mealybug (*P. manihoti*); coconut black headed caterpillar (*O. arenosella*)

(*Continued*)

TABLE 11.1 (*Continued*)
Legislative Provisions, List of Pest Outbreaks, Pests Under Surveillance and Pests with Control Measures in Asia-Pacific Countries

Country Legislative Provisions	Outbreaks	Surveillance Programme/ Pest-Free Status	Economically Important Pests Meriting Control Options
Timor-Leste Timor-Leste's legislation relating to phytosanitary restrictions 2016	Desert locust (*S. gregaria*)	Cotton blue disease (*Cotton leafroll dwarf virus*); desert locust (*S. gregaria*)	Banana bunchy top virus
Tonga Plant Quarantine Act 1988; Noxious Weed Act 2015	Banana weevil (*Cosmopolites sordidus*)	Bacterial crown rot (*E. papaya*)	Bacterial crown rot (*E. papaya*); banana aphid (*Pentalonia nigronervosa*); Mediterranean fruit fly (*Ceratitis capitata*); Melon aphid (*Aphis gossypii*); oriental fruit fly (*B. dorsalis*)
Vanuatu Plant Protection Amendment Act 2013	Taro leaf blight (*P. colocasiae*)	Fire ants; melon fly (*B. cucurbitae*); banana corm and leaf spot (*Marasmiellus stenophyllus*)	Bacterial soft rot (*E. chrysanthemi*); corm rot (*Athelia rolfsii*); taro leaf blight (*P. colocasiae*)
Vietnam Law on Plant Protection and Quarantine issued in 2001	Yellow-spined bamboo locust (*C. kiangsu*); grasshopper (*Hieroglyphus tonkinensis*); sawfly (*Shizocera* sp.)	Basic survey of pests on forest trees; National Programme for surveying and detecting early quarantine pests	Black pepper fungi (*Phytophthora parasitica*, *P. palmivora*; *Fusarium solani*; *Pythium* sp.; *Lasiodiplodia theobromae*; *R. solani*; *Rosellia* sp.)

Sources: FAO (2014), Hyun and Choi (2014), Kranthi (2016), Yongfan (2015), https://www.fao.org/asiapacific/apppc/en/, https://lrd.spc.int/plantprotectionorganisation-/pppo, https://www.nppc.gov.bt/wp-content/uploads/2016/07/4.-Chilli-pod-borer.pdf, https://www.rhs.org.uk/advice/profile?pid=457, http://www.mbdsnet.org/79105-2/, www.dpi.nsw.gov.au/__data/assets/pdf_file/0006/427515/Exotic-Pest-Alert-Tomato-potato-psyllid.pdf

*Relate to countries where pest-free area has been declared for certain pests as given under column 3.

11.2.2 Emerging Pests in Asia-Pacific

The term emerging pest has often been loosely used in literature. In the field of plant health, it means diseases or pests that are newly discovered, have increased in occurrence, or have spread to new locations (Anonymous 2010). The European Food Safety Association defines an emerging plant pest as a newly identified pest for which a significant probability of introduction and/or spread may occur, or an unexpected new or increased significant probability of introduction and/or spread of an already known plant pest, or a new or increased susceptibility of the host plants to a known plant pest (EFSA 2011). It may be noted that a pest can be emerging in a new location or country for the first time but might have caused outbreaks in other regions (e.g., BPH and hispine beetle). Similarly, a pest can always continue to emerge in many locations or countries but may not assume a severe enough proportion to cause an outbreak. A number of recently emerging pests affecting Asia-Pacific countries and their movement within the region, along with those having apparently restricted distributions, are given in Figure 11.1.

Hispine beetle (*Brontispa longissima*) that has spread fast with outbreaks in Southeast Asia poses a significant risk of spreading as an emerging pest in countries such as India, Sri Lanka and Bangladesh (Rethinam and Singh 2007; Takano et al. 2012).Coconut mite (*Aceria guerreronis*) has the potential to spread to other coconut countries in Asia, where 85% of the world's coconuts are grown (CABI 2016; Vincenzo 2016). Cassava mealybug (*Phenacoccus manihoti*), which has decreased cassava production in West Java to 30%–40%, now poses an acute risk to cassava production in Southeast Asia including the southern end of Karnataka in India, the eastern end of the NinhThuan province in Vietnam, and in West Timor in Indonesia (Parsa et al. 2012). Papaya mealybug (*Paracoccus marginatus*), which is native to Central America, was reported in 2002 in the Pacific Islands and in 2008 in Indonesia, India, and Sri Lanka. It has spread to South and Southeast Asian countries, posing the greatest threat to the rest of Asia-Pacific countries, including the Australian papaya industry (Meyerdirk et al. 2004; Muniappan et al. 2008, 2011; CABI 2016). Fire ant (*Solenopsis invicta*) in China has been introduced into parts of Australia and in some vulnerable ecosystems, including islands in the Pacific (New Zealand), and has the potential to colonize numerous other regions (McGlynn 1999; Korzukhin et al. 2001; CABI 2016). Wheat blast (*Magnaporthe oryzae) that recently emerged in Bangladesh from wheat imports from Brazil is the most recent episode of biosecurity threat to India and the Indian sub-continent* (Saharan et al. 2016; Callaway 2016; Malaker et al. 2016). Tomato leaf miner (*Tuta absoluta*) damage to tomatoes was first noticed in Pune and Bangalore during 2014 in India, and within a span of two years, the pest spread to nearly 1.5 million ha in other states. It has shown the potential to damage fruits to the extent of 100% (ICAR 2014). Tropical race 4 of Panama wilt of banana (*Fusarium oxysporum* f.sp. *cubense*) has now spread to many banana-growing countries in Asia-Pacific (Buddenhagen 2009; Ploetz 2015; Ordoñez et al. 2016). The Western flower thrips (*Frankliniell aoccidentalis*) emerging in Indonesia and India is responsible to transmit as many as five species of tospoviruses, of which *Tomato spotted wilt virus* alone is known to cause losses above US$ 1 billion (Tyagi and Kumar 2015).

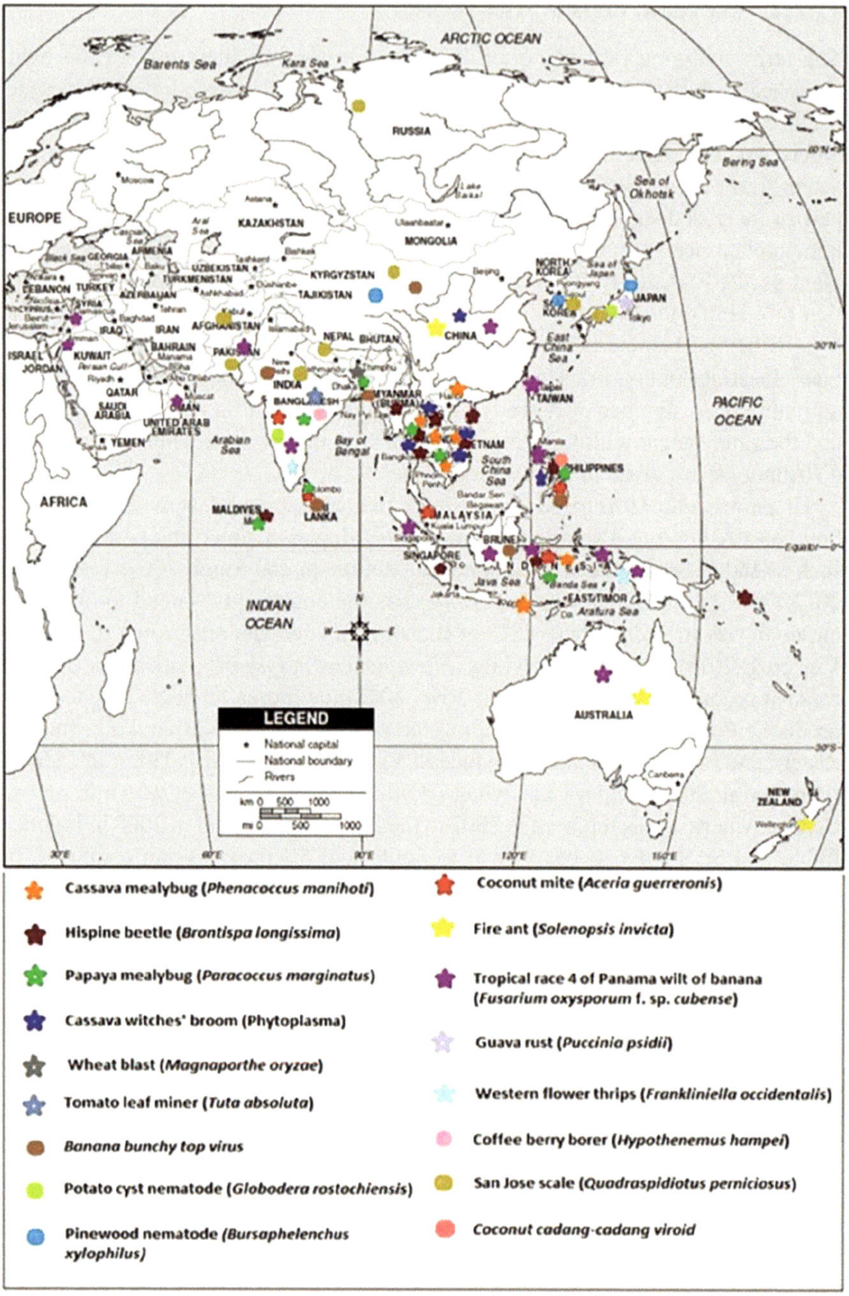

FIGURE 11.1 Distribution of important emerging pests in Asia-Pacific (pests indicated with circles in different colours are the pests with restricted distribution).

Among the emerging pests that have limited distribution in the region, *Coconut cadang-cadang viroid* (CCCVd) is found only in the Philippines, but if it spreads to the Asia-Pacific region, it may cause a huge loss to coconut farming (EPPO 2014). The banana bunchy top virus (BBTV), affecting bananas worldwide and causing a loss of up to 100%, was first recognized in Fiji (1889), and from there it was introduced to Sri Lanka and later spread to southern India and Bangladesh. BBTV is now also recorded in East and Southeast Asian countries, but its most disturbing effects in recent years have been seen in Pakistan, where the virus has almost destroyed the country's banana industry (CABI 2016). San Jose scale (*Quadraspidiotus perniciosus*) is indigenous to Eastern Asia, and from there, it has spread to many parts of the world. Most of the countries, such as Afghanistan, China (widespread), India (widespread), Japan, Republic of Korea, Nepal, Pakistan and Russia (far east), have the pest present, thereby destroying the orchards of many fruit crops (EPPO 2014). Potato cyst nematode (*Globodera rostochiensis*) has traversed from the Andes Mountains in South America all the way to the Asia-Pacific through Europe with infected potato tubers. In India, it is confined to the states of Kerala and Tamil Nadu, and in Japan, it is restricted to Hokkaido. Up to 80% of the crop can be lost when nematode populations are high (CABI 2016). Pinewood nematode (*Bursaphelenchus xylophilus*), native to North America, has been introduced into China, Japan and South Korea, and damaging the forests by causing the pine to wilt through the transport of infested wood materials, while the main vector for natural spread in China is the sawyer beetle, *Monochamus alternatus* (Xiang et al. 2015). The nematode has spread steadily, damaging pines on 87,000 ha and killing over 40 million pines. The direct economic loss is estimated to be US $3 billion from China alone (Yang and Wang 1989). Coffee berry borer is one of the economically most important pests of coffee (Vega 2008), causing an estimated loss of $500 million worldwide. First found in Uganda in 1908, it was reported in Asia during 1990 from a coffee-growing region in South India and has gradually spread to almost all the coffee growing areas, affecting the yield and quality of coffee and reducing the income of coffee growers (Jaramillo et al. 2006).

It may be seen from Figure 11.1 that during the last few years, about 12 important emerging plant pests have been reported in 14 countries in the region, some of which continue to spread further. Besides, some emerging pests with restricted distribution (Figure 11.1) were also observed. The emerging pests are, in fact, the alarm bells for the region, necessitating possible phytosanitary measures to contain their spread and minimize the potential losses caused by them.

11.3 REGIONAL EFFORTS IN ENSURING PLANT BIOSECURITY

Regional cooperation and coordination are critical in reducing the outbreaks of pests and their movement in the region. FAO, through its regional offices, undertakes plant protection activities related to integrated pest management (IPM), pesticide management, etc. and also facilitates the implementation of ISPMs of the International Plant Protection Convention (IPPC), which is an international agreement on plant health with 182 current signatories that aims to protect cultivated and wild plants by preventing the introduction and spread of pests (https://www.ippc.int/en/). The IPPC thus has a strong presence in each of the FAO regions, delivering its services through

Regional Plant Protection Organisations (RPPOs). Of the nine RPPOs of FAO, two are in the Asia-Pacific region, viz., Asia-Pacific Plant Protection Commission (APPPC) and Pacific Plant Protection Organization (PPPO). The APPPC, located at the FAO Regional Office in Bangkok, was established in 1956 and has 25 countries from Asia and the Pacific as members (http://www.fao.org/asiapacific/apppc/en/). The PPPO located at New Caledonia at the headquarters of the Secretariat of the Pacific Community was founded in October 1994 and has all 27 countries of Pacific Community as members (http://lrd.spc.int/plantprotectionorganisation-/pppo). All 25 member states of the APPPC and 27 of the PPPO vary greatly in size and level of economic development. Both APPPC and PPPO have a multi-facet responsibility and function that covers all the mandates of general plant protection of the member countries and also the mandate of IPPC for developing and harmonizing Regional Standards for Phytosanitary Measures (RSPMs). They work with National Plant Protection Organizations (NPPOs) to review the plant protection status, promote appropriate measures to prevent the introduction and spread of pests, develop and adopt RSPMs (Table 11.2, http://www.fao.org/asiapacific/apppc/en/), provide assistance to develop ISPMs, promote multi- and bilateral agreements, co-ordinate and arrange trainings, promote appropriate measures to control pests, including use of IPM, review status of and promote IPM and harmonize pesticide regulations. APPPC also promotes information exchange in the region through a website as a forum for the member countries to exchange information regularly about their plant quarantine

TABLE 11.2
Regional Standards for Phytosanitary Measures (RSPMs) Developed by APPPC

RSPM No. (Year)	Regional Standards
RSPM No. 1 (2004)	Guidelines for the development of the heat disinfestation treatments of fruit fly host commodities
RSPM No. 2 (2004)	Training requirements for plant quarantine inspectors
RSPM No. 3 (2005)	Requirements for the establishment and maintenance of pest-free areas for tephritid fruit flies
RSPM No. 4 (2005)	Guidelines for the confirmation of non-host status of fruit and vegetables to tephritid fruit flies
RSPM No. 5 (2007)	Guidelines for the application of emergency actions and the establishment of emergency measures
RSPM No. 6 (2007)	Guidelines for PRA on scale insects associated with commodities for human consumption
RSPM No. 7 (2014)	Guidelines for protection against South American Leaf Blight of rubber
RSPM No. 8 (2009)	Guidance on the operation of land border entry points for local trade
RSPM No. 9 (2014)	Approval of irradiation facilities
RSPM No. 10 (2014)	Approval of fumigation facilities

Source: https://www.fao.org/asiapacific/apppc/en/

and pest and pesticide management activities, as well as other issues related to plant protection. PPPO has prepared a Pacific Islands Pest List Database (PLD), which stores records of pests that are currently known to affect agriculture, forestry and the environment in Pacific Island Countries and Territories.

The FAO introduced IPM against cassava pink mealybug in Greater Meakong subregion countries to develop spread prevention strategies and ecological biocontrol options to manage the mealybug. In addition, FAO, through its Asia Regional IPM Programme, supported action-oriented research and development initiatives, most notably on late blight and leaf miner management as well as the promotion of bio-fumigation for the control of soil-borne diseases. Besides, the EU-funded FAO programme on IPM for cotton in Asia had six member countries: Bangladesh, China, India, Pakistan, the Philippines and Vietnam, where cotton is a chief cash crop and also faces huge losses due to many pests, including those of transboundary importance. An on-going Project on Capacity Development in Implementation of Plant Pest Surveillance and Information Management in Southeast Asian Countries has been financed by the Republic of Korea and involves Cambodia, Lao PDR, Myanmar, Nepal, Thailand and Vietnam (https://www.fao.org/fileadmin/templates/rap/files/meetings/2013/131129-report.pdf). Australia and New Zealand, being the major trading partners of Pacific Island Countries (PICs), have been assisting in encouraging increased trade with them. The PIC's limited ability to meet biosecurity requirements imposes a severe constraint on their capacity to increase export markets for currently important commodities and develop exports of a new range of products (https://lrd.spc.int/plantprotectionorganisation /pppo).

11.4 CHALLENGES IN ENSURING BIOSECURITY IN ASIA-PACIFIC

Ensuring a well-developed institutional mechanism for plant protection and quarantine: The challenges in managing pest outbreaks and transboundary pests in Asia-Pacific stem from the very fact that the member countries fall under different categories of development, i.e., developed, developing and least developed. This has a direct bearing on their resources and capacity, and hence their operational quality in dealing with the problem. Developed countries such as Australia, New Zealand and Japan have the best of regulatory provisions, resources, programmes, institutional infrastructure for plant protection and quarantine facilities with appropriate research back-ups to meet the challenges of biosecurity to a great extent. Also, there is a great variation among member countries in their organizational structures and national mandates, which may be more focused on biosecurity, whereas transboundary pest movement may be a component of plant protection to deal with but not the major focus. Thus, compliance with and implementation of different international standards as laid down by the IPPC are not at the same level within the region. This poses a huge challenge in harmonizing the quarantine norms to combat the transboundary movement of pests at the regional level.

Ensuring pest-free movement of seed and planting materials: The international trade in seed and planting materials in bulk and the exchange of germplasm are known to carry pests over long distances along with their movement (Khetarpal 2004; Khetarpal et al. 2009). The quarantine processing of germplasm material in most countries is often not as stringent as commercial material, despite the fact that germplasm can also be potential source of the spread of quarantine pests into the region. India has intercepted various quarantine pests in germplasm imports from outside the Asia-Pacific and also in imports from Asia-Pacific countries such as *Peronospora manshurica* on *Glycine max* from Indonesia, Japan, Malaysia, North Korea, South Korea, Taiwan, Thailand, and arabis mosaic virus on *Vigna radiata* from AVRDC (Taiwan), cherry leaf roll virus on *G. max* from AVRDC (Taiwan), Sri Lanka, Thailand, cowpea mottle virus on *V. unguiculata* from the Philippines, cowpea severe mosaic virus on *V. radiata* from Australia, tomato ringspot virus on *G. max* from AVRDC (Taiwan), Sri Lanka, Thailand, raspberry ringspot virus on *G. max* from AVRDC (Taiwan), pea enation mosaic virus on *Phaseolus vulgaris* from Nepal, etc.(Chalam and Khetarpal 2008; Chalam et al. 2008, 2012, 2014, 2017; Chalam and Maurya 2018; Khetarpal et al. 2001, 2006). If not intercepted, many of these quarantine pests would have entered and become emerging pests.

Ensuring the absence of important pests not yet introduced in Asia-Pacific: South American Leaf Blight (SALB) of rubber caused by *Microcyclus ulei* is one of the most important pests occurring on commercial commodities not yet occurring in Asia, particularly in Southeast Asia, where natural rubber is playing a prime role in the economy of the country, and there is a potential risk pathogen entry and establishment due to the favourable climatic conditions and rainfall similar to SALB endemic regions of the Amazon. Recently, APPPC has brought out a contingency plan for SALB, which provides background information on which to base a response. The plan provides a summary of information on the biology of the pest and the available control measures for the disease. It provides guidelines for steps to be undertaken and considered when developing a response plan for this pest. The response plan is operational and determines the resources that are needed (https://www.fao.org/asiapacific/apppc/en/). Mexican cotton boll weevil (*Anthonomus grandis*) is listed as an A1 quarantine pest by EPPO and has high quarantine significance for Asia-Pacific (CABI 2016). Since it exists in a subtropical region, it indicates that the cotton-growing areas in the region are at potential risk of invasion, and there is a need to be vigilant. On maize, which is an important food security crop in the region, ergot disease (which is endemic in central Mexico) caused by *Claviceps gigantean* has yet to be reported (CABI 2016).

Ensuring preparedness to mitigate uncertainties due to the impact of climate change: It is well known that new plant pests and diseases may emerge due to evolving selection and adaptation to new climatic situations. Migratory plant pests, in particular locusts, are totally dependent on rain, temperature and vegetation and their habitats change rapidly. The desert locust, like other locusts, can change its behaviour and physiology from solitary grasshoppers to gregarious stages that form swarms. Solitary desert locusts occur at low density in parts of India, Pakistan, Iran and Afghanistan. The outbreak area reaches from Mauritania to India and from southern Europe to Cameroon and Tanzania. Another key example of a migratory pest is the coffee berry borer. With climate change imminent, the coffee berry borer is expected to become an even greater threat (Jaramillo et al. 2011). Developing countries in the Asia-Pacific, especially India and Bangladesh, are likely to face the highest reductions in agricultural production in the world due to climate change. This will be through direct effects on crop production such as floods and drought and indirect effects through the spread of pests and diseases (http://legacy.crcplantbiosecurity.com.au/content/climate-change-asian-pacific-regions.html). An Asia-Pacific Network for Global Change project led by Cooperative Research Center for National Plant Biosecurity, Australia brought together Indian, Bangladesh and Australian policy experts and research scientists, comprising agro-meteorologists, crop modellers, entomologists and plant pathologists, demonstrating a global approach to addressing the risks posed by climate change to food security in both developing and developed countries (Luck et al. 2010).

Ensuring concerted efforts towards biosecurity by regional bodies: The regional bodies that exist and operate in various development activities in the region, such as Association of South East Association Nations, South Asian Association of Regional Cooperation, Bay of Bengal Initiative for Multisectoral Trade and Economic Cooperation, Asia-Pacific Economic Commission, Pacific Community, Pacific Agreement on Closer Economic Relations, Pacific Islands Trade and Investment Commission, Asian Development Bank and the global and regional donor agencies, need to prioritize plant biosecurity engagements, synergize their action plans for economic, food and environmental security and work closely with FAO, APPPC, PPPO and NPPOs.

11.5 CONCLUSIONS AND KEY ACTION PLANS

Recently, outbreaks of almost 90 pests occurred on important crops in Asia-Pacific, and a number of economically important emerging pests were also recorded. This implies that domestic markets and their needs are likely to be more imperative in the coming future than the desirability of export markets. The number of emerging pests is increasing in the region owing to the intensification of agriculture and the movement of pests with increasing trade from and within the region. Classification of a pest as causing outbreaks or as emerging is considered useful; for an outbreak, a rapid response strategy is needed by policymakers, whereas for an emerging pest, an immediate and extensive pest-specific surveillance programme and early detection protocols are called for. To ensure plant biosecurity, there is a great variation in the responses of different countries to pest outbreaks owing to variations in their economic growth and, hence, their institutional mechanisms and infrastructure. The developed countries Australia, New Zealand and Japan, which have the best possible biosecurity infrastructure, mechanisms and engagements, are investing substantially in the region, mainly because a lot of their imports are dependent on other developing and least developed countries and islands in the region. However, most countries in the Asia-Pacific region have insufficient enabling mechanisms, procedures and resources allocated to surveillance and monitoring, border control and inspections, expertise in risk assessment, diagnostic tools for early detections, expertise in diagnosis (taxonomy), data collection and access to information, tools for rapid response to entry, and establishment and spread of control measures at the source of the produce. FAO, through APPPC and PPPO, is coordinating with NPPOs to manage outbreaks and spreads of major pests and also facilitating the implementation of standards and guidelines for compliance with the SPS Agreement; however, the economic conditions of many countries do not allow them to sustain the global efforts after the project period is over. The role and responsibilities of national players, especially the NPPOs, thus become crucial. Attention from all policymakers, quarantine officials and researchers alike is needed to meet the challenges of biosecurity and ensure food, economic and environmental security in the Asia-Pacific.

The key action plans proposed are as follows:

- The Asia-Pacific countries need to ensure proper compliance with international and regional standards with special reference to pest risk analysis, pest surveillance and pest reporting. Some of the countries need to invest more in capacity building to develop infrastructure and expertise in various domains of SPS. Investments in early detection and control mechanisms will be critical to avoid higher costs of eradication and control. Good border control and rapid diagnostic tools for better surveillance of plant pests are needed.
- There is a dire need by countries for the strengthening of import and domestic quarantine programmes for pests with restricted distribution, with a focus on the Additional Declaration(s) in the Phytosanitary Certificate and provisions for post-entry quarantine monitoring disinfestation and disinfection treatments. The countries need to regularly update their regulated pest list and also their list of prohibited plant species from a given country based on the evolving scenario of emerging pests.
- Knowledge management on plant biosecurity at the regional level seems vital for easy access of information to researchers and other stakeholders so that needful innovative pest management strategies can be taken up well in advance (Khetarpal and Dashora 2014). There is a need for enhanced regional cooperation and coordination through information sharing on emerging pests in the area. A database on pests of exotic and indigenous origin needs to be prepared by the countries. The Pest and Disease Image Library (www.padil.gov.au/), a publicly available online database of Australia and New Zealand, is an example for other countries to emulate. Based on the country database, a regional database needs to be developed with the coordinated efforts of APPPC and PPPO.
- Interdisciplinary research in areas of biosecurity and biodiversity needs to be encouraged in relation to diagnostics, epidemiology, resistance breeding and the control of emerging pests. Development of Regional and National Plant Biosecurity Diagnostic Network as of Australia needs to be developed for other countries. The network may include a remote microscopy system, national reference collections, user-friendly portable devices for on-the-spot identification of known diseases that can detect multiple pests in the field, rapid, non-invasive detection of characteristic volatiles or electromagnetic radiation from infected plants, microarray technology, and ultimately, a national biosecurity chip for the diagnosis of all current threats to agriculture.

REFERENCES

Anonymous. Biological emergencies-high consequence plant diseases/pests and your crops. The Center for Food Security & Public Health, Iowa State University, USA. 2010. p2. https://www.prep4agthreats.org/Assets/Factsheets/High-Consequence-Plant-Diseases-Pests-and-Your-Crops.pdf.

Asghar, M.; Arshad, M.; Fiaz, M.; Suhail, A.; Sabir, A. M. A survey of rice farmers' farming practices posing threats to insect biodiversity of rice crop in the Punjab, Pakistan. *Int. J. Biodivers. Conserve*. 2013, 5, 647–654.

Buddenhagen, I. Understanding strain diversity in Fusariumoxysporum f. sp. cubenseand history of introduction of 'Tropical Race 4' to better manage banana production. *Acta Hortic*. 2009. 828, 193–204.

CABI. *Invasive Species Compendium*. Wallingford, UK: CAB International. 2016. www.cabi.org/isc.

Callaway, E. Devastating wheat fungus appears in Asia for first time. *Nature*. 2016. 532, 421–422.

Chalam, V. C.; Bhalla, S.; Singh, B.; Rajan. (eds). *Potential Quarantine Pests for India in Grain Legumes*. New Delhi, India: National Bureau of Plant Genetic Resources. 2012. p. 324.

Chalam, V. C.; Khetarpal, R. K. A critical appraisal of challenges in exclusion of plant viruses during transboundary movement of seeds. *Indian J. Virol*. 2008. 19, 139–149.

Chalam, V. C.; Maurya, A. K. Role of quarantine in ensuring biosecurity against transboundary plant viruses. *Agric. Res. J*. 2018. 55, 612–626.

Chalam, V. C.; Parakh, D. B.; Khetarpal, R. K.; Maurya, A. K.; Jain, A.; Singh, S. Interception of seed-transmitted viruses in cowpea and mungbean germplasm imported during 2003. *Indian J. Virol*. 2008. 19, 12 16.

Chalam, V. C.; Parakh, D. B.; Maurya, A. K. Role of viral diagnostics in quarantine for plant genetic resources and preparedness. *Indian J. Plant Genet. Resour*. 2017. 30, 271–285.

Chalam, V. C.; Parakh, D. B.; Maurya, A. K.; Singh, S.; Khetarpal, R. K. Biosecuring India from seed-transmitted viruses: The case of quarantine monitoring of legume germplasm imported during 2001–2010. *Indian J. Plant Prot*. 2014. 42, 270–279.

Cheng, J. Rice planthopper problems and relevant causes in China. In *Planthoppers: New Threats to the Sustainability of Intensive Rice Production Systems in Asia*. K. L. Heong; B. Hardy, (eds.). International Rice Research Institute, Los Baños, Philippines. 2009. pp. 157–178, p 460.

Dyck, V. A.; Thomas, B. The brown planthopper problem. In *Brown Planthopper: Threat to Rice Production in Asia*. International Rice Research Institute. 1979. pp. 3–17.

EFSA. Scientific Colloquium XVI – Identification of emerging plant health risks: Future challenges and approaches. Briefing notes for discussion groups. 9–10 June 2011. Parma, Italy.

EPPO. *PQR Database*. Paris, France: European and Mediterranean Plant Protection Organization. 2014. https://www.eppo.int/DATABASES/pqr/pqr.htm.

FAO. *Banana Market Review and Banana Statistics 2012–2013*. 2014. https://www.fao.org/docrep/i3627e/ i3627e/i3627e.pdf.

Flood, J. The importance of plant health to food security. *Food Secur*. 2010. 2, 215–231.

Heinrichs, E. A. The brown planthopper (*Nilaparvatalugen*). In *The Brown Planthopper: Threat to Rice Production in Asia*. S. Sadjad; S. Adisoemarto; M. A. Rifai, (eds.). International Rice Research Institute, Los Banos, Philippines. 1977. pp. 45–64.

Hodson, D. P. Shifting boundaries: Challenges for rust monitoring. *Euphytica*. 2011. 179, 93–104.

https://www.dpi.nsw.gov.au/__data/assets/pdf_file/0006/427515/Exotic-Pest-Alert-Tomato-potato-psyllid.pdf.

https://www.fao.org.
https://www.fao.org/asiapacific/apppc/en/.
https://www.fao.org/fileadmin/templates/rap/files/meetings/2013/131129-report.pdf.
https://legacy.crcplantbiosecurity.com.au/content/climate-change-asian-pacific-regions.html.
https://lrd.spc.int/plantprotectionorganisation-/pppo.
https://www.nppc.gov.bt/wp-content/uploads/2016/07/4.-Chilli-pod-borer.pdf.
https://www.rhs.org.uk/advice/profile?pid=457.
https://www.ippc.int/en/.
https://www.ippc.int/largefiles/adopted_ISPMs_previousversions/en/ISPM_05_2007_En_2007-07-26.pdf.

Hyun, I. H.; Choi, W. Phytophthora species, new threats to the plant health in Korea. *Plant Pathol. J.* 2014. 30, 331–342.

ICAR. *Tuta absoluta*: A new invasive pest alert. 2014. https://www.icar.org.in/en/node/8600.

Jaramillo, J.; Borgemeister, C.; Baker, P. Coffee berry borer *Hypothenemushampei* (Coleoptera: Curculionidae): Searching for sustainable control strategies. *Bull. Entomol. Res.* 2006. 96, 223–233.

Khetarpal, R. K. A critical appraisal of seed health certification and transboundary movement of seeds under WTO regime. *Indian Phytopathol.* 2004. 57, 408–421.

Khetarpal, R. K.; Dashora, K. Knowledge management for plant biosecurity in South Asia. *Sci. Res. Essays*. 2014. 9(24), 997–1002.

Khetarpal, R. K.; Gupta, K. Plant biosecurity in India: Status and strategy. *Asian Biotechnol. Dev. Rev*. 2007. 9, 83–107.

Khetarpal, R. K.; Lal, A.; Varaprasad, K. S.; Agarwal, P. C.; Bhalla, S.; Chalam, V. C.; Gupta, K. Quarantine for safe exchange of plant genetic resources. In *Hundred Years of Plant Genetic Resources Management in India*. A. K. Singh; K. Srinivasan; S. Saxena; B. S. Dhillon, (eds.). National Bureau of Plant Genetic Resources, New Delhi, India. 2006. pp. 83–108.

Khetarpal, R. K.; Mukhopadhyay, A. N.; Balaraman, N.; Vijayan, K. K.; Bandopadhyay, K. K.; Gupta, K.; Jain, R. K.; Kumar, J. Agricultural biosecurity. In *State of Indian Agriculture*. M. Rai; S. S. Acharya; S. M. Virmani; P. K. Aggarwal, (eds.). National Academy of Agricultural Sciences, New Delhi, India. 2009. pp. 151–174.

Khetarpal, R. K.; Singh, S.; Parakh, D. B.; Chalam, V. C. Viruses intercepted in exotic germplasm during 1991–2000 in quarantine. *Indian J. Plant Genet. Resour.* 2001. 14, 127–129.

Korzukhin, M. D.; Porter, S. D.; Thompson, L. C.; Wiley, S. Modeling temperature-dependent range limits for the fire ant *Solenopsisinvicta* (Hymenoptera: Formicidae) in the United States. *Environ. Entomol.* 2001. 30, 645–655.

Kranthi, K. R. Let not the whitefly see red. In *Cotton Statistics & News*. A. Singh, (ed.). Cotton Association of India. 2016. p. 12.

Lim, G. S.; Ooi, A. C.; Koh, A. K. Outbreak and control of the rice brown planthopper (NilaparvatalugensStal.) In TanjungKarang, Malaysia. Proc. Plant Prot. Conf. Malaysian Plant Protection Society, Kuala Lumpur, Malaysia. 1978. pp. 193–213.

Luck, J.; Asaduzzaman, M.; Banerjee, S.; Bhattacharya, I.; Coughlan, K.; Debnath, G. C.; De Boer, D.; Dutta, S.; Forbes, G.; Griffiths, W.; Hossain, D.; Huda, S.; Jagannathan, R.; Khan, S.; O'Leary, G.; Miah, G.; Saha, A.; Spooner-Hart, R. The Effects of Climate Change on Pests and Diseases of Major Food Crops in the Asia Pacific Region. Final Report for APN Project. Project Reference: ARCP2010-05CMY-LUCK. 2010. p. 73. https://www.apngcr.org/resources/files/original/1534fe7a80b1be6e9d00d2cd6934fae0.pdf.

Malaker, P. K.; Verma, N. C. D.; Tiwari, T. P.; Collis, W. J.; Duveiller, E.; Singh, P. K.; Joshi, A. K.; Singh, R. P.; Braun, H. J.; Peterson, G. L.; Pedley, K. F.; Ferma, M. L.; Valent, B. First report of wheat blast caused by Magnoportheoryzaepathotypetriticum in Bangladesh. *Plant Dis.* doi:10.1094/PDIS-05-16-0666-PDN.

Mani, M.; Joshi, S.; Kalyanasundaram, .M.; Shivaraju, C.; Krishnamoorthy, A.; Asokan, R.; Rebijith, K. B. A new invasive Jack Beardsley mealybug, *PseudococcusJackbeardsleyi* (Hemiptera: Pseudococcidae) on papaya in India. *Florida Entomol.* 2013. 96(1), 242–245. doi:10.1653/024.096.0135.

McGlynn, T. P. Non-native ants are smaller than related native ants. *Am. Naturalist.* 1999. 154, 690–699.

Meyerdirk, D. E.; Muniappan, R.; Warkentin, R.; Bamba, J.; Reddy, G. V. P. Biological control of the papaya mealybug, *Paracoccusmarginatus* (Hemiptera: Pseudococcidae) in Guam. *Plant Prot. Q.* 2004. 19, 110–114.

Mochida, O. Brown planthoppers reduce rice production. *Indones. J. Agric. Res. Dev.* 1979. 1, 2–7.

Muniappan, R.; Shepard, B. M.; Watson, G. W.; Carner, G. R.; Rauf, A.; Sartiami, D.; Hidayat, P.; Afun, J. C. K.; Goergen, G.; Rahman, A. K. M. Z. New records of invasive insects (Hemiptera: Sternorrhyncha) in Southern Asia and West Africa. *J. Agr. Urban Entomol.* 2011. 26(4), 167–174.

Muniappan, R.; Shepard, B. M.; Watson, G. W.; Carner, G. R.; Sartiami, D.; Rauf, A.; Hammig, M. D. First report of the papaya mealybug, *Paracoccusmarginatus* (Hemiptera: Pseudococcidae), in Indonesia and India. *J. Agr. Urban Entomol.* 2008. 25, 37–40.

Ordoñez, N.; García-Bastidas, F.; Laghari, H. B.; Akkary, M. Y.; Harfouche, E.; al Awar, B. N.; Kema G. H. J. First Report of *Fusariumoxysporum* f. sp. *cubense* Tropical Race 4 causing Panama disease in Cavendish bananas in Pakistan and Lebanon. *Plant Dis.* 2016. 100, 209.

Parsa, S.; Kondo, T.; Winotai, A. The cassava mealybug (*Phenacoccus manihoti*) in Asia: First records, potential distribution, and an identification key. *PLoS One.* 2012. 7(10), e47675. doi:10.1371/journal.pone.0047675.

Ploetz, R. C. Fusarium wilt of banana. *Phytopathology.* 2015. 105, 1512–1521. doi:10.1094/PHYTO-04-15-0101-RVW.

Poolprasert, P.; Jongjitvimol, T. Arthropod communities inhabiting organic rice agro-ecosystem. International Conference on Agricultural, Ecological and Medical Sciences, July 3–4, 2014, London, UK. 2014.

Rethinam, P.; Singh, S. P. Current status of the coconut beetle outbreaks in the Asia-Pacific region. In *Developing an Asia-Pacific Strategy for Forest Invasive Species: The Coconut Beetle Problem – Bidging Agriculture and Forestry.* S. Appanah; H. C. Sim; K. V. Sankaran, (eds.). Food and Agriculture Organization of the United Nations Regional Office for Asia and the Pacific, Bangkok. 2007. pp. 1–23.

Saharan, M. S.; Bhardwaj, S. C.; Chatrath, R.; Sharma, P.; Choudhary, A. K.; Gupta, R. K. Wheat blast disease – An overview. *J. Wheat Res.* 2016. 8, 1–5.

Sharma, S. B.; Wightman, J. *Vision Infinity for Food Security: Some Whys, Why Nots and Hows*! Springer. 2015. p. 98.

Shylesha, A. N.; Joshi, S. Occurrence of Madeira mealybug, *Phenacoccusmadeirensis* Green (Hemiptera: Pseudococcidae) on cotton in India and record of associated parasitoids. *J. Biol. Control.* 2012. 26, 272–273.

Singh, R. P.; Hodson, D. P.; Huerta-Espino, J.; Jin, Y.; Bhavani, S.; Njau, P.; Herrera-Foessel, S.; Singh, P. K.; Singh, S.; Govindan, V. The emergence of Ug99 races of the stem rust fungus is a threat to world wheat production. *Annu. Rev. Phytopathol.* 2011. 49, 465–481.

Soejitno, J. *Integrated Pest Management in Rice in Indonesia. A Success Story.* APAARI Publication No. 1999/2, Bangkok. 1999. p. 57.

Takano, S.; Takasu, K.; Ichiki, R. T.; Nakamura, S. Cold tolerance of the coconut hispine beetle, *Brontispalongissima* (Coleoptera: Chrysomelidae) in Japan. *Appl. Entomol. Zool.* 2012. 47, 173–180.

The Royal Society. Science Policy Centre Report – Reaping the benefits: Science and the sustainable intensification of global agriculture. 2009. https://royalsociety.org/~/media/Royal_Society_Content/policy/publications/2009/4294967719.pdf.

Tyagi, K.; Kumar, V. First report of western flower thrips, *Frankliniellaoccidentalis* (Pergande) (Thripidae: Thysanoptera) from India: A potential havoc to Indian agriculture. *HALTERES*. 2015. 6, 1–3.

Vega F. E. Coffee berry borer, Hypothenemushampei (Ferrari) (Coleoptera: Curculionidae: Scolytinae). In *Encyclopedia of Entomology*. J. L. Capinera, (ed.), 2nd ed. Springer, New York. 2008. pp. 959–960.

Vincenzo, V. *Handbook of Mites of Economic Plants: Identification, Bio-ecology and Control*. Wallingford, UK: CAB International. 2016. p. 832.

Waage, J. K.; Mumford, J. D. Agricultural biosecurity. *Phil. Trans. R. Soc. B*. 2008. 363(1492), 863–876.

Williams, D. J. *Mealybugs of Southern Asia*. Southdene SDN, Kuala Lumpur: The Natural History Museum. BHD, 2004. p. 896.

www.padil.gov.au/.

Xiang, Y.; Wu, X. Q.; Zhou, A. D. Bacterial diversity and community structure in the pine wood nematode *Bursaphelenchusxylophilus* and *B*. mucronatuswith different virulence by high-throughput sequencing of the 16S rDNA. *PLoS One*. 2015. 10(9), e0137386. doi:10.1371/journal.pone.0137386.

Yang, B.; Wang, Q. Distribution of the pine wood nematode in China and susceptibility of some Chinese and exotic pines to the nematode. *Can. J. For. Res*. 1989. 19, 1527–1530.

Yongfan, P. 27th Session of the Technical Consultation among Regional Plant Protection Organizations, 2–6 November 2015, Memphis, Tennessee, USA. *APPPC Update*. 2015.

12 Mycotoxins
Challenges and Preparedness for Reducing Impact on Human Health

Rashmi Aggarwal, Deeba Kamil, Bishnu Maya Bashyal, and Mahender S. Saharan
ICAR-IARI

12.1 INTRODUCTION

Many fungal species produce toxic chemical compounds known as mycotoxins, which are hazardous for human beings and animals and can cause diseases and death. These are low molecular weight metabolites largely produced by ascomycetous group of fungi such as *Aspergillus*, *Fusarium* and *Penicillium* (Liew and Mohd-Redzwan, 2018). When the toxic fungus grows on animals and produces disease, it is called as mycoses, whereas when toxic fungus when come into contact with dietary, respiratory, dermal and other exposures and produce diseases, it is called as mycotoxicoses, and the study of mycotoxicoses is called mycotoxicology (Bennett and Klich, 2003). Some fungal species produce mycotoxins, which inhibit the growth of bacteria (such as penicillin) and are called antibiotics. The mycotoxin term was coined in 1962 as the result of Turkey X disease, in which approx. 100,000 turkey poults died in London and England due to mycotoxins (Blout, 1961; Forgacs et al., 1962; Maggon et al., 1977). After the study of Turkey X disease, scientists observed that peanut seeds were also contaminated with aflatoxins secreted by the fungus *Aspergillus flavus*, and researchers found that the secondary metabolites produced by many moulds might be poisonous (Richard, 2007). Later, nearly 400 mycotoxins were identified, among them aflatoxins, ochratoxins, zearalenone, fumonisins and trich thecenes, have great significance with respect to human health (Ates et al., 2013).

On the basis of chemistry, mycotoxins are grouped into five major groups: asaflatoxins, ochratoxin, fumonisins, deoxynivalenol/nivalenol and andzearalenone. Based on mycotoxins found in food, the fungi have been divided into two groups: field fungi that attack the crop before harvest and storage fungi that occur only after harvest (Aggarwal et al., 2022). Mycotoxins can be produced in favourable conditions such as high temperatures, high moisture content and unhygienic conditions during transportation and storage. These lethal mycotoxins generally occur in animal feeds, crops and animal products. The maximum level of mycotoxins can

DOI: 10.1201/9781003403142-12

be found in concentrated animal feed materials. The fungi are transferred to the other growing crop by asexual spores, i.e., conidia, through wind or insects. Due to their harmful effects on plants and animals, it is necessary to remove or manage the development and production of mycotoxins. There are many strategies developed by scientists to manage mycotoxin production. Methodologies include breeding for resistance, chemical control and cultural control. Resistant plants provide an effective and environmentally friendly approach to managing mycotoxins and the fungi producing them.

12.2 MYCOTOXINS: GLOBAL CONCERN FOR FOOD SAFETY AND SECURITY

Mycotoxins, the low molecular mass (MW ~ 700 Da) secondary metabolites mainly produced by *Aspergillus, Penicillium* and *Fusarium*, are highly noxious substances in animals and humans. However, not all mycotoxins are classified as such; for example, penicillin is widely used as an antibiotic (Speight, 2012). The structural form of mycotoxins varies from simple four C compounds, e.g., moniliformin (MON), to complex substances such as phomopsins (Zain, 2011). Fungal proliferation and production of mycotoxins rise naturally due to environmental factors, especially during tropical conditions (Mohd-Redzwan et al., 2013). Besides, downstream processing, such as poor harvesting practices, improper storage and less than optimal conditions during transportation, processing and marketing can also contribute to the growth of fungi and increase the risk of the major food spoilage agent caused by mycotoxin production (Khazaeli et al., 2014). Due to the ubiquitous nature of fungi, mycotoxins have been increasingly attracting the concern of health organizations where their occurrence in foods cannot be ignored and already pose a risk to consumers (Jahanian, 2016; Aggarwal, 2021).

12.3 OCCURRENCE OF MYCOTOXICOSIS

It has been estimated that 25% of the world's crops, including nuts and cereals, are contaminated by mould and fungal growth, as reviewed by the United Nations Food and Agriculture Organization and the World Health Organization, which harm animal and human health (Pandya and Arade, 2016). Exposure to mycotoxins is mostly by ingestion but also occurs by the dermal and inhalation routes. The extent of adverse effects of mycotoxins on human or animal health mainly depends on the extent of exposure (dosage and period), type of mycotoxins, physiological and nutritional status, as well as possible synergistic effects of other chemicals to which the animals or humans are exposed (Gajecka et al., 2013). In 1960, interest in mycotoxins was initiated by the occurrence of Turkey X disease caused by AF, which killed more than 100,000 turkeys. Subsequently, it was found that AFs are carcinogenic and cause hepatocellular carcinoma (HCC) in animals and humans, and this has stimulated research on mycotoxins (Peraica et al., 1999). Since then, around 400 mycotoxins are known, but AFs, ochratoxins, zearalenone (ZEA), fumonisins (FBs) and trichothecenes are mostly focused on public health issues (Ates et al., 2013). Mycotoxin exposure is not only limited to pure mycotoxins but also masked mycotoxin, which forms when plants protect themselves by conjugating mycotoxins with biopolymers. In addition, some people are more susceptible to getting mycotoxicosis than others, and this is due to the pharmacogenetic variability where specific gene mutations such as cytochrome p450 (CYP 450) genes could either increase or decrease the metabolic activity (cytotoxicity) of the challenging mycotoxins (Sun et al., 2016). For instance, in both *in vivo* (Muhammad et al., 2017) and *in vitro* (Lewis et al., 2000) studies, CYPs' 1A2 and 3A4 appear as the most important enzymes that increased metabolism of AFB1 to its active form, AFB1–8,9epoxide and subsequently to AFB1-DNA adduct formation, in which the biomarker has been linked to the development of liver cancer (Ceccaroli et al., 2015). Chronic mycotoxicosis has a greater impact on human health. Mycotoxin can induce diverse and powerful toxic effects in test systems; some are carcinogenic, mutagenic, teratogenic, oestrogenic, haemorrhagic, immunotoxic, nephrotoxic, hepatotoxic, dermatoxic and neurotoxic (Milicevic et al., 2010). Frequently, mycotoxicosis remains unrecognized by medical professionals. Mycotoxicosis can be weighed when a disease appears in several people with no obvious connection to a known etiological agent, such as microorganisms (Viegas et al., 2015).

12.4 TYPES OF MYCOTOXINS AND ABOUT FUNGI PRODUCING THEM

Agriculture is the backbone of the Indian economy because 60%–70% of people depend on agriculture for their basic needs or employment. Mostly, every crop is infected with numerous types of diseases that are spread by fungi, bacteria and viruses. Nearly 25% of crops are affected by fungi worldwide. Mycotoxins are produced by fungi after infection with crops, and most cereal crops are infected with fungal infections (Table 12.1). Maize is one of the highly susceptible crops, whereas rice is the least susceptible to mycotoxins. The fungi associated with crops reduce their quality and quantity, and also spread different types of diseases to humans and animals.

12.5 CHEMISTRY AND BIOSYNTHESIS OF MYCOTOXINS

Mycotoxins are mainly toxic secondary metabolites that directly or indirectly impact human, plant and animal health (Kiessling, 1986). Some mycotoxins contaminate most cereal grains and leguminous crops, *viz.* aflatoxins, penicillic acid (PA), fumonisins, fusaric acid (FA), fusarin C, citrinin, patulin, ochratoxin, deoxynivalenol and zearalenone. Many cereal crops, such as barley, sorghum and wheat, can contain mycotoxins. The mycotoxigenic fungi involved in the human food chain belong mainly to three main genera: *Aspergillus*, *Penicillium* and *Fusarium*. The former two genera are commonly found as contaminants in foods during drying and storage, while the latter are plant pathogens that produce mycotoxins before or immediately after harvesting. Aflatoxins (AF) are a group of mycotoxins that are of greatest significance in foods and feeds and are produced mainly by *Aspergillus flavus* and *Aspergillus parasiticus*. The four main aflatoxins produced are B_1, B_2, G_1 and G_2. They are difuranocoumarin derivatives, with

TABLE 12.1
Major Mycotoxins and Fungi Producing Them

Mycotoxins	Fungi Associated
Aflatoxins	*Aspergillus flavus*, *A. parasiticus*
Ochratoxin A	*Aapergillus ochraceus*, *A. alliaceous*, *A. mellis*, *P. viridicatum*, *P. cycloplum*
Patulin	*Penicillium expansum*, *P. patulum*, *P. melini*, *P. equinum*, *Aspergillus clavatus*, *A. terreus*, etc.
Alternaria toxins (Tenuazonic acid, alternariol, alternariol methyl ether, altenuen and altertoxins)	*Alternaria* spp.
Deoxynivalenol (DON), 3-acetyl DON, 15-acetyl DON, nivalenol, fusarenone (FX), Zearalenone (ZEN)	*Fusariumculmorum*
DON, 15 – ADON, NIV, FX, ZEN	*F. graminearum*
T-2 toxin, HT-2 toxin, neosolaniol (NEO), diacetoxyscirpinol (DAS), FX, ZEN	*F. sporotrichioides*
T-2 toxin, HT-2 toxin, NIV, DAS, FX	*F. poae*
Fumonisins, moniliformin, Fusarin C	*F. verticilloides*
	F. proliferatum
Moniliformin, wortmannin, Fusaric acid	*F. oxysporum*
Sambutoxin	*F. sambucinum*

the B and G nomenclature deriving from the blue and green fluorescent colours produced under UV light on thin layer chromatography plates, with the subscript numbers indicating major and minor compounds, respectively. Aflatoxin B_1 (AFB_1) is widely regarded as the most potent liver carcinogen known for a wide variety of animal species, including humans (Dragan and Pitot, 1994). Aflatoxin M_1 and M_2 are hydroxylated derivatives of AFB_1 and AFB_2, which are formed and excreted in the milk of lactating animals, including humans, that have consumed aflatoxin-contaminated foods. A precursor of the aflatoxins, sterigmatocystin (ST), is a mycotoxin that is characterized by a xanthone moiety fused to a dihydrodifuran or tetrahydrofuran moiety; it is also acutely toxic and carcinogenic (Smith et al., 1994). The trichothecenes are a group of mycotoxins that are produced by several fungal genera, including *Fusarium*, *Trichoderma*, *Trichothecium*, *Stachybotrys*, *Cephalosporium* and *Myrothecium*. The trichothecenes are chemically very diverse, but are all tricyclic sesquiterpenes with a 12,13-epoxy-trichothec-9-ene ring. The fumonisins are another important group of mycotoxins produced primarily by the cereal pathogen *Fusarium moniliforme*. A number of other fungal species also produce fumonisins, including *F. proliferatum*, *F. anthophilum*, *F. dlamini* and *F. napiforme*.

The aflatoxin biosynthetic pathway is a well-understood pathway. Initially, acetate and malonyl CoA are converted to a hexanoyl starter unit by a fatty acid synthase, which is then extended by a polyketide synthase to norsolorinic acid, the first stable precursor in the pathway. The polyketide then undergoes approximately 12–17 enzymatic conversions through a series of pathway intermediates. While aflatoxins are produced only by certain strains of *A. parasiticus*, *A. flavus* and *A. nominus*, numerous ascomycetes and deuteromycetes, including *A. nidulans*, produce the mycotoxin sterigmatocystin (ST), the penultimate intermediate in the AF biosynthetic pathway. The ST pathway is believed to include at least 15 enzymatic activities involving each enzyme activity from the AF pathway, barring the penultimate steps involving the conversion of ST to AF. The trichothecene pathway begins with the cyclization of farnesyl pyrophosphate (FPP) to trichodiene by the enzyme trichodiene synthase. The subsequent pathway involves a number of oxygenations, isomerizations, cyclizations and esterifications leading from trichodiene to diacetoxyscirpenol, T-2 toxin and 3-acetyldeoxynivalenol. Fumonisins are thought to be synthesized through the condensation of the amino acid alanine into an acetate-derived precursor. Branched-chain methyl groups are added at C-12 and C-16 by an S-adenosyl methionine transferase. The subsequent biosynthetic steps involve oxygenation and esterification of the acetate-derived backbone (Sweeney and Dobson, 1999).

12.6 MYCOTOXINS: TYPES AND EFFECTS ON AGRICULTURE, HORTICULTURE AND FOOD

12.6.1 Aflatoxins

Aspergillus species are known to produce G_1 and G_2 aflatoxin of the G group, along with B aflatoxins. In 1960, aflatoxins were first identified after a serious epidemic of Turkey X disease, which was found to be associated with the diet of peanuts in Britain. *A. flavus* grows throughout the world and is found in food crops such as peanuts, maize and cottonseed in humid and warm climates. *A. parasiticus*is comparatively more limited in geographical distribution than *A. flavus* and generally infects peanuts.

Mode of action of aflatoxins: Aflatoxin inhibits chlorophyll a, chlorophyll b and protochlorophyllide biosynthesis. The level of protein and nucleic acid is also suppressed by aflatoxin in germinating seeds (Sinha, 1996). In addition, aflatoxin also inhibits the activity of chromatin-bound DNA-dependent polymerase. In the absence or unavailability of m-RNA, aflatoxin inhibits protein synthesis, whereas DNA synthesis is inhibited by binding to DNA during the replication process (Tripathi and Misra, 1983). Aflatoxins impair the cell cycle, promote apoptosis in cultured cells and lead to apoptosis through the accumulation of DNA damage in cells (Fadl-Allah et al., 2011).

12.6.2 Ochratoxin

Aspergillus and *Penicillium* are mainly two filamentous fungi known to produce ochratoxin. OTA is more complex to study than aflatoxin. The first OTA toxin was discovered in 1965 (Van der Merwe et al., 1965) and divided into three groups, *viz.* A, B and C (Heussner and Bingle, 2015). Among these groups, Ochratoxin A is dominant and applicable fungal toxin, which has the distinguishing feature of being chlorinated (Heussner and Bingle, 2015). Ochratoxin B (OTB) and ochratoxin C (OTC) are non-chlorinated and grouped in ethyl esters, which are comparatively less toxic than ochratoxin A (Bayman and Baker, 2006). Crops grown in a temperate climate are ideal for the production of OTA. In tropical climates, OTA affects mainly coffee and grapes, whereas cereals are mainly contaminated in temperate climates. *Penicillium verrucosum* produces OTA in cereal crops in cool temperate climates, while *Aspergillus* produces it in warm climates. Production of ochratoxin by *A. ochraceus* requires 31°C temperature, pH 3–10 and minimum water activity of 0.8, while 20°C, pH 6–7 and a minimum water activity of 0.86 are required by *P. verrucosum* (Reddy and Bhoola, 2010).

Mode of action of ochratoxin: Ringot et al. (2006) reported that OTA toxicity involves suppression of protein synthesis, prohibition of mitochondrial respiration, DNA damage, disruption of calcium homeostasis and interference in the metabolic system. Wang et al. (2012) studied the phytotoxic mechanisms of OTA during gene and protein expression in *Arabdiosis thaliana.* Some researchers summarized OTA phytotoxicity as when OTA comes in contact with *A. thaliana* leaves, it results in a hypersensitive response. OTA affects protein synthesis by changing the expression of genes and also triggering the membrane transport system. Salicylic acid, ethylene and jasmonic acid are the main mediator molecules causing the phytotoxicity of OTA. Wang et al. (2014) reported that OTA activates a defence response by enhancing the transcript level of enzymes to maintain the total glutathione content.

12.6.3 Penicillic Acid

The PA mycotoxin is mainly produced by the fungus *Penicillium puberulum, Penicillium roqueforti* mould and *Aspergillus flavus.* Penicillicmycotoxin was first isolated and named by Bainier, but it was synthesized for the first time by Ralph Raphael in 1947 during the Second World War. It is also the major product of the acid degradation of penicillin. PA has also been isolated from many *Penicillium* species, especially *P. aurantiogriseum, P. canescens, P. polonicum* and *P. radicicola* (Ismaiel and Papenbrock, 2015).

Mode of action of PA: The PA mycotoxin is activated when it interacts with the SH residues in enzymes and is readily inactivated by thiol enzymes (Ciegler, 1972). It has been found that the PA inhibits the thiol enzymes, i.e., alcohol dehydrogenase and lactic dehydrogenase (Ashoor and Chu, 1973). PA was measured in infected corn when its quantities reached more than 2% (Ciegler and Kurtzman, 1970). PA toxins increase during low temperature storage conditions and are a serious hazard for storage products (Kurtzman and Ciegler, 1970; Keromnes and Thouvenot, 1985).

12.6.4 Deoxynivalenol

Deoxynivalenol (DON), also called as vomitoxin, is mainly produced by the fungus *Fusarium graminearum*, but in many geographical regions, it is also produced by the fungus *F. culmorum* (Richard, 2000).

Mode of action of Deoxynivalenol: Deoxynivalenol are mycotoxins that play a major role in disease development. DON and 3-acetyldeoxynivalenol (3-ADON) inhibit seed germination and root development (Wakulinski, 1989). DON is important for the colonization of the *F. graminearum* into plant system. Fungus is allowed to spread because protein synthesis and the development of host enzymes are inhibited by DON (Snijders, 2004).

12.6.5 T-2 Toxin

The T-2 mycotoxin comes under Type A non-macrocyclic trichothecenes group of mycotoxins. *F. sporotrichioides* is the primary fungus that produces T-2 toxin along with some closely related HT-2 toxin and diacetoxyscirpenol mycotoxins of the same chemical class (CAST, 2003).

Mode of action of T-2 Toxin: T-2 and HT-2 toxins are produced by *F. sporotrichioides* known as a poor plant pathogen (Miller, 1994). In germinating wheat seedlings, a dose of 1 µg/mL of T-2 toxin affected root and shoot growth (Wakulinski, 1989). Among the trichothecenes, the T-2 toxin was tested as highly toxic for animal cells compared to other trichothecenes (Thompson and Wannemacher, 1986; Terse et al., 1993). Rahman et al. (1993) described that the concentration of T-2 toxin moderates cell division during prophase and metaphase. Mostly, T-2 toxin inhibits protein, DNA and RNA synthesis. It was also reported that these toxins decrease antibody levels, immune globulins and some hormones during cytokinesis.

12.6.6 Moniliformin

MON is a K/Na or sodium salt of 1-hydroxycyclobut-1-ene-3,4-dione produced by many *Fusarium* spp. but particularly *F. proliferatum* on the corn kernel (Vesonder et al., 2000).

Mode of action of MON: The hostile effect of MON can be seen on developing leaf, root and seedlings of wheat and corn (Wakulinski, 1989). Van Asch et al. (1992) reported that MON significantly reduced fresh mass accumulation at concentrations of 10 and 100 µg/mL.

12.6.7 Fumonisin

Fumonisin comes under the non-fluorescent mycotoxins grouped in FB1, FB2 and FB3, which are primarily produced by *Fusarium verticillioides* and *F. proliferatum*. The strains of Fusarium have variable toxin-producing abilities (CAST, 2003). Fumonisin mainly affects corn, but it is also found in sorghum and rice. The disease, mouldy corn toxicosis, was first reported in the horses fed with mainly maize from the USA (1891) and later reported in Argentina, China, Egypt, South Africa and Australia.

Mode of action of Fumonisin: Fumonisin mainly causes physiological impairment and sudden death in plants. Janardhan et al. (2011) reported that seeds of different pulse crops decreased their germinating ability when soaked in toxic fungal strains. Endosperm production was also inhibited by FB1 because it interferes metabolically with germination. FBl-treated maize callus cells showed alterations in cell ultrastructure (Van Asch, 1990). FB1 was also found to decrease the chlorophyll content (Abbas et al., 1993).

12.6.8 Fusarin and Fusaric Acid

Fusaric acid (FA) is a derivative of picolinic acid, which was first isolated from *Fusarium heterosporium* (Yabuta et al., 1934). FA was first reported to be involved in the pathogenesis of tomato wilt disease caused by *F. oxysporum* f.sp. *lycopersici* (Gaumann, 1957). Many members of the *Gibberella fujikuroi* species complex were found to produce fusarin C on maize, soybeans, wheat, rye and barley. Several *Fusarium* species also produce fusarin C, FA and fumonisins FB1, FB2 and FB3.

Mode of action of fusarins and FA: The action mechanism is still not fully understood. It is likely to inhibit the enzyme that converts dopamine to norepinephrine, i.e., dopamine beta-hydroxylase. It also inhibits DNA synthesis and cell proliferation. FA also acts as an inhibitor for quorum sensing. FA changes the membrane permeability, decreases mitochondrial activity, oxygen uptake and ATP synthesis and causes root growth inhibition (Bouizgarne et al., 2006). The *Fusarium* wilt index is positively correlated with the concentration of FA.

12.7 NEGATIVE EFFECTS OF MYCOTOXINS ON HUMANS

Mycotoxicoses, like all toxicological syndromes, can be categorized as acute or chronic. Acute toxicity generally has a rapid onset and an obvious toxic response, while chronic toxicity is characterized by low-dose exposure over a long period of time, resulting in cancer and other generally irreversible effects. Prior to the discovery and implementation of modern milling practices, *Fusarium* species had been implicated in several human outbreaks of mycotoxicoses. Cereal grains contaminated with *F. sporitrichoides* and *F. poae* were implicated in alimentary toxic aleukia in Russia from 1932 to 1947. Symptoms included mucous membrane hyperaemia, oesophageal pain, laryngitis, asphyxiation, gastroenteritis and vertigo (Lewis et al., 2005).

Aflatoxicosis is a toxic hepatitis leading to jaundice and, in severe cases, death. Repetitive incidents of this nature have occurred in Kenya (during 1981, 2001, 2004 and 2005), India and Malaysia (Shephard, 2004; Lewis et al., 2005). AFB_1 has been extensively linked to human primary liver cancer, in which it acts synergistically with HBV infection, and was classified by the International Agency for Research on Cancer (IARC) as a human carcinogen (Group 1 carcinogen) (IARC, 1993). This combination represents a heavy cancer burden in developing countries. A recent comparison of the estimated population risk between Kenya and France highlighted the greater burden that can be placed on developing countries (Shephard, 2006).

The largest risk of AF in humans is usually the result of chronic dietary exposure. Such dietary AF exposures have been associated with human hepatocellular carcinomas, which may be compounded by the hepatitis B virus. Approximately 250,000 deaths are caused by hepatocellular carcinomas in China and Sub-Saharan Africa annually and are attributed to risk factors such as a high daily intake of AF and a high incidence of hepatitis B (Wild et al., 1992). Aflatoxins have been found in the tissues of children suffering from Kwashiorkor and Reye's syndrome and are thought to be a contributing factor to these diseases. Reye's syndrome, which is characterized by encephalopathy and visceral deterioration, results in liver and kidney enlargement and cerebral oedema (Blunden et al., 1991). Aflatoxin has long been linked to

Kwashiorkor, a disease usually considered a form of protein energy malnutrition, although some characteristics of the disease are known to be among the pathological effects caused by aflatoxins in animals. Aflatoxin exposure was associated with reduced levels of secretory immunoglobulin A (IgA) in Gambian children (Turner et al., 2003).

Changes in differential subset distributions and functional alterations of specific lymphocyte subsets have been correlated with aflatoxin exposure in Ghanaian adults and indicate that aflatoxin could cause impairment of human cellular immunity that could decrease resistance to infections (Jiang et al., 2005). Of the other health risk factors, the morbidity and mortality associated with unsafe sex, unsafe water and indoor smoke arise from infectious diseases such as HIV/AIDS, infectious diarrhoea and lower respiratory tract infection. This may be due to immunological suppression associated with aflatoxin, and possibly DON could adversely affect all these outcomes. The modulating effect of aflatoxins in cases of zinc, iron and vitamin A deficiency in human health is less clear, but evidence from animal nutrition would suggest it could be significant (Williams et al., 2004). Fumonisins have been implicated in one incident of acute foodborne disease in India in which the occurrence of borborygmy, abdominal pain and diarrhoea was associated with the consumption of maize and sorghum contaminated with high levels of fumonisins. Fumonisin B1, the most abundant of the numerous fumonisin analogues, was classified by the IARC as a Group 2B carcinogen (possibly carcinogenic in humans) (IARC, 2002). Fumonisins, which inhibit the uptake of folic acid *via* the folate receptor, have also been implicated in the high incidence of neural tube defects in rural populations known to consume contaminated maize, such as the former Transkei region of South Africa and areas of Northern China (Marasas et al., 2004). Both DON and ZEN from toxic Fusaria have been linked to scabby grain toxicoses in the USA, China, Japan and Australia. Symptoms include nausea, vomiting and diarrhoea. Fumonisin B1 was associated with an illness outbreak in India with symptoms of acute onset of abdominal pain and diarrhoea. Fumonisins have also been implicated in oesophageal cancer in China (Yoshizawa et al., 1994). However, with limited causal relationships and the presence of several confounding factors, the data compiled by the International Agency for Research on Cancer were not conclusive for fumonisin carcinogenicity in humans (Castegnaro and Wild, 1995). Trichothecenes have been suggested as potential biological warfare agents. For example, T-2 toxin was implicated as the chemical agent of ‘yellow rain’ used against the Lao People’s Democratic Republic from 1975 through 1981 (Peraica et al., 1999). In an investigation of similar biological warfare agents in Cambodia from 1978 to 1981, T-2 toxin, DON, ZEN, nivalenol and DAS were isolated from water and leaf samples collected from the affected areas (Peraica et al., 1999). Clinical symptoms preceding death included vomiting, diarrhoea, haemorrhage, breathing difficulty, chest pain, blisters, headache, fatigue and dizziness. In addition to nephritic congestion, autopsy findings included necrosis of the lining of the stomach and upper small intestine, lungs and liver. It should be noted, however, that the origin of the samples of yellow rain is still a subject of debate. For example, one theory attributed the source of illnesses to unidentified endemic factors because the yellow rain was found to be a native bee faecal material devoid of mycotoxins (Seeley and Visscher, 1985).

12.8 DETECTION OF MYCOTOXINS

Confirmatory and quantitative approaches for the detection of mycotoxins mainly include thin layer chromatography (TLC), high-performance liquid chromatography (HPLC) and liquid chromatography coupled with mass spectrometry (LC–MS). PCR-based technologies make possible the rapid sorting of commodities based on fungal content, which can be used as a useful tool to reduce the exposure of food and feed matrices to mycotoxins (Niessen, 2007). The basic principle of PCR-based mycotoxigenic fungi detection technologies is to develop primers targeting the genes involved in the mycotoxin biosynthesis. Several PCR test systems for the detection of mycotoxinogenic fungi are considered using sets of primers oriented to the structural and regulatory genes involved in the biosynthesis of aflatoxins (*omt-1*, *norl* and *verl*), fumonisins (*fuml* and *fum13*), trichothecenes (*tril*, *tri3*, *tri5*, *tri6*, *tri12* and *tri13*), zearalenone (*PKS4*, *PKS13*, *ZEB1* and *ZEB2*) and ochratoxin (*pks* and *OTAnps*) (Table 12.2). Detection technologies that are based on DNA sequenced amplification include PCR, real-time PCR, nested PCR, RT-PCR, LAMP, *in situ* PCR, PCR-DGGE, co-operational PCR, multiplex PCR, DNA arrays, MCH-PCR and PCR-RFLP (Edwards et al., 2002; Elizaquível et al., 2014).

Immunochemical techniques rely on the specificity of binding between antibodies and antigens. The high affinity and specificity of antibodies for antigens have been used in the development of various immunochemical methods. The major immunochemical methods used in mycotoxins analysis comprise radioimmunoassay (RIA), enzyme-linked immunosorbent assay (ELISA), immunoaffinity column assay (ICA) and immunosensors. Lateral flow assaya (LFA) are based on a competitive format that is the same as microplate-based assays. Up to now, pAb- and mAb-based LFAs have already been developed for detecting mycotoxins, including AFB1 (Jia et al., 2021), AFM1 (Wang et al., 2020), DON (Geballa-Koukoula et al., 2020), OTA (Hao et al., 2021), ZEN (Chen et al., 2019), FB1 (Chen et al., 2021), T-2 toxin (Qie et al., 2019), cyclopiazonic acid (Li et al., 2020) and tenuazonic acid (Cai et al., 2021). Due to the complexity of the co-occurrence of mycotoxins, the demand for simultaneous detection of multiple mycotoxins is increasing. To date, a number of multiplex LFAs for mycotoxins have been successfully developed with good performance, which could simultaneously detect up to six mycotoxins (Charlermroj et al., 2021). A time-resolved fluorescence lateral flow assay based on two anti-idiotypic nanobodies for simultaneous detection of AFB_1 and ZEN in maize products was also developed (Charlermroj et al., 2021). Aptamer-based LFAs have also been designed for AFB_1, ZEN and OTA.

Biosensors are portable bioanalytical devices that incorporate biological recognition elements for binding target molecules and a signal transducer to convert the biorecognition event into a measurable signal. Based on various bioinspired recognition elements, mycotoxin sensors can be categorized into immunosensors, aptasensors, peptide-based sensors and MIPs-based sensors. Compared with other analytical methods mentioned above, it is much easier to realize real-time monitoring of reaction changes dynamically using biosensors, with the output of the results in digital formats. A variety of transducers have been explored for mycotoxin sensor development. Electrochemical, optical, mass-sensitive, calorimetric and magnetic

TABLE 12.2
Details of Gene-Specific Primers Used for the Detection of Mycotoxin

Mycotoxin	Target Gene	Primer Designation and Sequence	References
Trichothecene	tri5 Tr5F	5'-AGCGACTACAGGCTTCCCTC-3'	Doohan et al. (1999)
	Tr5R	5'-AAACCATCCAGTTCTCCATCTG-3'	
	tri5 Tox5-1	5'-GCTGCTCATCACTTTGCTCAG-3'	Niessen and
	Tox5-2	5'CTGATCTGGTCACGCTCATC-3'	Vogel (1998)
	tri5 HATriF	5'-CAGATGGAGAACTGGATGGT-3'	Edwards et al. (2001)
	HATriR	5'GCACAAGTGCCACGTGAC-3'	
	tri6 Tri6F	5'-CTCTTTGATCGTGTTGCGTC-3'	Bluhm et al. (2002)
	Tri6R	5'-CTTGTGTATCCGCCTATAGTGATC-3'	
DON	tri7	5'-GGCTTTACGACTCCTCAACAATGG-3'	Lee et al. (2001)
	GzTri7/f1	5'-AGAGCCCTGCGAAAGACTGGTGC-3'	
	GzTri7/r1		
NIV	tri7	5'-GGCTTTACGACTCCTCAACAATGG-3'	Lee et al. (2001)
	GzTri7/f1	5'-AGAGCCCTGCGAAAGACTGGTGC-3'	
	GzTri7/r1		
Fumonisin	Int1	5'-CCGAGTTACAACTCCCAAA-3'	Doohan et al. (1999)
	Int2	5'-ACAGAGTTTAGGGGTCCTCT-3'	
	Fum5F	5'-GTCGAGTTGTTGACCACTGCG-3'	
	Fum5R	5'-CGTATCGTCAGCATGATGTAGC-3'	
Patulin	IDH-1	5'-CAATGTGTCGTACTGTGCCC-3'	Paterson et al. (2000)
	IDH-2	5'-ACCTTCAGTCGCTGTCCTC-3'	
Aflatoxin	aflR660	5'-CGCGCTCCCAGTCCCCTTCATT-3'	Sweeney et al. (2000)
	aflR1249	5'-CTTGTTCCCCGAGATGACCA-3'	
	APA-450	5'-TATCTCCCCCCGGGCATCTCCCGG-3'	Schnerr et al. (2001)
	APA-1482	5'-CCGTCAGACAGCCACTGGACACGG-3'	
	ord1501	5'-TTAAGGCAGCGGAATACAAG-3'	Sweeney et al. (2000)
	ord2226	5'-GACGCCCAAAGCCGAACACAAA-3'	
	nortaq-1	5'-GTCCAAGCAACAGGCCAAGT-3'	Mayer et al. (2003)
	nortaq-2	5'-TCGTGCATGTTGGTGATGGT-3'	
	nor1	5'-ACCGCTACGCCGGCACTCTCGGCAC-3'	Geisen (1996)
	nor2	5'-GTTGGCCGCCAGCTTCGACACTCCG-3'	
	ver1	5'-GCCGCAGGCCGCGGAGAAAGTGGT-3'	Geisen (1996)
	ver2	5'-GGGGATATACTCCCGCGACACAGCC-3'	
	VER-496	5'-ATGTCGGATAATCACCGTTTAGATGGC-3'	Schnerr et al. (2001)
	VER-1391	5'-CGAAAAGCGCCACCATCCACCCCAATG-3'	
	omt1	5'-GTGGACGGACCTAGTCCGACATCAC-3'	Geisen (1996)
	omt2	5'-GTCGGCGCCACGCACTGGGTTGGGG-3'	
	OMT208	5'-GGCCCGGTTCCTTGGCTCCTAAGC-3'	Schnerr et al. (2001)
	OMT-1232	5'-CGCCCCAGTGAGACCCTTCCTCG-3'	

transducers stand out as the most important sensing platforms. There is an increasing tendency for the development of electrochemical immuno- and aptasensors for mycotoxin detection. In the past decade, numerous aptasensors towards several mycotoxins have been introduced, including OTA, AFB1, AFM1, PAT, FB1, ZEN and T-2

TABLE 12.3
Mycotoxin Detection Kits Commercially Available

Mycotoxins	Lateral Flow Assay	ELISA
Total AFs (AFB_1, B_2, G_1, G_2)	AgraStrip, Smart Strip, RIDA QUICK, QuickTox, QuickTox for QuickScan, ROSA AFQ-FAST, ROSA AFQ-WETS5, MycoTube, Rapid Test Strip, AgraQuant, Veratox HS	SENSISpec, I'screen AFLA, Screening Card, ELISA Kit, Plate Kit, Tube Kit,
DON	AgraStrip, Reveal Q+, Reveal Q+ MAX, Smart Strip, TotalTox Comb, ROSA FAST5, ROSA DONQ2, ROSA WET-S5, Rapid Test Strip, AgraQuant, Agri-Screen, Veratox HS, Celer	B ZERO, Plate Kit
FUM	AgraStrip, Smart Strip, QuickTox for QuickScan, TotalTox Comb, ROSA FAST5, Rapid Test Strip, Rapid Test Strip, Agri-Screen, Veratox, Veratox HS, Celer	B ZERO, Plate Kit
ZEN	Reveal Q+ MAX, Smart Strip, QuickTox for QuickScan, TotalTox Comb, ROSA FAST5, ROSA WET-S5, Rapid Test Strip, Veratox, Veratox, Celer	B ZERO, ELISA Kit, Plate Kit
OTA	AgraStrip, Smart Strip, Reveal Q+ MAX, QuickTox for QuickScan, Charm OCHRAQ-G, Rapid Test Strip, AgraQuant, Veratox, Veratox HS	B ZERO, RIDASCREEN, Screening Card, ELISA Kit
HT-2	Reveal Q+, Reveal Q+ MAX, ROSA, Charm OCHRAQ-G, Rapid Test Strip, Veratox	Plate Kit

toxin. However, ultrasensitive approaches are difficult to develop *via* simple aptasensor recognition. Therefore, a series of novel aptasensors with signal amplification and enhancement have been introduced for mycotoxins, which can meet the requirements of the low to maximum contamination level set by many countries and organizations.

A summary of the various commercially available mycotoxin detection kits for food and feed is tabulated in Table 12.3, where aflatoxins (AF), ochratoxins (OT), fumonisins (FUM), patulin (PAT), zearalenone (ZEA) and trichothecenes (TCTC), including deoxynivalenol (DON), are presented.

12.9 MANAGEMENT OF MYCOTOXINS

Owing to the negative effects of mycotoxins, various pre- and postharvest prevention and control strategies for several crops are continuously being developed. Due to the harmful effects of mycotoxins on animals, plants and humans, it is important to manage mycotoxin contamination in agricultural commodities. A number of physical and chemical approaches have been taken to reduce the effect of mycotoxins. The efficiency of these techniques depends on the level of contamination and the distribution of mycotoxins in the produce. A new approach is also applied for managing mycotoxins through transformations by using genetic engineering and biotechnological techniques (genetic transformation) that offer a specific, efficient and eco-friendly way for detoxification of mycotoxins that leaves neither toxic residue nor any undesirable by-product. However, the efficiency of these techniques depends on the level of contamination and the distribution of mycotoxins throughout the grain.

12.10 PRE-HARVEST STRATEGIES

Pre-harvest strategies, primarily cultural practices, can be adjusted to reduce the survival or spread of mycotoxin-producing fungi in the field or to avoid mycotoxin contamination of the crop through the timing of planting and harvest. Because most mycotoxin-producing fungi survive in crop residue, some reductions in contamination can be achieved through crop rotation, tillage, or other sanitation practices. The success of this strategy is limited; however, due to the ability of these fungi to spread by airborne spores over fairly long distances, cultural practices such as plant spacing, fertilization, weed control and irrigation can be altered to promote environmental conditions that are less favourable to infection by discouraging fungal development or reducing plant susceptibility. Properly timed fungicide applications can reduce infection and mycotoxin contamination, but disease control does not always result in reduced mycotoxin contamination. Biological control methods have been developed for Fusarium head blight of wheat and for *A. flavus* infection of cottonseed and other crops. For Fusarium head blight, microorganisms that are antagonistic to the *Fusarium* pathogens are applied to the wheat heads. The *A. flavus* biological control strategy involves the application of an atoxigenic strain of *A. flavus* to the soil surface, which then colonizes the crop and outcompetes the existing toxigenic strains.

12.11 POSTHARVEST STRATEGIES

For most of the agricultural products affected by mycotoxins, postharvest drying and maintaining proper storage conditions are key elements of a mycotoxin management plan. Fungal growth and mycotoxin production can be prevented when the moisture content of the product is held below a critical level. Optimal moisture contents have been determined for the storage of cereal grains, beans and nut crops. Control of storage temperature is also very important. These tactics are easily implemented in developed countries but can be very challenging in developing countries where adequate drying and storage facilities are not available.

Grain or nuts that are infected by mycotoxin-producing fungi often have visual symptoms or other physical characteristics that allow them to be separated from the healthy product. This can be a very effective way to reduce the mycotoxin level in a grain lot without removing a large proportion of the product. Methods include air-screen cleaners, optical sorters and density separation. In addition, the pericarp or hull of the grain often contains the most concentrated mycotoxins and de-hulling methods have also been shown to greatly reduce mycotoxin contamination. Finally, in some cases, whole lots of contaminated grain are diverted from use as human food or feed for sensitive livestock species and instead fed to less-sensitive livestock, such as mature beef cattle.

Detoxification of contaminated products can be achieved by thermal, biological, or chemical means. Thermal inactivation of mycotoxins is usually implemented as a stage in food or industrial processing. Most mycotoxins are relatively heat-stable, and it can be difficult to significantly reduce mycotoxin levels while maintaining product quality. Microorganisms have been identified that are capable of metabolizing certain mycotoxins. In particular, bovine rumen microorganisms have been shown to detoxify DON under some conditions, and a feed additive based on at least one of these organisms has been developed. Various chemical methods have been developed with some success in breaking down mycotoxins in postharvest products. The most well-known approach is the use of ammonia gas or ammonium sulphate to reduce aflatoxin levels in maize. This is a widely accepted practise internationally but is not used extensively in the U.S. Other approaches include solvent extraction of contaminants from feed and adding mycotoxin-binding agents to feed. In general, detoxification is one of the least preferred approaches to managing mycotoxins due to the expense and the limitations sometimes placed on the use of decontaminated grain.

The modern direction is towards non-invasive decontamination methods, which do not alter the chemical composition and the quality of the raw materials but which have a certain effectiveness in reducing the presence of mycotoxins. The modern direction is towards non-invasive decontamination methods, which do not alter the chemical composition and the quality of the raw materials but which have a certain effectiveness in reducing the presence of mycotoxins. Cold plasma technology reduces and degrades mycotoxins in/on food and feed materials, and it might be a sustainable method that requires lower energy input and investment. A recent innovation in the area of mycotoxin decontamination is the use of electro-magnetic radiation, which includes gamma radiation, pulsed light, radio frequency and microwaves. In recent years, nanomaterials have played a leading role in the photocatalytic degradation of mycotoxins. They are low-cost, environmentally friendly, and easily applied with no secondary pollution. Nanomaterials such as graphene/ZnO hybrids, Fe_2O_3 and titanium dioxide (TiO_2) have been used for mycotoxin photocatalytic degradation (Zhou et al., 2020).

The existing types of active packaging that could control mycotoxin formation are moisture regulators and antimicrobial active packaging. The former uses moisture absorbers; the latter are based on carbon dioxide, ethanol, preservatives, inorganic nanoparticles and natural extracts. Antioxidant and oxygen scavenger packaging can also reduce microbial growth while maintaining oxidative stability, as can thermal insulation packaging based on phase change material. In this direction, given the growing need for plastic reduction, bio-based packaging is also being developed to reduce foodborne pathogens. Nanofibers have acquired significant interest in the packaging industry. Nanotechnologies existing in packaging include biodegradable nano-sensors for temperature and moisture monitoring, nanoclays and nanofilms to prevent spoilage and oxygen adsorption, and nanoparticles for antimicrobial and antifungal topcoats (EFSA, 2014). Previously, many studies have shown that genetic transformations can be used to manage mycotoxins by using specific genes responsible for detoxifying a specific mycotoxin. For instance, detoxification of deoxynivalenol (DON) and zearalenone (ZEN) was achieved by the expression of the acetyltransferase gene (Tri101) and lactonohydrolase gene (Zh101), respectively (Okubura and Blechl, 2002; Takahashi-Ando et al., 2004).

A common approach used in food safety programmes is HACCP, or Hazard Analysis Critical Control Point. This approach has been recommended and implemented in relation to preventing mycotoxin contamination of food products. HACCP for mycotoxin prevention is a proactive food management system based on the control of processes and materials all along the supply chain. The approach is based on seven principles:

1. Identification of hazards; assessment of risk; description of control measures
2. Identification of the critical control points (CCP)
3. Establishment of critical limits
4. Establishment of procedures for monitoring CCPs
5. Establishment of corrective actions
6. Establishment of verification procedures
7. Documentation and record-keeping

HACCP programmes have been established for mycotoxin control in various crops, including maize, coffee, nut crops and apples, in a number of different countries and impact of mycotoxins on humans and animals.

12.12 CONCLUSION

Fungi cause human illness in different ways. Mycoses are the best-known diseases of fungal aetiology, but toxic secondary metabolites produced by saprophytic species are also an important health hazard. The term mycotoxin is an artificial rubric used to describe pharmacologically active mould metabolites characterized by vertebrate toxicity. They fall into several chemically unrelated classes, are produced in a strain-specific way and elicit some complicated and overlapping toxic activities in sensitive species that include carcinogenicity, inhibition of protein synthesis, immunosuppression, dermal irritation and other metabolic perturbations. Mycotoxins usually enter the body via ingestion of contaminated foods, but inhalation of toxigenic spores and direct dermal contact are also important routes.

It is difficult to prove that a disease is mycotoxicosis. Moulds may be present without producing any toxin. Thus, the demonstration of mould contamination is not the same thing as the demonstration of mycotoxin contamination. Moreover, even when mycotoxins are detected, it is not easy to show that they are the etiological agents in a given veterinary or human health problem. Nevertheless, there is sufficient evidence from animal models and human epidemiological data to conclude that mycotoxins pose an important danger to human and animal health, albeit one that is hard to pin down. The incidence of mycotoxicoses may be more common than suspected. It is easy to attribute the symptoms of acute mycotoxin poisoning to other causes; the opposite is true of the aetiology. It is not easy to prove that cancer and other chronic conditions are caused by mycotoxin exposure. In summary, in the absence of appropriate investigative criteria and reliable laboratory tests, the mycotoxicoses will remain diagnostically daunting diseases. Adaptation at higher trophic levels, such as mammalian species, has been crucial to survival as well.

12.13 FUTURE CHALLENGES

Mycotoxins are expected to become a greater concern for the feed and food supply chains in the future. Focused research on mycotoxin detection will help to manage them and reduce mycotoxin-related illnesses in the food and feed supply. Mycotoxins should be treated as an emerging concern, and we should try to keep them under control for their low risk to consumers. A number of tools to control their precursor fungi should be developed to save the lives of people.

REFERENCES

Abbas, H. K., Duke, S. O., & Tanaka, T. (1993). Phytotoxicity of fumonisins and rfzated compounds. *Journal of Toxicology: Toxin Reviews*, *12*(3), 225–251.

Aggarwal, R. (2021). Aflatoxin: Types and their effects on human health. Paper presented in ICAR-Industry Meet, January 4, 2021.

Aggarwal, R., Kamil, D., & Bashyal, B. M. (2022). Mycotoxins and their impact on human health. In: *Management of Postharvest Diseases and Value Addition of Horticulttural Crops* (Eds. Dinesh Singh, V. Devappa, S. Jahagirdar, H. R. Gautam, & Rashmi Aggarwal). Today and Tomorrow's Printers and Publishers, New Delhi, pp. 388.

Ashoor, S. H., & Chu, F. S. (1973). Inhibition of alcohol and lactic dehydrogenases by patulin and penicillic acid in vitro. *Food and Cosmetics Toxicology*, *11*(4), 617–624.

Ates, E., Mittendorf, K., Stroka, J., & Senyuva, H. (2013). Determination of Fusariummycotoxins in wheat, maize and animal feed using on-line clean-up with high resolution mass spectrometry. *Food Additives & Contaminants:* Part A, 30(1), 156–165. DOI:10.1080/19440049.2012.729162

Bayman, P., & Baker, J. L. (2006). Ochratoxins: A global perspective. *Mycopathologia*, 162, 215–223.

Bennett, J. W. & Klich, M. (2003). Mycotoxins. *Clinical Microbiology Reviews*, *16*, 497–516.

Blout, W. P. (1961). Turkey 'X' disease. *Turkeys*, *9*, 52.

Bluhm, B. H., Flaherty, J. E., Cousin, M. A., & Woloshuk, C. P. (2002). Multiplex polymerase chain reaction assay for the differential detection of trichothecene-and fumonisin-producing species of Fusarium in cornmeal. *Journal of Food Protection*, 65(12), 1955–1961.

Blunden, G., Roch, O. G., Rogers, D. J., Coker, R. D., Bradburn, N., & John, A. E. (1991). Mycotoxins in food. *Medical Laboratory Sciences*, *48*(4), 271–282.

Bouizgarne, B., El-Maarouf-Bouteau, H., Frankart, C., Reboutier, D., Madiona, K., Pennarun, A. M., … Bouteau, F. (2006). Early physiological responses of *Arabidopsis thaliana* cells to fusaric acid: Toxic and signalling effects. *New Phytologist*, *169*(1), 209–218.

Cai, P., Wang, R., Ling, S., & Wang, S. (2021). A high sensitive platinum-modified colloidal gold immunoassay for tenuazonic acid detection based on monoclonal IgG. *Food Chemistry*, *360*, 130021. DOI:10.1016/j.foodchem.2021.130021.

CAST. (2003). Mycotoxins: Risks in Plant, Animal and Human Systems. Report No. 139. Council for Agricultural Science and Technology, Ames, IA, USA.

Castegnaro, M., & Wild, C. P. (1995). IARC activities in mycotoxin research. *Natural Toxins*, *3*(4), 327–331.

Ceccaroli, C., Pulliero, A., Geretto, M., & Izzotti A. (2015). Molecular fingerprints of environmental carcinogens in human cancer. *Journal of Environmental Science and Health*, *33*, 188–228. DOI:10.1080/10590501.2015.1030491.

Charlermroj, R., Phuengwas, S., Makornwattana, M., Sooksimuang, T., Sahasithiwat, S., Panchan, W., Sukbangnop, W., Elliott, C. T., & Karoonuthaisiri, N. (2021). Development of a microarray lateral flow strip test using a luminescent organic compound for multiplex detection of five mycotoxins. *Talanta*, *233*, 122540.

Chen, X., Miao, X., Ma, T., Leng, Y., Hao, L., Duan, H., Yuan, J., Li, Y., Huang, X., & Xiong, Y. (2021). Gold nanobeads with enhanced absorbance for improved sensitivity in competitive lateral flow immunoassays. *Foods, 10*, 1488. DOI:10.3390/foods10071488.

Chen, Y., Fu, Q., Xie, J., Wang, H., & Tang, Y. (2019). Development of a high sensitivity quantum dot-based fluorescent quenching lateral flow assay for the detection of zearalenone. *Analytical and Bioanalytical Chemistry, 411*, 2169–2175. DOI:10.1007/s00216-019-01652-1.

Ciegler, A. (1972). Bioproduction of ochratoxin A and penicillic acid by members of the *Aspergillus* ochraceus group. Canadian Journal of Microbiology, *18*(5), 631–636.

Ciegler, A., & Kurtzman, C. P. (1970). Penicillic acid production by blue-eye fungi on various agricultural commodities. *Applied Microbiology, 20*(5), 761–764.

Doohan, F. M., Weston, G., Rezanoor, H. N., Parry, D. W., & Nicholson, P. (1999). Development and use of a reverse transcription-PCR assay to study expression of Tri5 by *Fusarium* species in vitro and in planta. *Applied and Environmental Microbiology, 65*(9), 3850–3854.

Dragan, Y. P. & Pitot, H. C. (1994). A£atoxin carcinogenesis in the context of the multistage nature of cancer. In: *The Toxicology of A£atoxins* (Eds. D. L. Eaton & J. D. Groopman). Academic Press, London, pp. 179–206.

Edwards, S. G., O'callaghan, J., & Dobson, A. D. (2002). PCR-based detection and quantification of mycotoxigenic fungi. *Mycological Research, 106*(9), 1005–1025.

Edwards, S. G., Pirgozliev, S. R., Hare, M. C., & Jenkinson, P. (2001). Quantification of trichothecene-producing Fusarium species in harvested grain by competitive PCR to determine efficacies of fungicides against fusarium head blight of winter wheat. *Applied and Environmental Microbiology, 67*, 1575–1580.

EFSA (2014). *Inventory of Nanotechnology Applications in the Agricultural, Feed and Food Sector*. Wiley, Hoboken, NJ.

Elizaquível, P., Aznar, R., & Sánchez, G. (2014). Recent developments in the use of viability dyes and quantitative PCR in the food microbiology field. Journal of Applied Microbiology, *116*(1), 1–13.

Fadl-Allah, E. M., Mahmoud, M. H., Abd El-Twab, M. H., & Helmey, R. K. (2011). Aflatoxin B1 induces chromosomal aberrations and 5S rDNA alterations in durum wheat. *Journal of the Association of Arab Universities for Basic and Applied Sciences, 10*(1), 8–14.

Forgacs, J., Koch, H., Carll, W. T., & White-Stevens, R. H. (1962). Mycotoxicoses I. Relationship of toxic fungi to moldy-feed toxicosis in poultry. *Avian Diseases, 6*(3), 363–380.

Gajecka, M., Stopa, E., Tarasiuk, M., Zielonka, Ł., & Gajecki, M. (2013). The expression of type-1 and type-2 nitric oxide synthase in selected tissues of the gastrointestinal tract during mixed mycotoxicosis. *Toxins, 5*, 2281–2292.

Gaumann, E. (1957). Fusaric acid as a wilt toxin. *Phytopathology, 47*(6), 342–357.

Geballa-Koukoula, A., Gerssen, A., & Nielen M.W.F. (2020). Direct analysis of lateral flow immunoassays for deoxynivalenol using electrospray ionization mass spectrometry. *Analytical and Bioanalytical Chemistry, 412*, 7547–7558. DOI:10.1007/s00216-020-02890-4.

Geisen, R. (1996). Multiplex polymerase chain reaction for the detection of potential aflatoxin and sterigmatocystin producing fungi. *Systematic and Applied Microbiology, 19*(3), 388–392.

Hao, L., Chen, J., Chen, X., Ma, T., Cai, X., Duan, H., Leng, Y., Huang, X., & Xiong Y. (2021). A novel magneto-gold nanohybrid-enhanced lateral flow immunoassay for ultrasensitive and rapid detection of ochratoxin A in grape juice. *Food Chemistry, 336*, 127710. DOI:10.1016/j.foodchem.2020.127710.

Heussner, A. H., & Bingle, L. E. (2015). Comparative ochratoxin toxicity: A review of the available data. *Toxins, 7*(10), 4253–4282.

IARC (1993). Monographs on the evaluation of carcinogenic risks to humans: Some naturally occurring substances: Food items and constituents, heterocyclic aromatic amines and mycotoxins. *Lyon, France: International Agency for Research on Cancer, 56*, 1–599.

IARC (2002). *Some Traditional Herbal Medicines, Some Mycotoxins, Naphthalene and Styrene. IARC MonogrEvalCarcinog Risks Hum*, vol. 82. Lyon, France: IARC.

Ismaiel, A. A., & Papenbrock, J. (2015). Mycotoxins: Producing fungi and mechanisms of phytotoxicity. *Agriculture*, *5*(3), 492–537.

Jahanian, E. (2016). Mycotoxin-induced toxicity; an updated mini-review on the current concepts. *Immunopathol. Persa*, 2, e11.

Janardhan, A., Subramanyam, D., Praveen, A., Reddi, M., & Narasimha, G. (2011). Aflatoxin impacts on germinating seeds. *Annals of Biological Research*, *2*(2), 180–188.

Jia, B., Liao, X., Sun, C., Fang, L., Zhou, L., & Kong, W. (2021). Development of a quantum dot nanobead-based fluorescent strip immunosensor for on-site detection of aflatoxin B1 in lotus seeds. *Food Chemistry*, *356*, 129614. DOI:10.1016/j.foodchem.129614.

Jiang, Y. I., Jolly, P. E., Ellis, W. O., Wang, J. S., Phillips, T. D., & Williams, J. H. (2005). Aflatoxin B1 albumin adduct levels and cellular immune status in Ghanaians. *International immunology*, *17*(6), 807–814.

Keromnes, J., & Thouvenot, D. (1985). Role of penicillic acid in the phytotoxicity of *Penicillium* cyclopium and Penicillium canescens to the germination of corn seeds. *Applied and Environmental Microbiology*, *49*(3), 660–663.

Khazaeli, P., Najafi, M. L., Bahaabadi, G. A., Shakeri, F., & Naghibzadehtahami, A. (2014). Evaluation of aflatoxin contamination in raw and roasted nuts in consumed Kerman and effect of roasting, packaging and storage conditions. *Life Science Journal*, *10*, 578–583.

Kiessling, K. H. (1986). Biochemical mechanism of action of mycotoxins. *Pure and Applied Chemistry*, *58*(2), 327–338.

Kurtzman, C. P., & Ciegler, A. (1970). Mycotoxin from a blue-eye mold of corn. *Applied Microbiology*, *20*(2), 204–207.

Lee, M. S., Chang, P. C., Shien, J. H., Cheng, M. C., & Shieh, H. K. (2001). Identification and subtyping of avian influenza viruses by reverse transcription-PCR. *Journal of Virological Methods*, 97(1–2), 13–22.

Lewis, C., Smith, J., Anderson, J., & Freshney, R. (2000). Increased cytotoxicity of food-borne mycotoxins toward human cell lines *in vitro* via enhanced cytochrome p450 expression using the MTT bioassay. *Mycopathologia* 148, 97–102. DOI:10.1023/A:1007130923558.

Lewis, L., Onsongo, M., Njapau, H., Schurz-Rogers, H., Luber, G., Kieszak, S., … Kenya Aflatoxicosis Investigation Group. (2005). Aflatoxin contamination of commercial maize products during an outbreak of acute aflatoxicosis in eastern and central Kenya. *Environmental Health Perspectives*, 113(12), 1763–1767.

Li, Y., Liu, L., Kuang, H., & Xu, C. (2020). Visible and eco-friendly immunoassays for the detection of cyclopiazonic acid in maize and rice. *Journal of Food Science*, *85*, 105–113. DOI:10.1111/1750-3841.14976.

Liew, W. P. P., & Mohd-Redzwan, S. (2018). Mycotoxin: Its impact on gut health and microbiota. *Frontiers in Cellular and Infection Microbiology*, 8, 60.

Maggon, K. K., Gupta, S. K., & Venkitasubramanian, T. A. (1977). Biosynthesis of aflatoxins. *Bacteriological Reviews*, *41*(4), 822–855.

Marasas, W. F., Riley, R. T., Hendricks, K. A., Stevens, V. L., Sadler, T. W., Gelineau-van Waes, J., … Merrill Jr, A. H. (2004). Fumonisins disrupt sphingolipid metabolism, folate transport, and neural tube development in embryo culture and in vivo: A potential risk factor for human neural tube defects among populations consuming fumonisin-contaminated maize. *The Journal of Nutrition*, 134(4), 711–716.

Mayer, Z., Bagnara, A., Färber, P., & Geisen, R. (2003). Quantification of the copy number of nor-1, a gene of the aflatoxin biosynthetic pathway by real-time PCR, and its correlation to the cfu of *Aspergillus flavus in foods. International Journal of Food Microbiology*, *82*(2), 143–151.

Milicevic, D. R., Skrinjar, M., & Baltic, T. (2010). Real and perceived risks for mycotoxin contamination in foods and feeds: Challenges for food safety control. *Toxins*, *2*(4), 572–592.

Miller, J. D. (1994). Epidemiology of *Fusarium* ear diseases of cereals. In: *Mycotoxins in Grain: Compounds Other than Aflatoxin* (Eds. J. D. Miller & H. L. Trenholm). St. Paul, MN: Eagan Press, pp. 19–36.

Mohd-Redzwan, S., Jamaluddin, R., Abd.-Mutalib, M. S., & Ahmad, Z. (2013). A mini review on aflatoxin exposure in Malaysia: Past, present and future. *Frontiers in Microbiology*, 4, 334.

Muhammad, I., Sun, X., Wang, H., Li, W., Wang, X., Cheng, P., … Hamid, S. (2017). Curcumin successfully inhibited the computationally identified CYP2A6 enzyme-mediated bioactivation of aflatoxin B1 in arbor acres broiler. *Frontiers in Pharmacology,* 8, 143.

Niessen, L. (2007). PCR-based diagnosis and quantification of mycotoxin producing fungi. *International Journal of Food Microbiology*, 119(1–2), 38–46.

Niessen, M. L., & Vogel, R. F. (1998). Group specific PCR-detection of potential trichothecene-producing Fusarium-species in pure cultures and cereal samples. *Systematic and Applied Microbiology*, *21*(4), 618–631.

Okubura, P. A., & Blechl, A. E. (2002). Engineering deoxynivalenol metabolism in wheat through the expression of a fungal trichotheceneacetyltransferase gene. *Theoretical and Applied Genetics*, *106*, 74–83

Pandya, J. P., & Arade, P. C. (2016). Mycotoxin: A devil of human, animal and crop health. *Advanced Life Sciences*, *5*, 3937–3941.

Paterson, D. L., Swindells, S., Mohr, J., Brester, M., Vergis, E. N., Squier, C., … Singh, N. (2000). Adherence to protease inhibitor therapy and outcomes in patients with HIV infection. *Annals of Internal Medicine*, *133*(1), 21–30.

Peraica, M., Radić, B., Lucic, A., & Pavlovic, M. (1999). Toxic effects of mycotoxins in humans. *Bulletin of the World Health Organization*, *77*(9), 754–766.

Qie, Z., Yan, W., Gao, Z., Meng, W., Xiao, R., & Wang, S. (2019). Ovalbumin antibody-based fluorometricimmunochromatographic lateral flow assay using CdSe/ZnS quantum dot beads as label for determination of T-2 toxin. *Microchimica Acta*, *186*, 816. DOI:10.1007/s00604-019-3964-x.

Rahman, S., El-Ayouty, Y. M., & Kamael, H. A. (1993). Characterization of heptapeptide toxins extracted from *Microcystisaeruginosa* (Egyptian isolate) comparison with some synthesized analogs. *International Journal of Peptide and Protein Research*, *41*(1), 1–7.

Reddy, L., & Bhoola, K. (2010). Ochratoxins-food contaminants: Impact on human health. *Toxins*, *2*(4), 771–779.

Richard, J. L. (2000). Mycotoxins-An overview. In: *Romer Labs' Guide to Mycotoxins* (Ed. J. L. Richard), vol. 1, Romer Labs, Union, MO, pp. 1–48.

Richard, J. L. (2007). Some major mycotoxins and their mycotoxicoses-An overview. *International Journal of Food Microbiology*, *119*(1–2), 3–10.

Ringot, D., Chango, A., Schneider, Y. J., & Larondelle, Y. (2006). Toxicokinetics and toxicodynamics of ochratoxin A, an update. *Chemico-Biological Interactions*, *159*(1), 18–46.

Schnerr, H., Niessen, L., & Vogel, R. F. (2001). Real time detection of the tri5 gene in *Fusarium* species by LightCycler(tm)-PCR using SYBR(r) Green I for continuous fluorescence monitoring. *International Journal of Food Microbiology*, *71*(1), 53–61.

Seeley, T. D., & Visscher, P. K. (1985). Survival of honeybees in cold climates: The critical timing of colony growth and reproduction. *Ecological Entomology*, *10*(1), 81–88.

Shephard, G. S. (2004). Mycotoxins worldwide: Current issues in Africa. In: *Meeting the Mycotoxin Menace* (Eds. D. Barug, H. Van Egmond, R. Lopez-Garcia, T. Van Ossenbruggen, & A. Visconti). Wageningen Academic, Wageningen, pp. 81–88.

Shephard, G. S. (2006). Mycotoxins in the context of food risks and nutrition issues. In: *The Mycotoxin Fact Book* (Eds. D. Barug, D. Bhatnagar, H. P. Van Egmond, J. W. Van der Kamp, A. Van Ossenbruggen, & A. Visconti). Wageningen Academic, Wageningen, pp. 21–36.

Sinha, K. K. (1996). *Mycotoxin Induced Physiological Responses in Crop Plants*. MD Publications Pvt. Ltd., New Delhi, India.

Smith, J. E., Lewis, C. W., Anderson, J. G., & Solomons, G. L. (1994). Mycotoxins in Human Health. Report EUR 16048 EN. European Commission, Directorate-General XII, Brussels.

Snijders, C. H. A. (2004). Resistance in wheat to *Fusarium* infection and trichothecene formation. *Toxicology Letters*, *153*(1), 37–46.

Speight, N. (2012). Mycotoxin-related illness. In: *Advancing Medicine with Food and Nutrients* (Ed. I. Kohlstadt), 2nd Edn. CRC Press; Taylor & Francis Group, Boca Raton, FL, pp. 821–850.

Sun, L. -H., Zhang, N. -Y., Zhu, M. -K., Zhao, L., Zhou, J. -C., & Qi, D.-S. (2016). Prevention of aflatoxin B1 hepatoxicity by dietary selenium is associated with inhibition of cytochrome P450 isozymes and up-regulation of 6 selenoprotein genes in chick liver. *Journal of Nutrition*, *146*, 655–661. DOI:10.3945/jn.115.224626.

Sweeney, M. J., & Dobson, A. D. W. (1999). Molecular biology of mycotoxin biosynthesis. *FEMS Microbiology Letters*, *175*, 149–163.

Sweeney, M. J., Pàmies, P., & Dobson, A. D. (2000). The use of reverse transcription-polymerase chain reaction (RT-PCR) for monitoring aflatoxin production in Aspergillus parasiticus 439. *International Journal of Food Microbiology*, *56*(1), 97–103.

Takahashi-Ando, N., Ohsato, S., Shibata, T., Hamamoto, H., Yamaguchi, I., & Kimura, M (2004). Metabolism of zearalenone by genetically modified organisms expressing the detoxification gene from Clonostachysrosea. *Applied and Environmental Microbiology*, *70*(6), 3239–3245.

Terse, P. S., Mahastha, M. S., Zurovac, O., Stringeffow, D., Marquardt, R. R., & Kemppainen, B. W. (1993). Comparison of in vitro and in vivo biological activation of mycotoxins. *Toxicon*, *31*, 913–919.

Thompson, W. L., & Wannemacher Jr, R. W. (1986). Structure-function relationships of 12, 13-epoxytrichothecene mycotoxins in cell culture: Comparison to whole animal lethality. *Toxicon*, *24*(10), 985–994.

Tripathi, R. K., & Misra, R. S. (1983). Mechanism of inhibition of maize seed germination by aflatoxin BI. In: *Mycotoxins in Food and Feed* (Eds. K. S. Bilgrami, T. Prasad, & K. K. Sinha). Allied Press, Bhagalpur, India, pp. 129–141.

Turner, P. C., Moore, S. E., Hall, A. J., Prentice, A. M., & Wild, C. P. (2003). Modification of immune function through exposure to dietary aflatoxin in Gambian children. *Environmental Health Perspectives*, *111*(2), 217–220.

Van Asch, M. A. J. (1990). *Studies on the resistance of wheat and maize to fungal pathogenesis* (Doctoral dissertation).

Van Asch, M. A. J., Rijkenberg, F. H. J., & Coutinho, T. A. (1992). Phytotoxicity of fumonisin B1, moniliformin, and T-2 toxin to corn callus cultures. *Phytopathology*, 82, 1330–1332.

Vander Merwe, K. J., Steyn, P. S., Fourie, L., Scott, D. B., & Theron, J. J. (1965). Ochratoxin A, a toxic metabolite produced by *Aspergillus* ochraceusWilh. *Nature*, *205*(4976), 1112–1113.

Vesonder, R. F., Wu, W., Weisleder, D., Gordon, S. H., Krick, T., Xie, W., Abbas, H. K., & McAlpin, C. E. (2000). Toxigenic strains of *Fusariummoniliforme* and *Fusariumproliferatum* isolated from dairy cattle feed produce fumonisins, moniliformin and a new C21H38N2O6 metabolite phytotoxic to Lemna minor L. *Journal of Natural Toxins*, 9(2), 103–112.

Viegas, C., Pinheiro, A. C., Sabino, R., Viegas, S., Brandao, J., & Verı́ssimo, C. (2015). *Environmental Mycology in Public Health: Fungi and Mycotoxins Risk Assessment and Management*. Academic Press; Elsevier, Cambridge, UK.

Wakulinski, W. (1989). Phytotoxicity of the secondary metabolites of fungi causin wheat head fusariosis (head blight). *Acta Physiologiae Plantarum*, 11(4), 301–306.

Wang, Y., Peng, X., Xu, W., Luo, Y., Zhao, W., Hao, J., Liang, Z., Zhang, Y., & Huang, K. (2012). Transcript and protein profiling analysis of OTA-induced cell death reveals the regulation of the toxicity response process in *Arabidopsis thaliana. Journal of Experimental Botany*, *63*(5), 2171–2187.

Wang, Y., Zhao, W., Hao, J., Xu, W., Luo, Y., Wu, W., Yang, Z., Liang, Z., & Huang, K. (2014). Changes in biosynthesis and metabolism of glutathione upon ochratoxinA stress in Arabidopsis thaliana. *Plant Physiology and Biochemistry*, *79*, 10–18.

Wang, J., Chen, Q., Jin, Y., Zhang, X., He, L., Zhang, W., & Chen, Y. (2020). Surface enhanced Raman scattering-based lateral flow immunosensor for sensitive detection of aflatoxin M1 in urine. *Analytica Chimica Acta*, *1128*, 184–192. DOI:10.1016/j.aca.2020.06.076.

Wild, C. P., Hudson, G. J., Sabbioni, G., Chapot, B., Hall, A. J., Wogan, G. N., … Groopman, J. D. (1992). Dietary intake of aflatoxins and the level of albumin-bound aflatoxin in peripheral blood in The Gambia, West Africa. *Cancer Epidemiology, Biomarkers & Prevention: A Publication of the American Association for Cancer Research, Cosponsored by the American Society of Preventive Oncology*, *1*(3), 229–234.

Williams, J. H., Phillips, T. D., Jolly, P. E., Stiles, J. K., Jolly, C. M., & Aggarwal, D. (2004). Human aflatoxicosis in developing countries: a review of toxicology, exposure, potential health consequences, and interventions. *The American Journal of Clinical Nutrition*, 80(5), 1106–1122.

Yabuta, T., Kobe, K., & Hayashi, T. (1934). Biochemical studies of the 'bakanae' fungus of rice. I. Fusarinic acid, a new product of the 'Bakanae' fungus. *Zeitschrift fur Pflanzenkrankheiten (P flanzenpathologie) und P flanzenschutz*, 10, 1059–1068.

Yoshizawa, T., Yamashita, A., & Luo, Y. (1994). Fumonisin occurrence in corn from high- and low-risk areas for human esophageal cancer in China. *Applied and Environmental Microbiology*, 60(5), 1626–1629.

Zain, M. E. (2011). Impact of mycotoxins on humans and animals. *Journal of Saudi Chemical Society*, *15*(2), 129–144.

Zhou, Y., Wu, S., Wang, F., Li, Q., He, C., Duan, N., & Wang, Z. (2020). Assessingthe toxicity in vitro of degradation products from deoxynivalenol photo-catalytic degradation by using upconversion nanoparticles@TiO2 composite. *Chemosphere*, *238*, 124648.

13 Pest Risk Analysis

A Tool to Facilitate Agriculture Trade of Economically Important Crops

V. Celia Chalam and Kavita Gupta
ICAR-National Bureau of Plant Genetic Resources

13.1 INTRODUCTION

Plant quarantine is a legislative measure to protect our agriculture from the onslaught of exotic pests. It is one of the most important strategies for exclusion of exotic pests that might inadvertently move along with movement of people and commodities, whereas pest risk is the probability, known or perceived, that a hazardous organism will enter, colonize and/or become established in an area where it is not known to occur. Quarantine services in most countries determine the entry status of imported agricultural commodities based on perceived risk.

Ever since the Sanitary and Phytosanitary Measure (SPS) Agreement has come into force, several countries are using it as a tool for negotiating imports and exports based on pest-risk analysis (PRA) and identification of pest-free areas (PFA), which are obligatory requirements as per Articles V & VI for facilitating trade among member countries under the Agreement. According to the Plant Quarantine (Regulation of Import into India) Order 2003, notified in November, 2003 and effective from 1st January, 2004, a PRA has become mandatory for issuing an import permit for all materials to be imported except those mentioned in Schedules IV, V and VI, which specify the list of materials either prohibited or restricted for import with additional declarations. (https://plantquarantineindia.nic.in/pqispub/pdffiles/pqorder2015.pdf).

The International Plant Protection Convention (IPPC) is one of the standard-setting organizations of the WTO for its Agreement on Application of Sanitary and Phytosanitary Measures (SPS Agreement) (https://www.wto.org/english/tratop_e/sps_e/spsagr_e.htm). It sets the standards for plant health through the development of International Standards for Phytosanitary Measures (ISPMs) (Khetarpal and Rajan, 1999). The IPPC, along with the two other standard-setting organizations, is aimed at helping governments discipline their judgement on SPS measures by emphasizing on scientific evidence and the principles involved in the analysis of risk in the international movement of plants and planting material (Khetarpal and Gupta, 2002;

DOI: 10.1201/9781003403142-13

Chalam et al., 2021). The PRA is a multi-stage process that includes the identification of hazards, likelihood of occurrence and economic losses a pest may cause under the prevailing conditions of the new locality. The IPPC, in its publications ISPM-2, ISPM-11 and ISPM-11 supplement on Guidelines for PRA (https://www.ippc.int/en/), has elaborated on the general requirements for a PRA. There are two ways of initiating a PRA – pathway-based and pest-based. The entire procedure is a blend of inductive, deductive and expert opinion that facilitates the identification and characterization of risk (Gupta and Rashid, 2020). The databases on pests, their biology and ecology, experience and expert judgement are used in interpreting the implications. This is a literature-based exercise, and databases are highly useful in risk analysis. One such database, the "Crop Protection Compendium," developed by CAB International has been found to be of great help in the preparation of PRAs, as it provides not only the procedure through various formats and questionnaires but also provides information and supports the scientists' arguments from the authentic pest database (https://www.cabi.org/cpc).

The preparation of PRA is an elaborate process requiring scientific input on various aspects of the pest and the associated risks involved in its inadvertent introduction.

13.2 DETAILS OF PEST RISK ANALYSIS

PRA consists of **risk assessment** (scientific estimation of the likelihood and magnitude of the establishment of a given pest risk) and **impact assessment** (estimation of the consequences of the establishment of a pest). Therefore, to ensure that imported commodities present a minimal pest risk or no pest risk to our agriculture and forestry, PRA must be conducted.

The requirement for a PRA consists of three stages, viz., initiation, risk assessment and risk management (Figure 13.1), whose compilation should be fully documented in the event of a review or whenever a dispute arises. The PRA should clearly document information sources and the rationale used in reaching a management decision regarding mitigation measures that have been taken or are to be taken.

The preparation of PRA is an elaborate process requiring scientific input on various aspects of the pest and the associated risks involved in its inadvertent introduction.

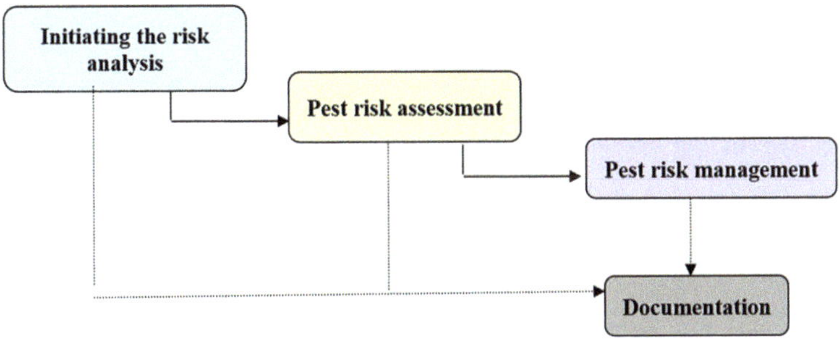

FIGURE 13.1 Different steps involved in pest risk analysis.

Under the IPPC, the term plant pest refers to all organisms harmful to plants or plant products, including other plants, bacteria, fungi, insects, other animals, mites, molluscs, nematodes and viruses. Pests can be either regulated or not, and the IPPC recognizes and defines two categories of regulated pests of plants: quarantine pests and regulated, non-quarantine pests. PRA assists with determining whether a pest fits either of these two categories and the strength of phytosanitary measures, if any, that should be taken in response to it.

QUARANTINE PEST

A quarantine pest is defined by the IPPC as "a pest of potential economic importance to the area endangered thereby and not yet present there, or present but not widely distributed and being officially controlled" (IPPC, 1997). In other words, it is any organism that is injurious or potentially injurious, directly or indirectly, to plants or plant products or by-products of plants and includes bacteria, fungi, insects, mites, molluscs, nematodes, other plants and viruses, not present in a specified area at risk or, if present, being controlled by an NPPO.

REGULATED NON-QUARANTINE PEST

A regulated non-quarantine pest (RNQP) is defined by the IPPC as "a non-quarantine pest whose presence in plants for planting affects the intended use of those plants with an economically unacceptable impact and which is therefore regulated within the territory of the importing contracting party". RNQPs are usually widely distributed in the country where they are regulated.

PRA is a process to answer the following questions:

Is the organism a pest?
What is the likelihood of introduction, establishment and spread?
How much economic (including environmental and social) damage (unacceptable impacts) does it cause?
What can be done to mitigate unacceptable impacts?

Consequently, any phytosanitary measures implemented by a country to mitigate against a particular pest should be technically justified by the PRA.

13.2.1 Pest-Initiated vs. Pathway-Initiated PRAs

There are two widely adopted approaches to conducting a PRA: one focused on a pathway and the other focused on a particular pest associated with one or more pathways. A commodity PRA is one type of pathway PRA. In the case of a pathway PRA, it is necessary to conduct PRAs on those pests associated with the pathway that are identified as potential quarantine pests. In the case of a pest PRA, consideration needs to be given to all possible pathways or commodities with which the pest may be associated (Nagarajan et al., 2002).

13.2.2 Stages in PRA

PRA consists of **risk assessment** (scientific estimation of the likelihood and magnitude of the establishment of a given pest risk) and **impact assessment** (estimation of the consequences of the establishment of a pest). Therefore, to ensure that imported commodities present a minimal pest risk or no pest risk to our agriculture and forestry, PRA must be conducted.

13.2.2.1 Initiating the Process

Initiating the process involves the identification of pests or pathways for which the PRA is needed. The list of pests is generated by any combination of databases, literature sources or expert consultation. Once the list of pests has been established, it is preferable to prioritize it by using expert judgement before the next step. According to the results obtained, it may or may not be necessary to conduct a risk assessment on all pests on the list.

13.2.2.2 Pest Risk Assessment

Pest risk assessment determines whether each pest identified as such or associated with a pathway is a quarantine pest, characterized in terms of likelihood of entry, establishment, spread and economic importance. In doing so, the PRA considers all aspects of each pest, and in particular, actual information about its geographical distribution, biology and economic importance. Expert judgement is then used to assess the establishment, spread and economic importance potential in the PRA area. Finally, the potential for introduction into the PRA area is characterized.

Pest risk

$$\text{Pest risk} = (\text{Probability of introduction}) \times (\text{Magnitude of impact})$$

Probability of introduction

$$\text{Probability of introduction} = (\text{Probability of entry}) \times (\text{Probability of establishment}) \times (\text{Probability of spread})$$

Magnitude of impact

$$\text{Magnitude of impact} = (\text{Economic impacts}) + (\text{Environmental impacts}) + (\text{Social impacts})$$

13.2.2.3 Pest Risk Management

Pest risk management involves developing, evaluating, comparing and selecting options for reducing the risk. A list of options for reducing risks to an acceptable level should be assembled. These options will primarily concern pathways and, in particular, the conditions for permitting the entry of commodities. The pest risk management to protect the endangered areas should be proportional to the risk identified in the pest risk assessment (Figure 13.2).

A PRA should be sufficiently documented so that when a review or dispute arises, the PRA will clearly state the sources of information and the rationales used in reaching a management decision regarding phytosanitary measures taken or to be taken. Complete and careful documentation for each PRA is necessary in order to implement an effective and efficient review process for the PRA. Documentation provides a valuable record of the process and is a useful tool for handling future PRA issues. A well-documented PRA is more easily communicated or updated when necessary and may also serve as a useful resource for future PRAs on related subjects.

The PRA process can stop during or after any one of the above stages, depending on the conclusions reached. Reasons to end the process can include the following:

- concerns about the risk no longer exist
- all associated risks are considered acceptable
- existing controls are deemed to be sufficient
- the activity generating the risk ceases
- the risk is not manageable

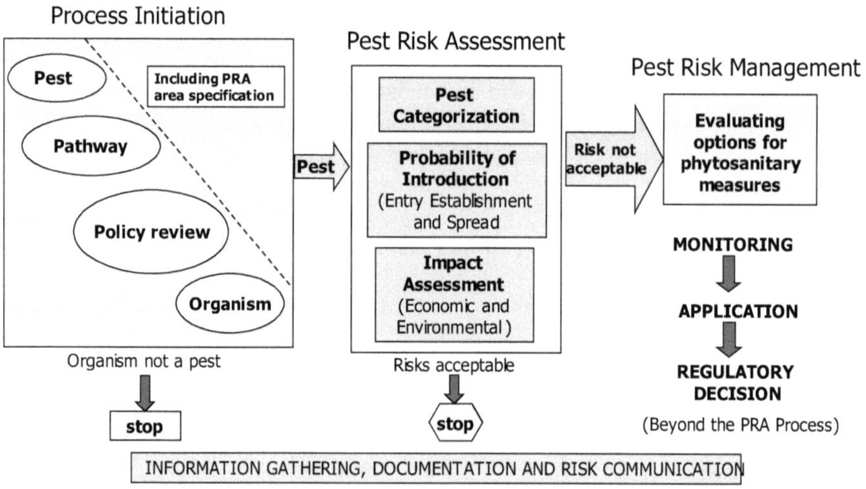

FIGURE 13.2 Pest risk analysis flow chart. (Adapted from ISPM No. 2, 2021.)

13.2.3 ISPMs on Pest Risk Analysis

The ISPMs that are most relevant to PRA are listed below.

13.2.3.1 ISPM 2 – Framework for Pest Risk Analysis

This standard was first adopted by the Twenty-eighth Session of the FAO Conference in November 1995 as *Guidelines for pest risk analysis*. This first revision was adopted by the Second Session of the Commission on Phytosanitary Measures in March 2007 as the present standard. It provides a framework that describes the PRA process within the scope of the IPPC. It introduces the three stages of PRA – initiation, pest risk assessment and pest risk management. The standard focuses on the initiation stage. Generic issues of information gathering, documentation, risk communication, uncertainty and consistency are addressed (https://www.fao.org/3/k0125e/k0125e.pdf).

The PRA process is a technical tool used for identifying appropriate phytosanitary measures. The PRA process may be used for organisms not previously recognized as pests (such as plants, biological control agents or other beneficial organisms, living modified organisms (LMOs)), recognized pests, pathways and a review of phytosanitary policy. The process consists of three stages: (1) initiation, (2) pest risk assessment and (3) pest risk management.

This standard provides detailed guidance on PRA Stage 1, summarizes PRA Stages 2 and 3 and addresses issues generic to the entire PRA process. For Stages 2 and 3, it refers to ISPM 3 (*Guidelines for the export, shipment, import and release of biological control agents and other beneficial organisms*), ISPM 11 (*Pest risk analysis for quarantine pests*) and ISPM 21 (*Pest risk analysis for regulated non-quarantine pests*) dealing with the PRA process.

The PRA process is initiated in Stage 1 with the identification of an organism or pathway that may be considered for pest risk assessment or as part of the review of existing phytosanitary measures, in relation to a defined PRA area. The first step is to determine or confirm whether or not the organism considered is a pest. If no pests are identified, the analysis need not continue. The analysis of pests identified in Stage 1 continues to Stages 2 and 3 using guidance provided in other standards. Information gathering, documentation and risk communication, as well as uncertainty and consistency, are issues common to all PRA stages.

13.2.3.2 ISPM 11 Pest Risk Analysis for Quarantine Pests

ISPM 11 (*Pest risk analysis for quarantine pests*) was adopted by the Third Session of the Interim Commission on Phytosanitary Measures in April 2001. In April 2003, the Fifth Session of the Interim Commission on Phytosanitary Measures adopted a supplement to ISPM 11 on analysis of environmental risk and agreed that it should be integrated into ISPM 11. This resulted in ISPM 11 Rev. 1 (*Pest risk analysis for quarantine pests including analysis of environmental risks*). In April 2004, the Sixth Session of the Interim Commission on Phytosanitary Measures adopted a supplement on PRA for LMOs and agreed that it should be integrated into ISPM 11 Rev. 1 (*Pest risk analysis for quarantine pests including analysis of environmental risks and living modified organisms*). The supplementary text on environmental risks is marked with "S1," and the supplementary text on LMOs is marked with "S2" (https://www.fao.org/3/j1302e/j1302e.pdf).

The Interim Commission on Phytosanitary Measures acknowledges the collaboration and support of the Secretariat of the Convention on Biological Diversity, as well as the participation of experts from Parties to the Convention, in the preparation of the supplements to ISPM 11.

Annex 4 on PRA for plants as quarantine pests, together with associated changes in the core text of the standard, was adopted by the Eighth Session of the Commission on Phytosanitary Measures in April 2013.

The standard provides details for the conduct of PRA to determine if pests are quarantine pests. It describes the integrated processes to be used for risk assessment as well as the selection of risk management options.

S1 It also includes details regarding the analysis of the risks of plant pests to the environment and biological diversity, including those risks affecting uncultivated/unmanaged plants, wild flora, habitats and ecosystems contained in the PRA area. Some explanatory comments on the scope of the IPPC in regard to environmental risks are given in Annex 1.

S2 It includes guidance on evaluating potential phytosanitary risks to plants and plant products posed by LMOs. This guidance does not alter the scope of ISPM 11 but is intended to clarify issues related to the PRA for LMOs. Some explanatory comments on the scope of the IPPC in regard to PRA for LMOs are given in Annex 2.

Specific guidance on conducting PRA for plants as quarantine pests is provided in Annex 4.

As per the standard, the objectives of a PRA are, for a specified area, to identify pests and/or pathways of quarantine concern and evaluate their risk, to identify endangered areas, and, if appropriate, to identify risk management options. PRA for quarantine pests follows a process defined by three stages:

> **Stage 1 (initiating the process)** involves identifying the pest(s) and pathways that are of quarantine concern and should be considered for risk analysis in relation to the identified PRA area.
> **Stage 2 (risk assessment)** begins with the categorization of individual pests to determine whether the criteria for a quarantine pest are satisfied. Risk assessment continues with an evaluation of the probability of pest entry, establishment and spread and of their potential economic consequences (including environmental consequences – S1).
> **Stage 3 (risk management)** involves identifying management options for reducing the risks identified at Stage 2. These are evaluated for efficacy, feasibility and impact in order to select those that are appropriate.

13.2.3.3 ISPM 21 – Pest Risk Analysis for Regulated Non-Quarantine Pests

This standard provides guidelines for conducting PRA for RNQPs. It describes the integrated processes to be used for risk assessment and the selection of risk management options to achieve a pest tolerance level (https://www.fao.org/3/y5722e/y5722e.pdf).

The objectives of a PRA for RNQPs are, for a specified PRA area, to identify pests associated with plants for planting, to evaluate their risk, and, if appropriate, to identify risk management options to achieve a tolerance level. PRA for RNQPs follows a process defined by three stages:

Stage 1 (initiating the process) involves identifying the pest(s) associated with the plants for planting that are not quarantine pests but which may be of regulatory concern and that should be considered for risk analysis in relation to the identified PRA area.
Stage 2 (risk assessment) begins with the categorization of individual pests associated with the plants for planting and their intended use to determine whether the criteria for an RNQP are satisfied. Risk assessment continues with an analysis to determine if the plants for planting are the main source of the pest infestation and if the economic impact(s) of the pest on the intended use of those plants for planting are unacceptable.
Stage 3 (risk management) involves identifying a pest tolerance level to avoid the unacceptable economic impact(s) identified at Stage 2 and management options to achieve that tolerance.

Although ISPMs are internationally agreed to and adopted, they are meant to be guidelines, and their use is not mandatory within the framework of the IPPC. In addition, their interpretation and application on a national level vary from country to country. This variation is illustrated by the many different national systems and procedures that exist for carrying out PRA. Countries often take slightly different approaches to implementing the standards while still staying true to their intent (Gupta and Khetarpal, 2004). A PRA conducted according to international standards can provide the basis for our national planners to allocate resources to problem areas posing the highest risk and, thereby, enable them to better protect our agriculture and environment from the entry and establishment of exotic pests.

13.3 SIGNIFICANCE OF PRA

13.3.1 As a Safeguard

The results of PRA and their application to operational procedures or the development of regulations or guidelines can be considered as a safeguard. Safeguards are designed to lower the risk of entry and establishment of a quarantine pest to an acceptable or tolerable level. Among the safeguards routinely used by quarantine services are regulations (exclusions), permits, phytosanitary certificates issued by the competent authority of the exporting country, inspection at the place of origin and/or upon entry, sample testing and treatment at origin and/or entry. PRA, if ready prior to import, would facilitate the smooth functioning of quarantine services. The quarantine workers, prior to receiving of any material, would be aware of the likelihood of the pests/pathogens that could be expected to be associated with it.

13.3.2 To Facilitate Export

A complete PRA prepared according to international standards will improve our agricultural export potential and create more awareness about plant health and the quality of the produce. If India meets international standards and still market access is refused, the burden of proof rests with the importing country, but to counteract such situations, India should have the requisite information about its territory. Indian products/produce would be more acceptable and sought after only if they met the international standard of quality and certification (Gupta et al., 2002).

13.3.3 Fixing Appropriate Level of Protection

PRA helps in fixing the appropriate level of protection (ALP). Knowing the risk involved in the import of a commodity from a specific country would enable us to fix our own ALP or appropriate levels of risk (ALR). This should harmonize with international standards and could be higher, if justified. This will automatically meet the requirements of consistency and transparency in decision-making and solving disputes, if any. However, at present, there are no guidelines available for fixing ALP, and member countries are free to decide their own levels of protection required, provided they are in harmony with the norms set by the IPPC or any other international standards setting organization.

13.3.4 Identification of PFAs

PRA is also a tool to identify areas of low pest prevalence or PFAs. Furthermore, this information could be useful in converting these low pest areas into PFAs by launching official control programmes for such areas. Official control of the peanut stripe virus of groundnut in India is one such example.

13.3.5 Developing Pest Management Options

PRA can be useful in developing alternate strategies that could be used for mitigation or pest management. Knowing the risk involved, decisions could be made for the type/kind of pest and level of IPM required for a given crop.

13.3.6 Dispute Settlement

Due to unpreparedness with the PRA, cases of dispute were pled against many countries in the WTO. To resolve these disputes (involving scientific or technical issues) in trade between two parties, a panel seeks advice from experts in the concerned fields. The panel is free to ask any party to supplement their arguments with scientific facts and figures. Finally, the decision is based on the PRA and PFA of the exporting country (Khetarpal and Gupta, 2002).

13.3.7 Harmonization

PRA will help in the revision of existing policies and establishing new policies that will protect our country's interests without restricting the international movement of people and commodities. In turn, our policies would more easily harmonize with international ones.

13.3.8 Research Prioritization and Fund Allocation

A PRA conducted according to international standards can provide the basis for managers to identify areas of new research or problems of the highest risk and enable them to better protect a country's agriculture and environment from the establishment of exotic pests of quarantine significance. The analysis of existing knowledge on pest/disease scenario would reveal critical gaps in our understanding and determine areas where research is needed. Well-defined outputs and endpoints will further sharpen research objectives. It will help in re-orientation of funding based on research prioritization. More private sector investment in research can be anticipated. Further establishment of new quarantine stations and upgrading of existing ones would place greater emphasis on the development of suitable modern techniques and skilled scientific manpower.

A PRA conducted according to international standards can provide the basis for our national planners to allocate resources to problem areas posing the highest risk and, thereby, enable them to better protect our agriculture and environment from the entry and establishment of exotic pests. The following are two case studies to elaborate on how PRA was used as a tool to facilitate safe trade.

13.4 PRA CASE STUDIES

13.4.1 Case Study 1: Soybean Import into India (1997)

The US has been attempting to gain market access into India for whole soybeans destined for crushing and processing facilities within India. In response to these market access requests, India expressed concern about the possible introduction of a number of seed-borne pathogens that could be introduced in soybean shipments imported.

To address these concerns, the US government prepared the following two documents and presented to the Indian government:

i. *Evaluation of potential phytosanitary and economic risks of the plant pathogens on US soybean imports by India.*
ii. *Seed transmitted pathogens of soybeans present in the US: Assessment of risks to India of soybean imports for crushing.*

The Indian government, on the other hand, argued that since a small amount of soybeans imported for crushing could be diverted for seed use, phytosanitary standards would be required for all imports. Recognizing this negligible impact on the pest introduction potential, the pest risk assessment and identification of mitigating measures should take into consideration the remote possibility of seed-borne pathogens being introduced into India following this pathway.

In response to these scientific presentations and inquiries from the US, the Indian government identified 14 outstanding diseases and one nematode as significant pests. To mitigate these pests, the Indian government suggested that soybeans "should be allowed to be imported into India in split form only." Recognizing the US industry's inability to comply with a splitting or cracking requirement, both logistically and financially, the US Department of Agriculture requested a technical exchange with Indian plant health specialists to discuss alternate measures to mitigate the pest introduction risks for these pathogens.

13.4.1.1 Application of International Protocols by the US

Utilizing the World Trade Organization (WTO) and Sanitary and Phytosanitary principles, regulatory decisions should be based on sound science and be technically justified. The US government prepared a document applying those principles and demonstrating that none of these outstanding soybean pests present a risk of new introductions into India and that mitigating measures can be identified for each of the 15 pests (Khetarpal et al., 2001).

According to International Plant Protection Commission (IPPC) standards, pests requiring regulatory responses must be categorized as quarantine pests by the importing country. Categorization of pests is accomplished by applying the IPPC definition of quarantine pests as follows: *A quarantine pest is a pest of potential economic importance to the area that is endangered, not yet present there, or present but not widely distributed and being officially controlled.*

Another important internationally accepted plant health principle is the completion of PRA for the importing country to evaluate pests of potential concern and identify appropriate introduction pathways. The US accomplished a preliminary PRA with a general evaluation by the Indian government, and a document was brought out that complemented the preliminary PRA as the qualitative PRA.

Following the above-mentioned IPPC protocols and definition of terms, the US government submitted that the identified soybean pests would not require regulatory action nor be considered quarantine pests if they met one or more of the following conditions:

- Pest already in India and not under official control
- Pest not likely to follow the introduction pathway
- Pest not transmitted by seed in soybeans and not likely to follow the introduction pathway
- Pest has no potential economic importance in India.

The US government thus summarized the appropriate scientific conclusions on the occurrence and biology of each pest necessary to evaluate the need for regulatory requirements. In each case, the statements were referenced by either scientific publications or internationally renowned plant health organizations, so that the conclusions are based on sound science and are technically justified.

Based on this analysis, the US government thus showed that all the soybean pests of concern for India meet one of the above four conditions.

However, the Indian government, on analysis of the US document, found certain anomalies and is in the process of framing an appropriate reply to counter certain statements, along with scientific and technical justifications supported by the publications. The document, which is being finalized by the Directorate of Plant Protection, Quarantine and Storage, would provide the following critical information:

1. Pathway analysis for quarantine pests of soybean from the US
2. Pest risk assessment for top ranking quarantine pests of soybean from the US
   - Probability of entry
   - Probability of establishment and spread
   - Overall estimate of probability of introduction
   - Economic impact assessment
   - Combined risk rating of probability of introduction and economic impact assessment
3. Pest risk management (reviewing several risk managements options viz., pre-entry, entry and post-entry to address the identified risks).
   - Review of existing phytosanitary requirements
   - Evaluation of risk management options

13.4.1.2 PRA – Lesson to Learn from the Soybean Case

- To meet the challenges of liberalization of trade posed by the WTO, there is a need to have a Central Unit under the Ministry of Agriculture, which could act as the nodal agency for the agreement. This would facilitate a periodic review of its implications and the necessary actions to mitigate their repercussions in order to eventually harmonize our trade policies.
- Preparation of a database on endemic pests would be the first step in this direction (Chalam et al., 2016). In the case of soybean imports, the qualitative pathway-initiated pest risk assessment was developed by the US, which in fact should have been the responsibility of the importing party, i.e., India. A proposal for feasible options for India for the development of an effective PRA system with a cost estimate will be presented in the next report.
- The PRA submitted by any other government or agency should be thoroughly reviewed for its scientific and technical justifications. This is clear from the fact that the PRA of soybean imports into India from the US had a number of anomalies that could be technically challenged.
- The international protocols and agreements do not recognize the internal problems of any member country. This would mean that the problems of pilferage of soybean seeds by farmers for sowing purposes, which otherwise are meant to be imported for crushing and processing, would not justify the introduction pathway for the pests. Hence, a compromising attitude needs to be adopted in such cases where imports are not meant for sowing or planting purposes.

13.4.2 Case Study II: Import of Wheat Grain from the US (2005–2006)

During 2005–2006, there was an import of 6 million tonnes of wheat from various countries, such as Australia, Canada and Ukraine, for which crop-based PRAs were conducted by the government. The wheat import conditions as per PQ Order 2003 and the specifications for the import of wheat for consumption/processing were as follows (https://plantquarantineindia.nic.in/pqispub/pdffiles/pqorder2015.pdf):

- **Additional declarations** required to be incorporated into Phytosanitary Certificate
  Freedom from:
  a. Granary weevil (Sitophilus granarius)
  b. Ergot (Claviceps purpurea)
  c. Dwarf bunt (Tilletia contraversa)
- **Special conditions of import include**
  Fumigation with methyl bromide at 32 gm/cu·m at 28°C and above under NAP and the treatment shall be endorsed on a phytosanitary certificate or by any other fumigant/substance in the manner approved by the Plant Protection Adviser for this purpose.
- **Besides, imports also required freedom from the following quarantine weed species (Schedule VIII of the PQ Order)**

1. *Allium vineale*	16. *Echinochloa crus-pavonis*
2. *Ambrosia maritime*	17. *Froelichia floridana*
3. *Ambrosia psilostachya*	18. *Helianthus californicus*
4. *Ambrosia trifida*	19. *Helianthus ciliaris*
5. *Apera-spica-venti*	20. *Heliotropium amplexicaule*
6. *Bromus secalinus*	21. *Leersia japonica*
7. *Cenchrus tribuloides*	22. *Matricaria perforatum*
8. *Centaurea diffusa*	23. *Polygonum cuspidatum*
9. *Centaurea maculosa*	24. *Proboscidea lovisianica*
10. *Centaurea solstitialis*	25. *Salsola vermiculata*
11. *Cichorium pumilum*	26. *Senecio jacobaea*
12. *Cichorium spinosum*	27 *Solanum carolinense*
13. *Cordia curassavica*	28. *Striga hermonthica*
14. *Cuscuta australis*	29. *Thesium australe*
15. *Cynoglossum officinale*	30. *Thesium humiale*
	31. *Viola arvensis*

Note: As on 30.6.2022, imports require freedom from 57 quarantine weed species.

The Ministry of Food, Government of India, started importing wheat into India during the years 2006–2007, mainly for Public Distribution System (PDS) throughout India. No single person could fulfil the entire demand. A total of 6,088,565 MTs of wheat was imported through various ports such as Chennai, Cochin, Tuticorin,

Mumbai, Vizag, Kakinada, Kandla, Mundra and Manalore from various countries, viz., Australia, Russia, Hungary, Ukraine, Canada, France, Kazakhstan, Romania, Bulgaria, the Czech Republic, the Netherlands and Argentina.

RELAXED PHYTOSANITARY NORMS FOR IMPORT OF WHEAT IN TO INDIA BASED ON RECOMMENDATIONS MADE BY TECHNICAL COMMITTEE OF EXPERTS

(valid from July 2006 to July 2007)

Wheat grains for consumption and processing are to be allowed to be imported by private importers as well as by STC on government accounts, in accordance with the following conditions:

i. The wheat consignment would be required to be certified to be free from granary weevil (*Sitophilus granarius*).
ii. The wheat consignment would be required to be certified to contain less than 0.01% Ergot (*Claviceps purpurea*) w/w.
iii The wheat consignment would be required to be certified to contain less than 0.005% Dwarf bunt (*Tilletia controversa*).
iv The wheat consignment would be required to be certified to contain less than the average of 100 quarantine weed seeds, as per Schedule VIII of this Order, per 200 kg of samples drawn as per the prescribed procedure.
v. The wheat consignment would be required to be fumigated with methyl bromide at 32 g/cu·m at 21°C and above for 24 hours under normal atmospheric pressure and the treatment endorsed on the phytosanitary certificate OR the wheat consignment would be required to be fumigated with aluminium phosphide at 40 g/1,000 cu ft for a period of at least 21 days prior to arrival into India, provided the ship hold has provision for post re-circulation of phosphine gas during the period of transit.

Ever since the revised phytosanitary conditions came into force, 47 out of the total 145 consignments imported from Australia, Russia, Canada, Ukraine, France, Romania, the Czech Republic, the Netherlands and Argentina were found to be infested/infected with quarantine pests of consignments imported from various countries. Besides, a large number of non-quarantine pests, especially weeds, were also intercepted from the wheat consignments.

13.4.2.1 Exotic Pests Intercepted in Imported Wheat during 2006–2007

Pest Risk Assessment	*Claviceps purpurea*	*Cenchrus tribuloides*	*Solanum carolinense*	*Viola arvensis*
Pest Identity	*Claviceps purpurea*	*Cenchrus tribuloides*	*Solanum carolinense*	*Viola arvensis*
Type	Fungal pathogen	Weed	Weed	Weed
Intercepted from	Australia, Russia, Canada, Ukraine, France, Romania, Czech Republic, The Netherlands, Argentina	Australia	Australia	Russia
Regulatory Status	Regulated pathogen	Regulated weed	Regulated weed	Regulated weed
Introduction Potential	Medium	High	High	High
Establishment Potential	Medium	High	High	High
Spread Potential	Medium	High	High	High
Potential Economic Consequences	Medium	High	High	High
Potential Environment Impact	Medium	Medium	Medium	Medium
Overall Risk Associated	**Medium (unacceptable)**	**Medium (unacceptable)**	**Medium (unacceptable)**	**Medium (unacceptable)**

Source: DAC&FW, Ministry of Agriculture and Farmers Welfare.

The PRA conducted revealed that the spread of serious exotic pests was unavoidable in cases where the imported material was distributed by the Food Corporation of India (FCI) under PDS or sold by private traders as follows:

Sclerotia of Ergot (*Claviceps purpurea*) intercepted from 39 consignments could survive under adverse conditions, remain viable in soil for up to 10 years and survive on wild hosts/ weeds prevalent in the wheat-growing areas.

Unrestricted distribution throughout the country via PDS with multiple rural outlets might also pose a serious threat to the wheat-growing areas by spreading infected grain and weed seeds.

Majority of weed seeds intercepted could easily spread to wheat-growing areas through air, water, transport, planting materials, etc., and in addition to agriculture, they may also pose serious health problems, as evidenced by the introduction of Parthenium seeds along with imported wheat from the US in the early 1950s.

Wheat is a staple food for a large population of India, and invasion through grains by any of the quarantine pests might result in a loss of food for a large population. Hence, using PRA as a decision-making tool, India could technically justify its decision to import wheat from specific countries based on a rational risk analysis prior to import and avoid the potential threat posed by the exotic pests to its biosecurity (Sarathbabu et al., 2010).

13.5 STATUS OF PRA IN INDIA AND PERSPECTIVES

The developed countries, such as Australia, New Zealand, the UK and the US, are well equipped with databases and multivariate analytical models to interpret the observations and results through quantitative approaches. At the same time, some of the developing countries, like Jamaica, are serious about deciding and equipping themselves with PRAs for pests that can attack their native crops like coffee, bananas, citrus and sugarcane. These countries obtain information on pest status, distribution and host range from the exporting countries and prepare an informal analysis with the help of their experts (Khetarpal et al., 2009). This quantitative way of reaching a conclusion demands significant resources. The PRAs prepared for the import of commodities by the exporting countries are often biased, as can be exemplified by the PRA prepared by US scientists for soybean and wheat imports into India from the US.

India needs to channelize its scientific manpower and resources by networking the three concerned organizations, viz., Directorate of Plant Protection Quarantine and Storage of Department of Agriculture and Cooperation (DAC), Indian Council of Agricultural Research of Department of Agricultural Research and Education and State Agricultural Universities. In this direction, the DAC has established a National Unit on Pest Risk Analysis at the National Institute of Plant Health Management, Hyderabad. This unit apart from preparing/coordinating the development of PRAs also prioritizes the commodities for export for which PRAs have to be developed on priority (Khetarpal and Gupta, 2006). In order to prepare an updated list of endemic pests, authentic data on the survey/surveillance of the country as well as literature are indispensable.

The Plant Quarantine (Regulation of Import into India) Order 2003 came into force on January 1, 2004. This was important to fulfil India's legal obligations under the WTO Agreements. Till June 2021, 90 amendments to the Plant Quarantine Order 2003 have been notified to the WTO, revising quarantine pest lists, incorporation of ISPM-15 compliance, recognition of irradiation treatment, PFAs and cold treatments for fruit flies to allow the import of fresh fruits, and revising the lists of crops under Schedules V and VI to include 17 and 700 crops/commodities, respectively. The Directorate of Plant Protection Quarantine and Storage under the DAC&FW, which is responsible for all phytosanitation-related issues, has till date developed PRAs for about 732 crop and plant species (https://plantquarantineindia.nic.in/pqispub/pdffiles/pqorder2015.pdf). ICAR-National Bureau of Plant Genetic Resources on the other hand is contributing to this activity by preparing generic PRA to facilitate the import of germplasm. To date, PRA for 140 crop species has been submitted to DAC&FW for gazette notification under the PQ Order. Additionally, PRA for about 90 crop species has been completed to facilitate the import of many new species not listed in PQ Order 2003.

The research areas that need emphasis for generating data have been identified. The need for developing national standards for surveys, surveillance and PFAs was also highlighted (Rajan et al., 2005). The scientific community needs to further identify the focused areas of research to complement the national efforts. To carry out such a gigantic task, all the stakeholders need to contribute in the national interest. An enormous amount of work within a fixed time frame needs to be carried out to meet the challenge of an open market under the WTO regime. The plant protection specialists have a major role to play in this regard in promoting and facilitating exports and imports in the interest of the country.

REFERENCES

Chalam VC; S Bhalla; K Gupta; B Singh; Z Khan; SC Dubey. *Generic pest risk analysis: import of transgenic soybean.* ICAR-National Bureau of Plant Genetic Resources, New Delhi, 2016, p. 146. https://www.nbpgr.ernet.in/Downloadfile.aspx?EntryId=7319.

Chalam VC; K Gupta; R Sharma; VD Sharma; AK Maurya. Pest Risk Analysis and Plant Quarantine Regulations. In: *Emerging Trends in Plant Pathology*, eds: KP Singh; S Jahagirdar; BK Sarma, Springer, Singapore, 2021. https://doi.org/10.1007/978-981-15-6275-4_29.

Gupta Kavita; RK Khetarpal. Concept of regulated pests, their risk analysis and the Indian scenario. *Annual Review of Plant Pathology* 2004, Vol 3, 409–441.

Gupta Kavita; Irfan Rashid. Training manual on pest risk assessment developed for training under TCP/NEP/3702 on *Building capacities to improve and sustain forest health to enhance the resilience of forests and livelihoods of forest dependent communities in Nepal*, November 23–28, 2020, 100 p.

Gupta Kavita; BM Pandey; RK Khetarpal. Pest risk analysis and its significance to India in WTO Regime. *Seed Tech News* 2002, Vol 32(2&3), 266–269.

https://plantquarantineindia.nic.in/pqispub/pdffiles/pqorder2015.pdf.

https://www.fao.org/3/j1302e/j1302e.pdf

https://www.fao.org/3/k0125e/k0125e.pdf

https://www.fao.org/3/y5722e/y5722e.pdf

https://www.ippc.int/en/core-activities/standards-setting/ispms/

https://www.wto.org/english/tratop_e/sps_e/spsagr_e.htm

https://www.cabi.org/cpc

Khetarpal RK; Kavita Gupta. Implications of sanitary and phytosanitary agreement of WTO on plant protection in India. *Annual Review of Plant Pathology* 2002, Vol I, 1–26.

Khetarpal RK; Kavita Gupta. Policy Issues in Plant Protection. In: *One Hundred Years of Plant Pathology in India-An Overview,* eds: SS Chahal; RK Khetarpal; TS Thind, Indian Society of Mycology and Plant Pathology, Scientific Publishers, Jodhpur, India, 2006, pp. 381–401.

Khetarpal RK; Rajan. Sanitary and phytosanitary agreement of WTO. *Seed Tech News.* 1999, 29(4): 51–53.

Khetarpal RK; Rajan; Kavita Gupta. Sanitary and Phytosanitary (SPS) measures: A case study with soybean. In Proceedings of XXXI Annual Workshop-2001 of All India Co-ordinated Research Project on Soybean. Marathwada Agricultural University, Parbhani (Maharashtra). March 12–14, 2001, pp. 58–62.

Khetarpal RK; N Balaraman; SK Bandopadhyay; Kavita Gupta; RK Jain; J Kumar; AN Mukhopadhyay; KK Vijayan. Status of Agricultural Biosecurity. In: *State of Indian Agriculture,* eds: M Rai; SS Acharya; SM Virmani; PK Aggarwal, National Academy of Agricultural Sciences, New Delhi, 2009, pp. 151–174.

Nagarajan S; RP Singh; J Kumar; AK Sharma. Pest risk analysis for shipping wheat from Karnal bunt (*Tilletia indica*) infected areas to disease free destinations. Abstract. In 54th Annual Meeting and National Symposium on Crop Protection & WTO-An Indian Perspective. January 23–25, 2002, p. 2.

Rajan; Kavita Gupta; RK Khetarpal. Sanitary and Phytosanitary Issues for Trade in Horticultural Crops. In: *Challenging Problems in Horticultural & Forest Pathology*, eds: RC Sharma; JS Sharma, Indus Publishing Co., New Delhi, 2005, pp. 422–443.

Sarathbabu B; KS Varaprasad; Kavita Gupta; RK Khetarpal. Evolution of National Plant Quarantine System in India in Tune with International Phytosanitary Regime. In: *Sustainable Crop Protection Strategies* Volume I, eds: HR Sardana; OM Bambawale; D Prasa, Daya Publishers, Delhi, 2010, pp. 13–29.

14 Safeguarding Global Plant Resources from Ever-Growing Disease Threats

Shashi Bhooshan Sharma
The World BioProtection Forum and Harry Butler Institute, Murdoch University

14.1 INTRODUCTION

Planet Earth, when compared with billions of other planets, is, as far as we know, unique in that it harbours life. About 9 million species of diverse organisms, including humans, call this planet home (Mora et al., 2011). Of all these organisms, plants have the innate ability to capture energy from the Sun, a distant celestial object 150 million km away, and transform it into earthly products. One of these products is oxygen, which is vital for the life of all aerobic organisms such as humans and livestock. Plants are also integral to maintaining temperature and water balances on the planet and are a key player in maintaining planetary health. Our global food systems are dependent on the plants we eat and on organisms that eat plants. Over 80% of our food is derived from plants – even more if we include animal-based protein sourced from herbivores. It may not be easy to assign a monetary value to a wide range of benefits and ecosystem services provided by global plant resources; however, it would not be surprising if even a quick, back-of-the-envelope type calculation finds the value to be in trillions of dollars. Plants have a positive influence and a special niche in the development of human civilisation. Humans depend on plants for their survival. Life would be impossible on Earth in the absence of plants.

14.1.1 Impact of Humans on Global Plant Resources

Humans have not adequately reciprocated the positive influences and contributions of plants to human society. Many plants have suffered and continue to suffer due to inadvertent, and sometimes deliberate, human actions. Human population growth and associated infrastructure needs are resulting in the destruction of habitats for several plant species. Humans have caused mass destruction of all types of plants, particularly forest trees, for short-term gains while neglecting the long-term benefits of standing trees and other plants. Human actions have contributed to the extinction of many plant species and threatened the existence of other species. Unfortunately, humans with a present population of over 7 billion, which is just about 0.01% of all

 DOI: 10.1201/9781003403142-14

living organisms on this planet, are continuing to inflict enormous harm to plants in both managed and natural environments worldwide (Quinney, 2020).

One of the key causes of harm, which some global planners in their wisdom may endeavour to justify as collateral damage, is to meet the ever-growing needs and wants of human populations worldwide for food, medicine, clothing, building materials, heat and a wide range of consumables such as paper. Human activities are largely responsible for the international and inter-continental spread of many organisms that are dependent on plant species for their survival and well-being. This human-caused or anthropogenic introduction and movement is even resulting in some organisms crossing an evolutionary boundary, such as ecological separation (Cunningham et al., 2003; Anderson et al., 2004). Global plant resources are now exposed to a wide range of organisms that are harming their health and well-being.

14.1.2 Plant Health and Plant Disease

There is as yet no proper and universally accepted definition of plant health. The United Nations Food and Agriculture Organisation (FAO) designated 2020 as the International Year of Plant Health (IYPH). It would have been prudent to develop a succinct definition of 'plant health' during the IYPH, preferably in line with the definition of 'human health' as adopted by the United Nations World Health Organisation. For the purpose of this article, plant health is defined as a state of complete physical, physiological and genetic well-being, not merely the absence of disease or damage (Sharma, 2020a). Plant health is impacted by biological as well as non-biological stresses such as nutrition deficiency, toxicity, changing climates, degraded soils and inefficient waste disposal. The focus of this article is on a subset of plant health, that is, plant diseases or organisms harmful to plant health, specifically how plant health is influenced by the infestations and infections of disease-causing organisms such as bacteria, insects, fungi, nematodes, and viruses. Some of these organisms have affinity and dependence for survival on the same plant species on which humans are dependent for their food, fibre and many other requirements. Consequently, human dislike for these organisms is obvious, and it is evident from human branding of these organisms as 'enemies' or 'harmful organisms' or 'pests' or 'pathogens' or 'disease-causing organisms' or 'biological threats', etc. These terms have been interchangeably used in the article.

All these so-called harmful organisms have the potential to cause a reduction in plant biomass and yield, deterioration of plant products in post-harvest storage and transport, a reduction in the quality of produce, contamination of produce with toxins, a predisposition to attack by other organisms and vulnerability to abiotic stresses such as drought and nutritional deficiency. They may kill the host in some cases. These organisms not only impact plant health, but they also adversely impact access to markets for the affected plants and their products, product safety, quality, productivity and sustainability. According to a FAO report, harmful organisms to plants are responsible for losses of about 40% of food crops globally and for trade losses in agricultural products worth over US $220 billion each year (IPPC Secretariat, 2021a). This estimate covers only food crops, and the overall adverse impact would be significantly higher if losses caused to all plant species were considered. It should be noted that despite the application of 3 million metric tons of pesticides worldwide, which is about a US $40 billion expense, and the use of various biological and other non-chemical controls, between 35% and 42% of potential crop production is destroyed by harmful organisms. The value of this crop loss is estimated to be US $2,000 billion per year, yet there is still a US $4 return per dollar invested in pesticide control. Without pesticides and non-chemical controls, the damage inflicted by harmful organisms would be more severe than it is at present; crop losses may increase from 40% to 70% and would have serious adverse impacts on the world food supply and economy (Pimentel, 2009).

14.2 HUMAN-FACILITATED DISTRIBUTION OF HARMFUL ORGANISMS

Humans are responsible for exposing global plant resources to a wide array of organisms that are detrimental to their health and well-being. The era of globalisation, world trade and rapid advancements in transport technology have transformed the global economy and facilitated the quick international movement of goods and people (Keller et al., 2011). Over 1 billion international travellers and 18 million containers making over 200 million trips by sea alone in a year provide massive opportunities for organisms harmful to plants to move to different geographical regions (Shashi Sharma and John Wightman, 2015). A survey of dust and dirt samples taken from shipping container surfaces arriving in New Zealand ports revealed the presence of live nematodes in 81% of samples from 347 containers, although only 4% were plant parasitic nematodes (Watson, 2004). This is still equivalent to the potential arrival of plant parasitic nematodes on more than 10,000 containers entering New Zealand every year. There are countries that receive millions of containers, and a 4% infestation rate means that hundreds of thousands of containers bring in harmful organisms. In Australia, more than 90% of sea containers arriving in the country are not inspected for biosecurity risks because they are profiled or assessed as posing negligible or very low biosecurity risk (Inspector General of Biosecurity, 2019). Other potential carriers of harmful organisms, such as billions of plants that are produced in Central America and moved to North America and large numbers of plants produced in Africa and Asia and exported to Europe, have the potential to provide excellent opportunities for these organisms to migrate to new regions and find host plants (Santini et al., 2018). For example, *Xylella fastidiosa* arrived in Italy from Costa Rica on coffee plants in 2008. Coffee plants are latent carriers of this bacterium. It was transmitted to olives and, in due course, led to the death of several millions of olive trees, with significant environmental, social, political and economic consequences (Sicard et al., 2021).

Globalisation has opened up markets across continents, and the availability of agricultural produce is no longer dependent on local production. This market opportunity has stimulated the expansion of production areas for some key crops, which has resulted in the emergence of new plant pathogens, and some pathogens have become major production constraints. Expansion of soybean (*Glycine max*) cultivation from the USA to Brazil and Argentina has led to the emergence of Bean golden mosaic virus as a major threat to the production of beans in Brazil. In Argentina, Bean dwarf mosaic virus has emerged as a serious constraint. Both of these viruses are transmitted by the whitefly *Bemisia tabaci*, which thrives on soybean plants. The intensive cultivation of soybeans in Brazil and Argentina has provided excellent opportunity for large increases in the populations of whiteflies (Anderson et al., 2004). Human activities have contributed to the interspecific hybridisation of some organisms that are not naturally sympatric and resulted in the emergence of new strains of plant disease-causing organisms (Brasier and Duncan, 1999). This new scenario of perpetually enhanced exposure of plants to harmful organisms is causing serious concerns and affecting our ability to safeguard the health and well-being of global plant resources.

14.3 EMERGENCE OF NEW HARMFUL ORGANISMS TO PLANTS

There are rapid increases in the incidence, geographical distribution, plant damage-causing ability and host ranges of harmful organisms in plants. New plant damage-causing organisms and new vectors of harmful organisms are being found in different countries. Furthermore, changing climate and weather patterns are influencing the suitability of geographical regions and seasons for different species of plants and other organisms (Bebber, 2015; Velasquez et al., 2018; Priyadarsini et al., 2020). There are clear indications that climate change has already expanded the host range and geographical distribution of some pests and may further increase the risk of pest introduction to new areas. (IPPC Secretariat, 2021b). These changes are contributing to the emergence of new organisms or new strains that were hitherto considered benign to plant health. New interactions between harmful organisms and plants are contributing to expanding the range of harmful organisms and their vectors in plants. Needless to say, without human assistance, these organisms had limited abilities to move across large distances, find suitable host plants and mates and increase in large numbers. Trade leads to continental jumps in distribution rather than local spread through natural means. New locations may lack natural enemies, and introduced pathogens can spread unchecked as they are generally introduced to new areas without the competitors or enemies that keep them in check at home. Introduction of wheat blast into Bangladesh in infected seed (Inoue et al., 2017), incidence of cassava mosaic virus, which does not naturally occur on cassava in South America, on cassava in Africa following uptake of cassava as a major food crop there (Legg et al., 2015), emergence of a new race of wheat stem rust pathogen in Ethiopia with a very high combination of virulence genes, and a new race in Sicily with the ability to overcome host resistance in wheat cultivars (Hei et al., 2020) are just some of the reports of emerging plant pathogens on different crops and regions. Migration of pathogens or host plants into new geographic regions and hybridisation between species or individuals are resulting in the expansion of host ranges of pathogens. Hybridisation between a recently introduced exotic pathogen and a resident pathogen may allow rapid adaptation to new hosts or environments. Interspecific hybridisation and mutations within existing pathogen lineages can lead to the emergence of new pathogens, as in the case of the emergence of a pathogen of Solanum species, *Phytophthora andina*, in South America and the alder pathogen *Phytophthora alni* in the United Kingdom (Gladieux et al., 2014, Martin et al., 2016). Emergence of vector-transmitted diseases that lack vector specificity, as in the case of *X. fastidiosa*, poses significant challenges in the management of the pathogen and its economic impact. There is a high likelihood that *X. fastidiosa* isolates, when reaching a novel environment, will be transmitted by an endemic vector species (Almeida and Nunney, 2016). Zebra chip disease, caused by the phloem-limited bacterium *Candidatus* Liberibacter solanacearum, is transmitted by the potato-tomato psyllid *Bactericera cockerelli* Šulc. The disease has spread across North America, South America and New Zealand, and there is serious apprehension that native species of psyllids may acquire and transmit the bacterium in different countries.

Incidents of vector-borne microbial threats are rapidly increasing. Xylella disease, caused by *X. fastidiosa*, is vectored by several insect species, and, depending on the host, the disease is known by a range of common names, including Pierce's disease (in grapevine), alfalfa dwarf disease (in lucerne), phony disease (in peach), leaf scald (in plum), quick decline (in olive), and variegated chlorosis (in citrus). Other biological threats of growing concern include species of thrips and the orthotospoviruses (formally tospoviruses) they transmit to horticultural crops. Orthotospoviruses reduce yields, quality and marketability of infected plants and, in the worst-case scenario, cause near-complete crop failures. These are serious threats with expanding host ranges, geographic distributions and vectors. A study conducted at the University of Exeter, United Kingdom, has identified the biological threats likely to be the most invasive in the coming years. These include three species of tropical root knot nematodes, *Meloidogyne* spp., root parasites of thousands of plant species, *Blumeria graminis*, a fungus that causes powdery mildew on wheat and other cereals, and the citrus tristeza virus. Over 2,400 microorganisms occur in the seeds of 383 genera of plants (McGee, 1997), and it is estimated that up to one-third of plant viruses might be seedborne in at least one of their hosts (Stace-Smith and Hamilton, 1988).

Sometimes, introductions of these harmful organisms may remain dormant, pending the arrival of a facilitator vector organism before the plant disease can emerge and spread. The citrus tristeza virus became an economically important disease in South America after the introduction of its efficient aphid vector, *Toxoptera citricidus*, from Asia. Within two decades of the introduction of the vector, over 6 million citrus trees were destroyed in Sao Paulo alone in Brazil (Bar-Joseph et al., 1989). Pierce's disease, caused by *X. fastidiosa*, was present in California, USA, for over a century. The arrival of a new vector, the blue-green sharpshooter *Graphocephala atropunctata*, in 1997 led to the rapid spread of Pierce's disease in grapevines and caused over US $6 million in South California alone in 1999 (Anderson et al., 2004). In many cases, plants introduced for food or other commercial purposes are attacked by endemic species. The bacterium *Ralstonia solanacearum* causes the moko disease of bananas, which evolved in *Heliconia* species in Central America (Anderson et al., 2004).

14.4 NEW GLOBAL PLANT HEALTH RISK SCENARIO HAS EMERGED

In the early years of our civilisation, when human movement was localised, any organisms identified as causing damage to cultivated plants were controlled by local cultural practices such as rogueing, ploughing, fallowing and crop rotations. Subsequently, widespread use of major gene resistances, applications of pesticides and integrated pest management practices have been applied to manage plant damage and its impact on society (He et al., 2016). The elevated frequency of the introduction of harmful organisms has threatened our ability to safeguard plant resources. In Australia alone, over 18,000 vessels, 1.8 million sea cargo consignments, 41 million air cargo consignments, 152 million international mail items and 21 million passengers arrive in a year, and these numbers are growing annually, consequently increasing the risk of the introduction of harmful organisms. Between 2012 and 2017, the number of interceptions of potentially harmful organisms rose from 25,281 to 37,014 (Inspector General of Biosecurity, 2019).

The COVID-19 pandemic has taught us that harmful organisms do not respect the arbitrary lines on a map that humans consider to be national boundaries. Once harmful organisms have moved from their native region and established themselves in another country, they require ongoing management to minimise their adverse impacts. As these adverse impacts are intergenerational and irreversible, there are significant economic imposts on present and future generations. Pimentel et al. (2001) calculated that worldwide establishments of exotic biological threats were already costing over US $1.4 trillion per annum. These worldwide increases in the emergence of new biological threats to global food value chains, forests and natural biodiversity, and consequent intergenerational and irreversible adverse impacts on human populations and the natural environment, are inevitable and need urgent corrective actions (Shashi Sharma and John Wightman, 2015). There is a rise in reports of emerging infectious diseases caused by pathogens that have increased incidence, changed pathogenesis or are newly evolved or newly discovered (Anderson et al., 2004). Plant resources in all countries are at enhanced risk from these harmful organisms. Stem rust of wheat (Ug99 strain) in Africa, bacterial wilt of bananas in Uganda, cassava mosaic virus in East Africa, citrus greening in the USA, and Myrtle rust in Australia are only some of the examples of newly-detected pathogens in the past 20 years. The number of named plant viruses has increased from less than 380 species in 1991 to more than 900 species in 2011 (McKirdy et al., 2012).

Border inspections to mitigate the risk of these harmful organisms entering the country are generally more efficient at detecting insect infestations than pathogen infections. Between 2012 and 2018, the pest incidents database in Australia recorded more than 104,000 identified pests in air cargo and 35,800 identified pests in sea cargo. Of these pests, 76% involved invertebrate, mainly insect, species. Between 2007 and 2016, plant pathogens accounted for most incursions. Most plant-related responses were caused by incursions of diseases or viruses. Generally, incursions of disease-causing organisms such as viruses, bacteria and fungi are more difficult to eradicate than those of invertebrate organisms such as insects, snails and mites. Incursions of *Pseudomonas syringae* pv. *actinidiae* biovar 3, Panama disease (*Fusarium oxysporum* f.sp. *cubense* Tropical Race 4), Cucumber green-mottle mosaic virus, Melon necrotic spot virus, and Myrtle rust (*Austropuccinia psidii*) were found to be non-eradicable in Australia (Inspector General of Biosecurity, 2019). Consequently, the present and future generations of Australians will live with these established organisms and continually pay a price for keeping them under control.

This Australian example is an indicator of a global scenario. We cannot afford the perpetual increases in the number of established harmful organisms in any country.

As an example, the Mediterranean fruit fly, *Ceratitis capitata*, was introduced into Western Australia in 1895, and even now, about 127 years later, it continues to be a serious horticultural pest in Western Australia. It attacks and damages a wide range of cultivated fruits and some fruiting vegetables and causes millions of dollars of losses. It has seriously impacted production and market access for the horticulture industry in Western Australia. If it is not managed by using chemical and other control measures, some fruit industries will be devastated (Mumford, 2002). This fruit fly will continue to be an economic impost on present and future generations of Western Australians. This is the same story for all other harmful organisms that have established themselves in different regions of the world.

There is a need for a renewed perspective on safeguarding global plant resources from harmful organisms. The COVID-19 pandemic has highlighted the fact that disease epidemics and pandemics are aided in their spread across countries and continents by human activities. New initiatives and arrangements are needed to safeguard our plant resources from the worldwide migration of harmful organisms.

14.5 RENEWED GLOBAL APPROACHES FOR PLANT HEALTH MANAGEMENT NEEDED

Historical evidence reminds us that empires and global systems have been disrupted by the only force against which political genius and military might be impotent – epidemic diseases (Schepin Oleg and Waldemar Yermakov, 1991). Incidence and spread of potato late blight, a disease caused by *Phytophthora infestans*, led to widespread famine in Ireland as a result of the failure of the potato crop, which was a staple food. This disease, HEERB-1H, originated in Mexico and is presently estimated to cause US $6 billion of damage per year globally (Ramanujan, 2019). What if this late blight disease-causing organism had remained confined to its native region in Mexico? This organism on its own cannot migrate to other countries. Why did we expose potato crops all over the world to this organism? Who is responsible for the ongoing US $6 billion loss to the global community due to this organism? Is this the price potato plants and people worldwide are compelled to pay for the sake of globalisation?

We should not wait for a global plant pest or disease pandemic situation to emerge before we devise approaches and tools to protect our at-risk plant resources. Policymakers, planners, scientists and the global community must consider the potential for long-term intergenerational and irreversible damages from invading organisms. Many harmful organisms to plants may cause dramatic damages or losses, but they can collectively cause sizable economic losses. Living with these harmful organisms is expensive and is increasingly becoming unaffordable. There are emerging concerns that the widespread use of antimicrobials, including fungicides, is increasing antimicrobial resistance (Miller et al., 2022). This poses a significant threat to the health and well-being of people and livestock.

If incursions and spread of these organisms are not prevented, then not only present generations but future generations will also be made to bear the cost of managing them. There appears to be a lurking belief that the introduction and spread of exotic organisms can be managed by developing appropriate management strategies and approaches, as has been historically done. Living with harmful organisms appears realistic, but prevention is cheaper and should always be the first priority. This approach to living with harmful organisms and managing them must change to actively prevent the migration of potentially harmful organisms. This preferred option will include any measures that prevent the spread and introduction of harmful organisms from their endemic regions and the capacity to eradicate any likely invasions of these organisms. Eradication must always be the preferred option if it is technically and operationally feasible. Economic feasibility should be given lesser importance or at least assessed in the context of the long-term adverse economic impacts if the invading organism establishes. It would even be valuable to investigate the technical and operational feasibility of eradicating already-established harmful organisms and consider eradication options. Depending on the existing spread and available control measures, the eradication of established harmful organisms is most likely a long-term strategy. New technologies may shift technical and economic feasibility towards successfully eradicating established harmful organisms and warrant enhanced research funding for their development. The current global expenditure of over US$1.9 trillion for arms and ammunition to kill other humans and cause mass destruction must be diverted to research and managing the real threats, such as the current and emerging harmful organisms to plants, livestock and humans (Sharma, 2020c).

A global approach to plant health management is needed that promotes actions to contain organisms known to be harmful to plants in their native locations and countries. The basic philosophy of this approach is the understanding that most organisms that are harmful to plants can travel only short distances without the aid of some other agent, such as humans.

14.5.1 Status of Global Plant Health and Disease Management

Presently, FAO has some initiatives, such as the Plant Production and Protection Division and the International Plant Protection Convention (IPPC), that have global plant health mandates. The vision of IPPC is to protect the world's plant resources from pests (https://www.ippc.int/en/). IPPC is recognised by the World Trade Organisation (WTO) Agreement on the Application of Sanitary and Phytosanitary Measures (the SPS Agreement) as the international standard-setting body for plant health. IPPC as such is a multilateral treaty with more than 180 contracting parties, and it aims to ensure that the importation of a plant or plant product does not expose the importing country or region to unacceptable pest risk. The governing body for IPPC is the Commission on Phytosanitary Measures, which guides effective and globally harmonised action to prevent the spread and introduction of plant pests. The scope of IPPC extends beyond the protection of cultivated plants to the protection of natural flora and plant products and applies not only to international trade in plants and plant products but also to carriers of plant pests, including containers, ships, packaging and storage facilities.

Though IPPC claims to have a global plant health mandate and has made significant contributions to mitigating the risks of spreading harmful organisms via international trade, its mandate, framework and resources are inadequate to lead, manage and operate as a global plant health manager. The IPPC work programme has heavy reliance on the National Plant Protection Organisations (NPPO) of all the member countries for completing their functions and delivering outcomes competently. However, IPPC has no authority to ensure that NPPOs are resourced adequately to deliver targeted outcomes (Sharma, 2020a). There is no proper global coordination of work on established and endemic harmful organisms. It is extremely important to have specific programmes for endemic harmful organisms to ensure that they do not move out of their native location and country. There are many highly damaging harmful organisms that have established across countries and continents over the years and are among the key reasons for widespread and significant damage to cultivated plants and other plant species in the natural environment.

The present state of global plant health management is inadequate to safeguard global plant resources from the growing threats of harmful organisms. A global plant health management system is needed that promotes and coordinates actions to contain potentially harmful organisms in plants in their native regions. A global organisation that enables worldwide implementation of a biosecurity continuum of risk prevention and response preparedness is needed. This biosecurity continuum approach comprises an integrated framework of risk management policies and operations to reduce exposure to the risk of harmful organisms.

14.5.2 Plant Biosecurity and Biosecurity Continuum Approach

The intent of plant biosecurity is to safeguard plant resources from biological threats (Sharma et al., 2009; McKirdy et al., 2014). Biological threats include the full spectrum of potentially harmful organisms that are naturally occurring, introduced by accident, introduced by negligence, or introduced deliberately as biological weapons (Sharma, 2019). The biosecurity continuum is a set of biosecurity systems that work together to mitigate the risk of these biological threats and prevent systemic gaps that could lead to their introduction. Australia has greatly benefited from adopting the biosecurity continuum approach – a biosecurity arrangement where pre-border, border and post-border biosecurity systems work in tandem. Dodd et al. (2020) reported that the total benefits from assets vulnerable to biosecurity hazards in Australia were about A$251 billion per annum, or about A$5.7 trillion over 50 years. In the absence of a biosecurity system, the likely impact of newly introduced harmful organisms was estimated to cause about A$672 billion in damages to these assets over the next 50 years. The net present value of Australia's biosecurity system was assessed to be A$314 billion, at an average return on investment of 30:1.

Biosecurity continuum is an integrated approach and a framework of risk management actions taken at a time and place when they are most effective in mitigating the risk of harmful organisms. It includes actions pre-border (off-shore, such as surveys to underpin pest status, good production practices, treatment and inspection, and certification), at the border (such as inspection, detection, treatment, re-export and destruction) and post-border (such as diagnostics, surveillance, eradication, containment and management). This approach necessitates implementation of appropriate biosecurity strategies pre-border, border and post-border and their deployment by a range of people and organisations.

14.5.3 Pre-Border Biosecurity

The underlying philosophy of pre-border activities is that 'prevention is better than cure' – preventing entry of harmful organisms is a much more cost-effective strategy than reacting to the establishment of these organisms (Sequeira and Griffin, 2014). Pre-border risk reduction actions are generally implemented by the exporting countries to prevent potentially harmful organisms from reaching the borders of other countries. These actions include off-shore inspections to verify the phytosanitary status of the goods for export against the requirements of the importing country, the implementation of phytosanitary measures at production sites or post-harvest to address the risks of harmful organisms, and storage and transportation to minimise the likelihood of infestation by the harmful organisms. Science-based biosecurity policies, biosecurity legislation, regulations, capacity to undertake risk analysis and apply risk mitigation treatments are some of the important components of a pre-border biosecurity system. The effectiveness of pre-border biosecurity systems depends on the availability of adequate plant health risk management capability (trained human resources) and capacity (infrastructure) to reduce the likelihood of transporting harmful organisms.

The IPPC and NPPOs play an important role in establishing systems and processes to improve implementation of phytosanitary programmes by the exporting countries and to ensure the phytosanitary requirements of the importing country are met. All member countries of the IPPC and the WTO have obligations to notify incursions of exotic harmful organisms and report the status of endemic and established harmful organisms to plants and plant products. The pre-border biosecurity system can be further improved by the global sharing of information and knowledge on the risks and pathways identified as potential threats. This will help to expedite the development, adoption and enforcement of risk management measures and prevent the spread of harmful organisms. New technology initiatives to further improve reporting and sharing of the status of harmful organisms and to develop decision-making and decision-support systems for risk management are needed to enhance the efficiency and effectiveness of the pre-border biosecurity system.

14.5.4 Border Biosecurity

Border biosecurity activities include, but are not limited to, inspection of passengers, goods, vessels and international mail; audit activities; and on-arrival or post-entry quarantine at airports, seaports, mail centres and other points of entry into the country. These activities enable the interception of harmful organisms, if any, and the application of disinfestation treatments. Potential carriers of organisms harmful to plants, which include all plants and plant materials, may require a phytosanitary certificate issued by the NPPO of the exporting country. Importing countries verify, usually by inspecting imported goods at the border or at the point of entry into the country, that import requirements have been met and no regulated pests are present. These inspections provide opportunity to intercept material with any infestations of harmful organisms.

Often, potential carriers of these organisms are classified based on their risk profile. Imports of some potential carriers, including plants and plant materials, can be prohibited by a country pending the development and validation of effective risk mitigation treatments. The border biosecurity system includes legislative and regulatory measures and associated activities designed to minimise the entry and spread of harmful organisms (Whattam et al., 2014). However, biosecurity inspections of the increasing volume of potential carriers of harmful organisms entering a country pose several challenges in inspecting large quantities of containerised cargo and, increasingly, the risks associated with the containers themselves. In most cases, a representative sample of the imported goods is inspected. For example, in Australia, most cargo arrives in air or sea containers. The air cargo, such as fresh produce and flowers, is considered high biosecurity risk and is inspected at high levels. However, only about 22% of arriving sea containers are sent for biosecurity consideration based on the risk profiling of importer or broker declarations. Of those referred, a further 14% are released without further intervention after assessing import documents. Over 80% of the 21 million incoming travellers in 2017–2018 were released unassessed or screened, 14% were screened, and 7% (1.4 million) had their bags, personal items or clothing inspected. This yielded 220,000 instances where biosecurity risk material was destroyed, treated or exported (Inspector General of Biosecurity, 2019).

When visual inspections are not possible or are inadequate, technology-enabled inspections such as the use of electronic noses or imaging technology to detect pests and detector dog programmes are deployed to complement the visual inspections. Biosecurity inspection and risk management at the point of entry (e.g., airplanes, ships, road transport, boats), biosecurity threat identification (diagnostics), biosecurity risk categorisation, legislation (prescribing entry conditions and penalties and sanctions for non-compliance as a deterrent to deliberate illegal imports of risk materials), permitted organism list, biosecurity risk treatments (e.g., fumigation, irradiation, chemical, cold or heat treatment), biosecurity approved arrangements and sites that manage the entry of biosecurity risk material, biosecurity breach management, and post-entry quarantine facilities are some of the important components of border biosecurity system.

14.5.5 Post-Border Biosecurity

The post-border biosecurity measures are designed to keep a vigilant eye on and then report any new harmful organisms. These measures apply best practice monitoring procedures and tools for early detection and reporting of new or exotic harmful organisms. Early detection and reporting are important to achieve cost-effective eradication, containment or control outcomes. Post-border biosecurity measures are essentially to prevent the establishment of harmful organisms that enter a country.

Application of measures to achieve zero risk of introduction of harmful organisms is neither tenable nor acceptable under the WTO agreement. As such, the risks posed by any potential carrier of harmful organisms are mitigated to the extent that the risk is acceptable, which means some residual risk remains. This residual risk is managed by post-border biosecurity activities, which include monitoring and surveillance for exotic harmful organisms underpinned by diagnostic capacity and emergency response preparedness. Every country is expected to have a list of high-priority harmful organisms that are not present or wanted, as their incursions are considered to have unacceptable social, environmental and economic impacts. Targeted surveys for these organisms, including their potential carriers and pathways, should be regularly conducted. A comprehensive national surveillance and diagnostic system is needed in every country for early detection of any new introductions and to provide accurate, up-to-date, timely and verifiable records of endemic and established harmful organisms.

If an outbreak of a priority harmful organism is identified, it is extremely important that a rapid and effective response be implemented, which can be expedited by pre-prepared contingency plans and response strategies. These plans and strategies must clearly identify actions, arrangements and responsibilities of all stakeholders in the event of an outbreak. They can also focus on the tools that will be needed and ensure they can be accessed and deployed quickly. It is helpful to undertake exercises to road test the plans during peacetime and refine them to ensure they are fit for purpose in the event of an outbreak. Such emergency incident response or incursion management facilitates decision-making with limited information in stressful and time-poor situations about important matters that can have significant consequences for producers, communities and industries. Every country must develop adequate capacity, trained capability, policies, legislation, infrastructure, arrangements and processes to enable quick response and manage incursions of organisms harmful to plants. Australia and New Zealand have developed emergency response agreements between the government and plant industries that enable decision-making, communication and cost sharing in the event of a pest incursion. These also provide a useful forum for response preparedness, including biosecurity planning and response exercises.

If eradication or containment is deemed unfeasible, a transition to a management plan should be prepared to ensure minimal impact on production and markets for the host plants, environment, economy and community. Every country needs a national management framework to ensure that nationally significant harmful organisms to plant resources established in the country are contained or cost-effectively managed by rational adaptation of resistance, avoidance, elimination and remediation strategies individually and collectively (He et al., 2016). For example, in Australia, Myrtle rust, an exotic fungal disease caused by *Austropuccinia psidii*, was detected on the central coast of New South Wales on 23 April 2010 on *Agronis flexuosa* and other hosts. On 30 April, the heads of the Australian Government and state agencies responsible for biosecurity took the decision that it was not technically feasible to eradicate Myrtle rust from Australia. On 2 July, the Australian Government implemented an interim response plan to investigate whether containment and suppression might lead to the possible eradication of the fungus. On 23 December, Australia confirmed that eradication was not technically feasible, despite efforts under the interim response plan. It was agreed that there was a need for ongoing arrangements for the management of the pathogen due to the potential for further impact on the natural environment, the community and affected industries. This investment provided new information and tools that enabled industries and communities to mitigate the impacts of the disease in urban, primary production and natural environments (Plant Health Australia, 2011). Consequently, the Myrtle rust is contained in Eastern Australia, and options are available to mitigate its effect in the infested regions. Such transitions to management actions are needed when eradication is deemed to be unfeasible.

14.6 SOME IMPORTANT CONSIDERATIONS AND WAY FORWARD

14.6.1 Understand the Perspective of So-Called Harmful Organisms

'Know your enemy' and 'think like your enemy' are some of the ancient words of wisdom for winning wars (Thomas Cleary, 1988). For winning wars against the harmful organisms in plants, I have often wondered what these organisms, which we have labelled as biological threats, pests, diseases, pathogens and harmful organisms, think, and say about us! I believe if we somehow were to develop the uncanny ability to know what they think and say about us, we may be in for a surprise.

We may find them saying, 'We really like this animal called a human – our fellow companion on this planet has taken us along to places where we could never even dream of reaching on our own. Humans have created opportunities for us to enter new environments, providing us with new hosts and new mates'. We may even find these organisms talking about their similarities with us in food preferences and likings. However, what these organisms would find perplexing is that humans take them everywhere, create conditions for them to thrive and proliferate, and then try hard to poison and kill them. We could easily imagine them asking, 'Why do you take us around, giving us plentiful food and environments we can thrive in? We were happy where we were leading a simple life. Why did you bring us here, and pamper us so nicely, only to then spend so much time and energy trying to kill us? We were on this planet for millions of years, much before you (humans) appeared, and we have honed our survival skills. Perhaps you should focus more on your own actions and behaviour than on us, so that we can live in harmony'. These same organisms might similarly complain that humans have recently projected one of their fellow organisms as a monster and named it COVID-19, which is being blamed for almost everything going wrong in the lives of humans. This organism would have remained where humans found it – it had no ambition to roam around the world – human actions are responsible for its spread.

This perspective is thought-provoking and one of the takeaways is that humans must develop a better knack for problem prevention. Plant disease management is essentially a problem-solving approach. It encompasses actions taken after the disease has manifested. On the other hand, problem prevention is the ability to see potential problems arising and dispel them before they manifest. The biosecurity continuum approach is predominantly a problem-preventing approach, and it needs as much attention, if not more, as the management of established harmful organisms.

14.6.2 Supplement Globalisation with Glocalisation

There are further opportunities to enhance the effectiveness of a biosecurity continuum approach by exploring options to minimise the dispersal of harmful organisms through globalisation-related activities. While the COVID-19 pandemic, on the one hand, has exposed the weaknesses of globalisation in exacerbating the dispersal of harmful organisms, on the other, it has highlighted the importance of localisation, that is, local production for local consumption, using local resources and trade. Locally produced commodities are less likely to introduce harmful organisms. It is much more energy efficient to produce and trade locally, with a lesser risk of spreading harmful organisms. Possibly the solution lies in 'localisation at a global scale' – let us call it 'glocalisation', which must be allowed to flourish side by side with need-based globalisation as a part of strategy to prevent the dispersal of exotic harmful organisms (Shashi Sharma and John Wightman, 2015).

14.6.3 Technology-Enabled Disease Management System

New technologies through smartphone applications, electronic sensors for detection of harmful organisms, and technology-assisted capacity to quickly analyse disparate information appear very promising in enhancing our ability to better understand the harmful organisms and design effective mitigation measures as a part of biosecurity continuum actions. These promising technologies are helping to increase data sharing, improve the timeliness and accuracy of reporting of outbreaks globally, improve global plant disease reporting, enhance international shared access to data for disease surveillance, enable real-time data sharing, and aid in the exchange of epidemiological information among countries. We live in an era of digitisation, and the ongoing automation of traditional practices is providing excellent opportunities and novel options to reform our plant health management system. Wide use of smart technologies, large-scale machine-to-machine communication, and the internet of things have the potential to be among the major contributors to enhancing the efficiency and effectiveness of the global plant health management system. Artificial intelligence may enable the integration of knowledge, methods, and expertise from different disciplines and enhance the feasibility of interdisciplinary teams across multiple disciplines and scales working together to tackle emerging plant diseases. Predictive data for incursions of harmful organisms such as transportation and trade networks, geography, weather and climate parameters, and early pathogen detection data from DNA sequences, incidence records, disease sensors, citizen science, data mining, and web scraping tools and disease identification applications can be input into models to develop dynamic disease surveillance networks, which can be used to model and predict future outbreak hotspots on the landscape level (Ristaino et al., 2021). Already, the Program for Monitoring Emerging Diseases (ProMED) has developed capacity and experience in rapidly disseminating written reports of global outbreak information on infectious diseases of plants via email and reaching subscribers in over 185 countries. Environmental data and land-surface temperature and rainfall are being combined with models of pest life cycles to predict and reduce the impact of fall armyworm in corn in East Africa (Early et al., 2018). The wheat rust early warning systems in Ethiopia are using weather forecast data in combination with spore dispersal models and ground surveillance data (Allen-Sader et al., 2019).

14.6.4 A Global Approach to Plant Health Management

A globally coordinated approach to plant health management, implemented at the farm level, will help safeguard plant resources from harmful organisms. The objective is to safeguard each and every farm by applying basic biosecurity procedures, where possible. Farm biosecurity protocols may be difficult to implement at the small-scale farming enterprise level, where village-level biosecurity may be considered as a basic unit for implementing biosecurity measures. Farm biosecurity entails all measures required to keep plants on the farm free from harmful organisms. Biosecure farms have systems in place to prevent the introduction of harmful organisms (McKirdy et al., 2014; Shashi Sharma and John Wightman, 2015). Adoption of farm biosecurity protocols on a global scale will make a significant difference in mitigating the risk of exotic harmful organisms.

The need for a comprehensive global plant health management strategy and framework cannot be overemphasised. The present FAO initiatives, including the IPPC, are limited in scope, capacity and funds to be even considered as somewhat equal to those of the World Health Organisation, which is mandated to safeguard human health. There is no clarity on how we will safeguard the health of our global plant resources to feed and provide a clean and green environment and sustain the growing human population, which is likely to increase to about 10 billion in the foreseeable future. In the 20th century, there was heavy reliance on chemical poisons, the so-called pesticides, to protect plants. In the 21st century, this will not be a preferred option, and reliance will be on other strategies, including biologicals (Sharma, 2021). Organisations such as FAO and IPPC have a crucial global role in facilitating discoveries of new biologicals or inventions of eco-friendly new pest and disease management options that are as effective, or even more effective, as chemical pesticides in controlling harmful organisms. The FAO and IPPC frameworks are mainly inward-looking, whereas there is a need for looking outward and upward to establish linkages and partnerships with new emerging arrangements and structures required to be able to meet the needs and provide services to our evergrowing population. Though it may be beyond the scope of this article, it is worth mentioning that it is time to review the specialised agencies of the United Nations and assess the need to create an overarching Planetary Health Organisation underpinned by specialised organisations with global mandates on Plant Health, Animal Health, Human Health, Air Health, Soil Health, Water Health and Planetary Health Emergencies (Sharma, 2020b). This is not wishful thinking but a compelling need for the sustainability of our species on this planet and establishing much-needed new global arrangements for safeguarding global plant resources from harmful organisms.

REFERENCES

Allen-Sader, C.; William Thurston, W.; Meyer, M., et al. An early warning system to predict and mitigate wheat rust diseases in Ethiopia. *Environ. Res. Lett.* 2019, 14 (11), 115004. https://doi.org/10.1088/1748-9326/ab4034.

Almeida, R. P. P.; Nunney, L. Emerging insect-transmitted plant diseases: the bacterium *Xylella fastidiosa* as a case study. National Academies of Sciences, Engineering, and Medicine 2016. Global Health Impacts of Vector-Borne Diseases: Workshop Summary. Washington, DC: The National Academies Press. https://doi.org/10.17226/21792.

Anderson, P. K.; Cunningham, A. A.; Patel, N. G., et al. Emerging infectious disease of plants: pathogen pollution, climate change and agrotechnology drivers. *Trends Ecol. Evol.* 2004, 19 (10), 537–555.

Bar-Joseph, M.; Marcus, R.; Lee, R. F. The continuous challenge of citrus tristeza virus control. *Annu. Rev. Phytopathol.* 1989, 27, 291–316.

Bebber, D. P. Range-expanding pests and pathogens in a warming world. *Annu. Rev. Phytopathol.* 2015, 53, 335–356.
Bhattacharya, S. Deadly new wheat disease threatens Europe's crops. *Nature* 2017, 542, 145–146. https://doi.org/10.1038/nature.2017.21424.
Brasier, C. M.; Duncan, J. M. Origin of a new Phytophthora pathogen through interspecific hybridization. *Proc. Natl. Acad. Sci. USA* 1999, 96, 5878–5883.
Cleary, T. *Translation of Sun Tzu's Art of War*. Shambhala, Boston, 1988, 172 pages.
Cunningham, A. A.; Daszak, P.; Rodriguez, J. P. Pathogen pollution: defining a parasitological threat to biodiversity conservation. *J. Parasitol.* 2003, 89, S78–S83.
Dodd, A.; Stoeckl, N.; Baumgartner, J; Kompas, T. Key Result Summary: Valuing Australia's Biosecurity System. CEBRA Project 170713. Centre of Excellence for Biosecurity Risk Analysis (CEBRA) 2020, The University of Melbourne, Melbourne.
Early, R.; González-Moreno, P.; Murphy, S. T., et al. Forecasting the global extent of invasion of the cereal pest *Spodoptera frugiperda*, the fall armyworm. *NeoBiota* 2018, 40, 25–50.
Gladieux, P.; Feurtey, A.; Hood, M. E., et al. The population biology of fungal invasions. *Mol. Ecol.* 2014, 24, 1969–1986. https://doi.org/10.1111/mec.13028.
He, D.; Zhan, J.; Xie, L. Problems, challenges and future of plant disease management: from an ecological point of view. *J. Integr. Agric.* 2016, 15 (4), 705–715.
Hei, N. B.; Tsegaab, T.; Getaneh, W., et al. First report of *Puccinia graminis* f. sp. *tritici* race TTKTT in Ethiopia. *Plant Dis.* 2020, 103, 982.
Inoue, Y.; Vy, T. T. P.; Yoshida, K., et al. Evolution of the wheat blast fungus through functional losses in a host specificity determinant. *Science* 2017, 357 (6346), 80–83. https://doi.org/10.1126/science.aam9654.
Inspector-General of Biosecurity. *Pest and disease interceptions and incursions in Australia*. Department of Agriculture and Water Resources 2019, pp. 70, Canberra. ISBN: 978-1-76003-164-0
IPPC Secretariat. *International year of plant health - final report. Protecting plants, protecting life*. 2021a. FAO on behalf of the Secretariat of the International Plant Protection Convention, Rome. https://doi.org/10.4060/cb7056en.
IPPC Secretariat. *Scientific review of the impact of climate change on plant pests - a global challenge to prevent and mitigate plant pest risks in agriculture, forestry and ecosystems*. 2021b. FAO on behalf of the IPPC Secretariat, Rome. https://doi.org/10.4060/cb4769en.
Keller, R. P.; Geist, J.; Jeschke, J. M., et al. Invasive species in Europe: ecology, status, and policy. *Environ. Sci. Eur.* 2011, 23. https://doi.org/10.1186/2190-4715-23-23.
Legg, J. P.; Kumar, P. L.; Makeshkumar, T., et al. Cassava virus diseases: biology, epidemiology, and management. *Adv. Virus Res.* 2015, 91, 85–142.
Martin, M. D.; Vieira, F. G.; Ho, S. Y. W., et al. Genomic characterization of a South American *Phytophthora* hybrid mandates reassessment of the geographic origins of *Phytophthora infestans*. *Mol. Biol. Evol.* 2016, 33 (2), 478–491. https://doi.org/10.1093/molbev/msv241.
McGee, D. C. *Plant Pathogens and the Worldwide Movement of Seeds* (New York: APS Press, 1997), 109 pp.
McKirdy, S.; Rodoni, B.; Moran, J.; et al. Microbial threat- a growing challenge for plant biosecurity. *Microbiol. Aust.* 2012, 33, 12–14.
McKirdy, S. J.; Sharma, S. B.; Bayliss, K. L. Quarantine and Biosecurity. In *Encyclopedia of Agriculture and Food Systems*. Van Alfen, N. K., Ed. Academic Press, London, UK, 2014, pp. 11–20.
Miller, S. A.; Ferreira, J. P.; LeJeune, J. T. Antimicrobial use and resistance in plant agriculture: a one health perspective. *Agriculture* 2022, 12, 289. https://doi.org/10.3390/agriculture12020289.
Mora, C.; Tittensor, D. P.; Adl, S., et al. How many species are there on earth and in the ocean? *PLoS Biol.* 2011, 9 (8), e1001127. https://doi.org/10.1371/journal.pbio.1001127.

Mumford, J. D. Economic analysis of area-wide fruit fly management. In Proceedings of 6th International Fruit fly Symposium, 6–10 May 2002, Stellenbosch, South Africa, pp. 189–193.

Oleg, S. P.; Yermakov, W. V. *International Quarantine* (Connecticut: International University Press Inc., 1991), 344 pages. Translated from Russian.

Pimentel, D. Pesticides and Pest Control. In *Integrated Pest Management: Innovation-Development Process*. Peshin, R., Dhawan, A. K., Eds. Springer Science+Business Media B.V., Dordecht, 2009, pp. 83–87. https://doi.org/10.1007/978-1-4020-8992-3_3.

Pimentel, D.; McNair, S.; Janecka, J.; et al. Economic and environmental threat of alien plant, animal and microbe invasions. *Agric. Ecosyst. Environ.* 2001, 84, 1–20.

Plant Health Australia. Plant for transition to management of myrtle rust. Plant Health Australia Document, 2011, 1–15 pages. https://www.planthealthaustralia.com.au/wp-content/uploads/2018/10/Plan-for-Transition-to-Management-of-Myrtle-Rust.pdf

Priyadarsini, S. L.; Suresh, M.; Huisingh, D. What can we learn from previous pandemics to reduce the frequency of emerging infectious diseases like COVID-19? *Global Transitions* 2020, 2, 202–220. https://doi.org/10.1016/j.glt.2020.09.003.

Quinney, M. 5 reasons why biodiversity matters - to human health, the economy and your wellbeing. *World Economic Forum*. May 2020. https://www.weforum.org/agenda/2020/05/5-reasons-why-biodiversity-matters-human-health-economies-business-wellbeing-coronavirus-covid19-animals-nature-ecosystems/

Ramanujan, K. Irish potato famine pathogen originated in Mexico 2014. *Cornell Chronicle*. https://news.cornell.edu/stories/2014/06d. \P/irish-potato-famine-pathogen-originated-mexico.

Ristaino, J. B.; Anderson, P. K.; Bebber, D. P. The persistent threat of emerging plant disease pandemics to global food security *PNAS* 2021, 118 (23), e2022239118. https://doi.org/10.1073/pnas.2022239118.

Santini, A.; Liebhold, A.; Migliorini, D., et al. Tracing the role of human civilization in the globalization of plant pathogens. *ISME J.* 2018, 12, 647–652. https://doi.org/10.1038/s41396-017-0013-9.

Sequeira, R.; Griffin, R. The Biosecurity Continuum and Trade: Pre-border Operations. In *The Handbook of Plant Biosecurity*. Gordh, G., McKirdy, S., Eds. Springer Science+Business Media, Dordrecht, 2014, pp. 119–148. https://doi.org/10.1007/978-94-007-7365-3_5.

Sharma, S. B. Role of Biosecurity in Achieving Global Food Security. In *Applied Zoological Research for Sustainable Agriculture & Food Security*. Srivastava, C. P., Singh, R. N., Raju, S. V. S., and Raghuraman, M., Eds. Department of Entomology and Agricultural Zoology Institute of Agricultural Sciences, Banaras Hindu University, Varanasi, India, 2019, pp. 1–16.

Sharma, S. Global plant health management needs transformation. June 2020a. https://www.linkedin.com/pulse/global-plant-health-management-needs-transformation-shashi-sharma-phd/

Sharma, S. Global plant health management needs transformation - part 2. August 2020b. https://www.linkedin.com/pulse/global-plant-health-management-needs-transformation-part-sharma-phd/

Sharma, S. What Covid 19 is telling us - part 10: Spending $1.9 trillion on arms to kill humans is insane. November 2020c. https://www.linkedin.com/pulse/what-covid-19-telling-us-part-10-spending-trillion-arms-sharma-phd/?trackingId=xWhxxkALVTXHUUnih%2BHWMg%3D%3D

Sharma, S. Broken global pest and disease management system needs fixing: bioprotectants require a booster dose as a way forward. 2021. https://news.agropages.com/News/NewsDetail---40011.htm

Sharma, S. B.; Wightman, J. W. *Vision Infinity for Food Security: Some Whys, Why Nots and Hows!* Springer Briefs in Agriculture. Springer Cham Heidelberg, New York, Dordrecht and London, 2015, pp. 98. https://doi.org/10.1007/978-3-319-23249-2.

Sharma, S. B.; Varaprasad, K. S.; You, M. Biosecurity - an integral part of global food security strategy. *Indian J. Plant Prot.* 2009, 36, 165–172.

Sicard, A.; Saponari, M.; Vanhove, M., et al. Introduction and adaptation of an emerging pathogen to olive trees in Italy. *Microb. Genom.* 2021, 7, 000735. https://doi.org/10.1099/mgen.0.000735.

Stace-Smith, R.; Hamilton, R. I. Inoculum thresholds of seedborne pathogens: viruses. *Phytopathology* 1988, 78, 875–880.

Velasquez, A. C.; Castroverde, C. D. M.; He, S. Y. Plant-pathogen warfare under changing climate conditions. *Curr. Biol.* 2018, 28 (10), R619–R634. https://doi.org/10.1016/j.cub.2018.03.054.

Watson, R. N. Internal biosecurity - a realistic objective for plant nematodes. *NZ Plant Prot.* 2004, 57, 151–155.

Whattam, M.; Clover, G., Firko, M.; Kalaris, T. The Biosecurity Continuum and Trade: Border Operations. In *The Handbook of Plant Biosecurity*. Gordh, G., McKirdy, S., Eds. Springer Science+Business Media, Dordrecht, 2014, pp. 149–188. https://doi.org/10.1007/978-94-007-7365-3_6.

Index

Note: Bold page numbers refer to tables; *italic* page numbers refer to figures.